Uri M. Kupferschmidt
The Diffusion of "Small" Western Technologies in the Middle East

Studies on Modern Orient

—
Volume 44

Uri M. Kupferschmidt

The Diffusion of "Small" Western Technologies in the Middle East

Invention, Use, and Need in the 19th and 20th Centuries

DE GRUYTER

ISBN 978-3-11-077719-2
e-ISBN (PDF) 978-3-11-077722-2
e-ISBN (EPUB) 978-3-11-077730-7

Library of Congress Control Number: 2023939392

Bibliographic information published by the Deutsche Nationalbibliothek
The Deutsche Nationalbibliothek lists this publication in the Deutsche Nationalbibliografie; detailed bibliographic data are available on the internet at http://dnb.dnb.de.

© 2023 Walter de Gruyter GmbH, Berlin/Boston
Cover image: The typewriting class around 1908 in the School of Commerce at the Syrian Protestant College (in 1920 renamed American Universty of Beirut): "…all of these pupils are alike in one characteristic, namely the remarkable quickness with which they master the skilled use of the writing machine" (Remington Notes, vol 1/10 [1908], New York Public Library)
Printing and binding: CPI books GmbH, Leck

www.degruyter.com

Preface

The essays in this volume represent an aspect of the social and cultural history of the Middle East during the 19th and 20th centuries as I used to research and teach it.

Three of the essays, here chapters, have been published over the past two decades, the first being "The Social History of the Sewing Machine in the Middle East" in *Die Welt des Islams*, vol 44 (2004). At the time, my research on that topic led to writing "Who Needed Department Stores in Egypt? From Orosdi-Back to Omar Effendi" in *Middle Eastern Studies*, vol. 34 (2007). It derived from my more extensive book on that chain of department stores, *The Orosdi-Back Saga: European Department Stores and Middle Eastern Consumers* which appeared the same year at the Ottoman Bank Archives and Research Centre in Istanbul as winner of its bi-annual research prize. The chapter "On the Diffusion of 'Small' Western Technologies and Consumer Goods in the Middle East during the Era of the First Modern Globalization" appeared in Liat Kozma et al.(eds), *A Global Middle East, Mobility, Materiality and Culture in the Modern Age, 1880–1940* (London: Tauris 2015) and was in fact a development of my study on the impact of the sewing machine. I wish to thank the above publishers for their kind permission to reprint these articles in this volume. No changes have been made in the original texts, but some revisions have been inserted into the extensive Introduction to this volume.

Without giving up my interest in diffusion and consumerism, I continued the story of the sewing machine in the Middle East. Building on the research agenda which I had sketched out in my article of 2015, the last chapters elaborate on three further specific cases, namely typewriters, eyeglasses, and pianos in the Middle East. On the whole, overlapping and inconsistencies with my earlier findings have been unavoidable, but I hope that the volume in its entirety proves that the study of small technologies and their differential diffusion in the countries of the Middle East offers a wide range of scholarly challenges for even more research.

I have been very fortunate that close colleagues and friends, each in their field of expertise, but with similar research interests, have been kind enough to read and comment upon some of the chapters or parts of them when they were still in draft. I hereby gratefully acknowledge all their suggestions and additions. Often their help and learned advice were invaluable. They saved me from mistakes and misunderstandings, but where such remain, I alone bear responsibility for them. My profound thanks go in particular to Robert Adelson, Ami Ayalon, Farid Benfeghoul, Gad Gilbar, Henri Obstfeld, Ruth Roded, Yuval Shaked, Relli Shechter, Peter Stearns, and Fruma Zachs.

But they are not alone. Over the years I have been in contact with many more colleagues and friends on these subject matters, – asking questions, trying out ideas, checking references or verifying sources, and exchanging information. It is a long list, and I am greatly indebted to each of them, perhaps more than they realize. This is exactly what makes historical research for me a fascinating and pleasant challenge.

The latter academic and non-academic experts were, in alphabetical order: Reema Abusaleh, Amnon Acho, Mafalda Ade, Hélène Alexander, Ersin Altin, Reuven Aviv, Canan Balan, Rifat Bali, Henri Barda, Yuval Ben-Bassat, Eyal Sagui Bizawe, Dick Bruggeman, Marina Bugaev, Assaf Dar, Arndt Engelhardt, Amer Hanna Fatuhi, Carter Findley, Willem Floor, Wolfgang Förster, Peter Gaboda, Sonia Gergis, Nile Green, Ekmeleddin İhsanoğlu, Grace Hummel, Yavuz Köse, Evren Kutlay, Robert Labaree, Mordechay Lewy, Brinkley Messick, Titus Nemeth, David Ezra Okonşar, Filiz Özbaş, Anthony Phillipson, Moshe Porat, Markus Purkhart, Uriel Rozen, Manfred Sauer, Selim Sednaoui, Stephen Sheehi, Miri Shefer-Mossensohn, Qustandi Shomali, Lorans Tanatar Baruh, Peter Thoergersen, Tasha Vorderstrasse, and Keren Zdafee. Some more are mentioned in the footnotes. Please forgive me if, unintentionally, I have omitted some names but be assured that I have appreciated every bit of help and advice.

Last but not least, I wish to thank Marion Lupu, a dear colleague at the University of Haifa, for correcting and improving the manuscript – not the first time that I have enjoyed working with her on a book –, as well as Dr. Sophie Wagenhofer, Katharina Zühlke, Antonia Pohl, and Anett Rehner of De Gruyter for their friendly advice, constant encouragement, and great patience, to see this project through to its final publication. I am equally grateful to my wife Tamar for her support, patience and understanding, bearing my trying out of ideas on her over the years that I was writing the various chapters. Not to forget Zwi, Naama, and Aran, who each have also contributed to the project.

A note on transliteration

Owing to the use of sources from different periods and areas, and the reprinting of earlier published chapters, I have chosen a simplified form of transliteration from Arabic and Ottoman Turkish. For the rest, I have maintained the Arabic differentiation between the *'ayn* and the *alif* and marked diacritics and long vowels only where the transliteration might raise doubts. For Ottoman Turkish, modern orthography is used. But the Arabic *afandi*, and Turkish *efendi*, are rendered here in the common western form *effendi*. However, we retained inconsistencies such as Sultan Abdülhamid, and President 'Abd al-Nasir, or Midhat Pasha instead of Midhat Paşa. Most cities come under their present names, e.g., Istanbul, not Constantinople, Izmir, not Smyrna, with the exception of Salonika instead of Thessaloniki. It is hoped that readers familiar with the era, the region, and the subject, will recognize the names, terms, and words without difficulty. A last point concerns the name of the Orosdi-Back department stores, here written with a hyphen, although it was originally written without one.

Contents

Introduction —— 1
        On invention, use, and need —— 1
        Orosdi-Back: from a trading company to a chain of department stores —— 9
        Technologies as consumer goods, big and small —— 22
        Invention and in-use —— 27
        Diffusion: the Everett Rogers model —— 29
        "Differential diffusion" —— 30
        Delays and resistance —— 34
        Leapfrogging —— 38
        Four case studies —— 39
        In conclusion —— 44

1 Who needed department stores in Egypt? —— 46
1.1   From Orosdi-Back to Omar Effendi —— 46
1.2   Postscript —— 69

2 On the diffusion of "Small" Western technologies and consumer goods in the Middle East —— 70
2.1   "Big" and "small" —— 72
2.2   Diffusion and the process of acquisition —— 74
2.3   About use and need —— 75
2.4   The sewing machine —— 78
2.5   The typewriter —— 80
2.6   Photographic cameras —— 83
2.7   The piano —— 85
2.8   Incandescent light bulbs and electric household implements —— 90
2.9   The automobile —— 93
2.10 Some conclusions —— 96

3 The social history of the sewing machine in the Middle East —— 97
3.1   Marketing and consumption in the Middle East —— 101
3.2   Use of the sewing machine in industry —— 106
3.3   Home industry —— 107
3.4   Gender aspects —— 108
3.5   The sewing machine as a tool of development —— 111

3.6	Ready-made clothing ——	**112**
3.7	Conclusions ——	**114**

4 Leapfrogging from typewriter to personal computer in the Middle East —— 116

4.1	Introduction ——	**116**
4.2	The typewriter's genealogy ——	**117**
4.3	Techno-linguistic adaptation ——	**120**
4.4	Typography and calligraphy ——	**128**
4.5	From foreign use to local business ——	**130**
4.6	Typing (dactylography) and shorthand (stenography) as ancillary skills —— **132**	
4.7	From early adopters to an early majority: another time chasm ——	**134**
4.8	No local typewriter celebrities ——	**138**
4.9	Transitions: QWERTY and Turkey, and de-colonization in Egypt ——	**139**
4.10	The slow emergence of indigenous female typists and office workers —— **142**	
4.11	Leapfrogging to the personal computer ——	**148**

5 From Ibn Haitham to Abou Naddara and Al-Misri Effendi: eyeglasses in the Middle East —— 151

5.1	Evolution in Europe ——	**153**
5.2	Why did the Middle East not have them before Europe? ——	**157**
5.3	A small technology known to select elites ——	**159**
5.4	From "knowledge" to "persuasion" ——	**166**
5.5	Lesser prevalence of visual disabilities? ——	**167**
5.6	The use and the marketing of eyeglasses in the Middle East ——	**169**
5.7	Photographic evidence ——	**175**
5.8	Ophthalmology ——	**176**
5.9	Iconic figures ——	**180**
5.10	Some conclusions ——	**185**

6 Did the piano have a chance in the Middle East? —— 187

6.1	A short history of the piano ——	**189**
6.2	*Maqam*s and notation ——	**191**
6.3	*Alla Turca* and other orientalisms ——	**193**
6.4	When did modern pianos begin to reach the Middle East? ——	**194**
6.5	Missionaries, colonial officers, and diplomats ——	**202**
6.6	Trickling down? ——	**206**
6.7	Further diffusion ——	**211**

6.8	Statistics —— **214**	
6.9	An "oriental piano" for the Middle East? —— **216**	
6.10	The Arab Music Congress of Cairo in 1932 —— **221**	
6.11	Piano education —— **227**	
6.12	Edward Said —— **229**	
6.13	An acoustic "glass bottom"? —— **231**	
6.14	Conclusions —— **235**	

Bibliography —— 240
        Books and articles —— **240**
        Some periodicals —— **260**
        Diverse yearbooks and reports —— **260**
        Archives —— **261**
        Selected online sources —— **261**

List of Figures —— 263

Index of proper names —— 265

Index of geographical names —— 269

Index of subjects —— 272

Introduction

On invention, use, and need

The second half of the nineteenth century and the first decades of the twentieth century, sometimes called the age of "the first modern globalization"[1], saw an era of growing industrial mass production, of rising urbanization. and of a significant expansion of international free trade.[2] With a consensus on *laissez-faire* – itself a marker of western Imperialism –, industrial entrepreneurs, seizing on unfolding commercial opportunities, appeared on the global scene, and successfully created markets abroad for their new mass-produced goods.[3] Mass-produced, ready-to-wear clothing (called "prêt-à-porter" in France and "confection" in central Europe) – that is fashionable garments for women, and also for men – may have been pioneered in the United States, but European exporters to the Balkans and the Middle East had a relative advantage due to their geographical proximity, and possibly also as far as fashion and taste were concerned.

European merchants had a long tradition of trading with the countries of the Middle East, as emerging Austrian, French, British, and German industrialists began exporting new consumer goods. They were joined by large American industrial corporations. After successfully marketing durable consumer goods in their domestic markets, Singer, Remington, Edison, Kodak, and others, driven by capitalist motives, sought additional outlets around the globe, and in some cases set up

---

[1] Liat Kozma, Cyrus Schayegh, and Avner Wishnitzer (eds.), *A. Global Middle East, Mobility, Materiality and Culture in the Modern Age, 1880–1940* (London: Tauris 2005), pp. 1–8 *et passim*.
[2] "Western" here as an overarching general term for Europe and North America. Negative effects have been emphasized by many researchers, for instance, Daniel R. Headrick, *Tentacles of Progress: Technology Transfer in the Age of Imperialism, 1850–1940* (New York: Oxford University Press 1988), who reviewed mainly the impact of steam shipping, railways, telegraph networks, urban planning and infrastructures, large agricultural projects, and the like. He argued that such transfers from the West led to colonial segregation and failed to develop local economies or to raise per capita incomes. The diffusion of western products was advanced by the impact of all these, from faster maritime and riverine transport, the telegraph (the "first internet" as Tom Standage has called it), to print culture. Though printing presses had preceded steam engines by centuries, the production of Ottoman, Arab and Persian printing rose sharply from the 19th century onward.
[3] After the 1500s, the transfer of new technologies between the Islamic world and Europe is thought to have become reversed. With few exceptions, major technological innovations and newly developed mass-manufactured products originated in Britain, France, Belgium, and subsequently Austria, Germany, and the United States.

https://doi.org/10.1515/9783110777222-001

plants in Europe.⁴ At the time, the mental association of commercial expansion with military and commercial strategic conquests was not uncommon.⁵ It could be called a form of economic or soft imperialism.

This brings us to the relational nature of consumption which, in its oldest economic appearance, is conceived as "what one produces, is wanted or needed by somebody".⁶ Obviously, in the beginning, not the entire Middle Eastern population wanted or needed the new goods, simply because they were not aware of their existence or of their utility. But they arrived faster than ever before and became known to growing segments of the literate population due to the press.

Whether we call these new industrial goods "novelties", "*nouveautés*" (which in French meant clothing more often than not), or new inventions, their advent at first expressed itself among the upper urban sections of Middle Eastern populations, often minorities (as well as foreign residents whose presence also gradually rose). For many consumers, there was no immediate existential need to adopt these, but a craving to belong to what they identified as a more advanced world. This became visible in the way they dressed, lived, spent their leisure times, in what they read in what language, and which objects they installed in their palaces, homes, offices, or workshops. Such consumption goods were not imposed but voluntarily acquired, sometimes with no reason other than display and social prestige.

The study of consumerism takes simple consumption one or more steps further: A social, cultural, and behavioral order of obtaining an object "with the intention of use, as opposed to resale, regardless of the manner of acquisition", or even "patterns of buying, selling, and using goods that consider associated meanings such as status, sociability, enjoyment, taste, fashion, aesthetics, consumer choice, agency, and ethics, as well as anxiety and manipulation".⁷ With new turns in social and cultural historical research since the 1980s, the more profound

---

4 E.g., Singer in Clydebank, Glasgow; Kodak in Harrow, U.K., and Berlin; Ford in Berlin.
5 Robert B. Davies, "'Peacefully Working to Conquer the World': The Singer Manufacturing Company in Foreign Markets 1854–1889", *Business History Review* 43/3 (1969), the quote is from George Baldwin Woodruff, Singer's representative in London, who likened his agents to "a living moving army of irresistible power, peacefully working to conquer the world", and probably dates from 1882, see p. 304.
6 The biased division between alleged thriftless spending (or "shopping") by women and production ascribed solely to men, has long been abandoned.
7 The former is the definition of Suraiya Faroqhi, *Stories of Ottoman Men and Women, Establishing Status, Establishing Control* (Istanbul: Eren 2002), p. 38. The latter much broader definition comes from H. Hazel Hahn, "Consumer Culture and Advertising" in Michael Saler (ed.) *The Fin-de-Siècle World* (London: Routledge 2015), p. 392.

study of consumerism has reversed the erstwhile emphasis on top-down entrepreneurial marketing to local interests and behaviors, – in short, needs and wants.[8]

Consumer culture, globally speaking, is often perceived as one comprehensive phenomenon, but it assumed various forms in diverse countries, in graded urban-rural and other social, or ethnic and regional contexts. As so often happens, trends in the study of western history generated comparative questions among historians of the Ottoman lands, including Egypt, the Arab successor states, and also Iran. These insights challenged me with regard to the fine tuning of the local effects of consumerism. Part of my initial inspiration came from Donald Quataert's edited volume, *Consumption Studies and the History of the Ottoman Empire* (1999)[9]; from several books and collections by Soraiya Faroqhi in the following years[10]; and from Yavuz Köse's early articles and his subsequent comprehensive *Westlicher Konsum am Bosporus* which was published in 2010.[11] Together with books and articles by other prominent scholars, these opened the field widely to an investigation into parallels – and differences – between Europe and America, and the Middle East.

But not only new consumer products reached the Middle East, so did innovative commercial strategies, and new sales practices and styles. Indeed, many researchers have associated consumerism – maybe over-emphasized in relation to other stores – with department stores as the epitome of modern commerce.[12] The first of these had appeared in mid-19th century France, Britain, and the USA.[13] Michael Miller, an American social historian, published in 1994 his iconic

---

[8] See a landmark study by the social historian, Peter N. Stearns, *Consumerism in World History, The Global Transformation of Desire* (New York and London: Routledge, first ed, 2001, second ed. 2006). In separate chapters on the globalization of western-driven consumerism outside Europe and America, Stearns also discusses the Islamic Middle East but since his book, more detailed studies have been published. It is worth also looking at Frank Trentmann, *Empire of Things* (London: Penguin 2016).
[9] Donald Quataert (ed.), *Consumption Studies and the History of the Ottoman Empire, 1550–1922* (Albany, SUNY 2000).
[10] Soraiya Faroqhi has an impressive list of publications many of which pertain in one way or another to the history of consumption in the Ottoman Empire.
[11] Yavuz Köse, *Westlicher Konsum am Bosporus, Warenhäuser, Nestlé & Co.im Späten Osmanischen Reich (1855–1923)*, (München: Oldenbourg Verlag 2010).
[12] Over-emphasized because in the larger cities, luxury specialty stores emerged alongside, while at some places, relatively new goods such as umbrellas, prams, clocks, or eyeglasses, could also be found in traditional *suq*s, as Toufoul Abou-Hodeib, *A Taste for Home, the Ottoman Middle Class in Beirut* (Palo Alto: Stanford University Press 2017), p. 149, claims for Beirut. It proves how difficult it is to map out the process of diffusion.
[13] I have tabulated their early history in Europe and America in my *The Orosdi-Back Saga, European Department Stores and Middle Eastern Customers* (Istanbul: Ottoman Bank Archives and Re-

book on the Parisian department store, *Au Bon Marché*, which in a way inspired my own study of Orosdi-Back in the Middle East.[14]

At some point, it struck me that the book by the Marxist historian, Eric Hobsbawm, *The Age of Empire*, with its emphasis on the Industrial Revolution and its consequences for production, labor, and class transitions, referred to department stores. His book opens with a peculiar reference to his maternal uncle whose family name he did not spell out in the text. This was Albert Mayer who owned a chain of department stores in the Middle East, one of them in Alexandria, the place of his birth:

> Uncle Albert had built up a chain of stores in the Levant – Constantinople, Smyrna, Aleppo, Alexandria. In the early twentieth century there was plenty of business to be done in the Ottoman Empire and the Middle East, and Austria had long been central Europe's business window on the Orient. Egypt was both a living museum, suitable for cultural self-improvement, and a sophisticated community of the cosmopolitan European middle class, with whom communication was easily possible by means of the French language.[15]

Albert Mayer belonged to a category of Central-European Jewish entrepreneurs, generally entrepreneurial families, who founded department stores in Istanbul, Cairo, and other major cities of the Middle East. Among them we also find Stein, Tiring, and Stross. We will discuss Orosdi-Back in detail.

Indeed, before too long, after their appearance in Europe and America, similar department stores also appeared in the larger cities of the Middle East, from the expanding mega-cities such as Istanbul, Salonika, and Izmir, to Cairo, Alexandria, Beirut, to some middle-range cities.

In Istanbul, and other Ottoman cities, famous names were Bon Marché (the owners, the Bortoli Brothers, had no connection to the famous Parisian store),

---

search Centre 2007), p. 54. That book partly overlaps with Chapter One here but discusses the wider activities of the said firm all over the countries of the Middle East.

14 Michael Miller, *The Bon Marché: Bourgeois Culture and the Department Store, 1869–1920* (Princeton: Princeton University Press 1994). Other studies followed suit, e.g. Geoffrey Crossick and Serge Jaumain (eds.), *Cathedrals of Consumption, European Department Stores, 1850–1939*, (London: Routledge 1999). See also Amr Tawfik Kamal, "Empires and Emporia: Fictions of the Department Store in the Modern Mediterranean (PhD Thesis, University of Michigan 2013).

15 *The Age of Empire 1875–1914* (London: Weidenfeld and Nicolson 1987), pp. 1–2, and p. 29; see also his *Interesting Times* (London 2002), p. 3. Hobsbawm may have felt the involvement of his family in this capitalist branch embarrassing, but he was definitely interested in my initial research questions on department stores. Adelheid Mayer and Elmar Samsinger's book *Fast Wie Geschichten Aus 1001 Nacht, Die Jüdischen Textilkaufleute Mayer Zwischen Europa und dem Orient* (Vienna: Mandelbaum 2025) had not yet appeared. In our correspondence I should have discussed Hobsbawm's definition of the "cosmopolitan European middle class".

Baker, and Carlmann & Blumberg, while in Cairo, Cicurel, Chemla, Hannaux, Ades, Levi-Benzion, and Gattegno were some of the outstanding names. Many were Jews of Mediterranean background. Exceptions in Cairo were a few such as Sednaoui and Davis Bryan, respectively of Syrian-Christian and of Welsh origin.[16] Their erstwhile commercial prominence can still be gauged from the commercial guides of the time, as well as their erstwhile premises. Most may have been forgotten, or their reputation was suppressed in the later era of nationalizations; but some were redressed or imitated to serve as outlets for local industrial products.

Remarkably, one discerns in recent years a renewed interest in the history of these once famous department stores in Cairo, Beirut, Baghdad, or even Adana and other cities.[17] Part of this re-emerges in personal memoirs, which reflect the nostalgia of erstwhile Jewish and other minorities, and not only them, but another part of it ensues from a new interest in local urban or architectural heritages, as well as in the development of consumerism and gender in the late Ottoman Empire, in particular among younger Turkish scholars.[18] Besides, the internet occasionally reveals collectibles for sale, including postcards, souvenirs, publicity items, wrapping paper with the name of the firm, dinner coins for the personnel, and many invoices and envelopes (not only owing to the value of the post stamps).[19]

Quite of few of these new outlets called themselves in the *lingua franca* of the dominant classes, *grands magasins*.[20] The firm itself reported to its shareholders in 1908: "Nous sommes ici dans le domaine des grands bazars, mais il nous semble que nous pouvons en parler en même temps que des grands magasins".[21] Ironical-

---

[16] Cf. Baghdad where Orosdi-Back was rivaled by the Hasso Brothers' store who were Seventh Day Adventists.
[17] On the renewed, relatively intensive interest in Adana's economic development see Ahmet Nadir Işayağ, *Orosdi-Back Efsanesi (Horozdibeği) Adana* (Ankara: Akedemisyen Kitabevi 2019). Though relating the local story of the Adana branch, he relied heavily on my book and its title. See further Ela Yılmaz, "Osmanlı Topraklarında Uluslararası bir Kuruluş: Orosdi-Back ve Adana Mağazası", *Ankara Üniversitesi Sosyal Bilimler Dergisi*, vol. 13/1 (2022), pp. 28–44.
[18] Orosdi-Back in Beirut and Baghdad are regularly referred to as landmarks, not unlike famous department stores in Paris, London, or New York. University theses in Turkey are increasingly accessible on the internet.
[19] However, visual materials, and even advertisements, of some of the smaller branches are lacking.
[20] Though our research bears no less an inclination toward French sources, and to the assumed supremacy of French among higher circles, it does not mean that Ottoman Turkish, or Arabic, were absent.
[21] Note the use of the word *bazar*, of Persian origin, for an open market, which entered the French language in the beginning of the eighteenth century. In Paris, a department store with the name Bazar Napoléon had opened in 1856, renamed Le Bazar de l'Hotel de Ville in 1870

ly, the origin of the term *makhzan* for storehouse was Arabic and had much earlier entered French and other west-European languages via the Italian. While the derived plural *mağazalar* was equally used in Turkish, it was more remarkable that the word *bonmarşeler* (from Au Bon Marché) entered that language.

Yavuz Köse has aptly coined the notion of "vertical bazaars of modernity" to distinguish them from the traditional flat outlays of bazaars, called *suq*s in Arabic or *çarşı*s in Turkish.[22] These innovative stores, originally western outlets for ready-made clothing (a novelty at the time, more expensive and exclusive than handmade clothes) –, gradually assumed "glocal" forms of western modernity. Architecturally, the conspicuous Orosdi-Back, Tiring, and Sednaoui's main buildings in Cairo are hardly less impressive than Au Printemps in Paris which they sought to resemble; the former's building in Beirut (opened in 1900, coinciding with Sultan Abdülhamid's silver jubilee) had the first elevators in town (Fig.1), and they were maybe also the first in Baghdad to have them. In Philippopoli (Plovdiv), Bulgaria, they opened the allegedly first building made of concrete. In Tunis, a large accurate clock was erected in front of their premises.

Different sales floors were not the only innovation: It was a new practice of commerce which comprised more than architectural features, namely free access, orderly arrangement of goods, glass showcases (*vitrines*) and large display windows on the street side.[23] Above all, the imported new wares were advertised as "just arrived" *nouveautés*, and sold by a large contingent of trained personnel, including women, conversant in one or more foreign languages. The usual haggling of the bazaar was replaced by fixed prices, the sort of assumed "western" orderliness, though a closer examination also reveals flexibility in the form of special offers, holiday sales, lotteries (*tombolas*) and other bargains.

In Europe and America, department stores, and the new phenomenon of "shopping" were bound up with each other. However, analytical references to its evolution in the Middle East are few. By the late nineteenth century, in Istanbul, elite women could indeed be seen shopping regularly in the Beyoğlu area of Istan-

---

(today the upscale BHV). It was not the only one to use this word in the sense of a large modern establishment.

22 Yavuz Köse, "Vertical Bazaars of Modernity: Western Department Stores and their Staff in Istanbul (1889–1921) in supplement to *International Review of Social History*, vol. 14 (2009), pp. 91–114.

23 Gökhan Akçura, "Göz Avlama Sanatı, Türkiye Vitrin Tarihine Giriş", https://manifold.press/goz-avlama-sanati (accessed 25 June 2022). By 1949, Omar Effendi in Cairo still advertised "Visitez nos Étalages" (Visit our Windows) on cinema tickets.

Fig. 1: Postcard of the Orosdi-Back building at the Rue de la Douane, Old Port of Beirut (opened 1900 with display windows). At the time it was the largest department store in the Eastern Mediterranean, sensational due to its three floors and elevators. Two of the copulas on the facade had clocks. Later, the store would relocate closer to the center, near the Riyadh al-Sulh Square (author's collection).

bul.[24] The extant diary of a public servant named Said Bey indicates that he went out to buy much of his clothing himself, but that his wife purchased hers at the Bon Marché department store. It would seem that most of the time, the genders shopped separately, women often in small groups. Segregation thus remained more common than in the West, and this must still have been the norm for a long time all over the region.[25] The gender aspects, both of women shopping and of paying cash on the spot, or on credit, deserve more future study, as does the increasing number of saleswomen in the various outlets.[26]

---

24 Faroqhi, *Stories*, p. 52. See also Ersin Altin, *Rationalizing Everyday Life in Late Nineteenth Century Istanbul, c.1900* (PhD Thesis (digital), New Jersey Institute of Technology 2013), pp. 9–10, 136, and 194–195.
25 Together, upper class women were allowed to visit holy tombs, cemeteries, and increasingly to undertake excursions at times of festivals, but free "shopping" as a rule developed late.
26 Paul Dumont, "Said Bey – The Everyday Life of an Istanbul Townsman at the Beginning of the Twentieth Century" in Albert Hourani et al. (eds), *The Modern Middle East* (London: Tauris 1993), pp. 277, 279, 282–282, and 284; Melis Hafez, *Inventing Laziness: The Culture of Productivity in Late*

There is inevitably a question as to who in the Middle East really "needed" the new western mass products. Evidently, the owners of these department stores (and other big importers) felt that it was they who perceived the chances to sell their products and increase their profits. After all, they were traders and capitalists, and this did not differ in essence from their home bases. The expansion of mass-production of standardized goods became linked, in one way or another, to the evolution of local class and communal aspects. [27] Where new consumer goods became known to a segment of the population and were considered attractive enough, they were wanted as practical or social necessities, or "must have" entities.

Our literature shows a certain bias in the relative magnification of *bonmarşeler* and *grands magasins* (*mağazalar*) – with all their French connotations. Along the main modern streets of a number of big cities, single specialized stores also built a reputation on imported wares, including the latest Parisian or Viennese fashions, styles, and taste. But they also offered prestigious brands in foodstuffs and drinks, pharmaceuticals, cosmetics, furniture, musical instruments, optics, and electrical appliances.[28] Some foreign manufacturers had their own local agents.[29] In the large cities, the most prominent modern shopping streets, such as the Grande Rue de Pera and Yüksek Kaldırım in Istanbul, or Fuad al-Awwal and 'Imad al-Din streets in Cairo, and similar shopping areas elsewhere, were lined with first class retail stores.[30] Here, importers and specialized stores and

---

Ottoman Society (Cambridge: Cambridge UP 2021), pp. 148, 168, and 175. Closer study of novels and diaries has been proven to potentially yield more information.

27 More recently, on the Middle East, Relli Shechter (ed.), *Transitions in Domestic and Family Life in the Modern Middle East* (New York 2003). Another highly relevant study is Nancy Young Reynolds, *Commodity Communities: Interweaving of Market Cultures, Consumption Practices, and Social Power in Egypt, 1907–1961* (PhD Thesis, Stanford University 2003), reworked as *A City Consumed, Urban Commerce, the Cairo Fire, and the Politics of Decolonization in Egypt* (Stanford: Stanford UP 2012).

28 Ernest Giraud, *La France à Constantinople* (Constantinople 1907), pp. 4–23. Giraud was the president of the French Chamber of Commerce. David Todd, *A Velvet Empire, French Informal Imperialism in the Nineteenth Century* (Princeton: Princeton University Press 2021), discusses at length the export of French luxury goods, with a certain emphasis on champagne, which is presumably among the less wanted products in the Middle East.

29 Among many other foreign traders was the English firm, Nowill & Co, importers of metal wares, later of Remington typewriters, Gillette razors, and Milners safes, which worked from the Alime Han in lower Galata, Sidney E.P. Nowill, *Constantinople and Istanbul, 72 Years of Life in Turkey* (Leicestershire: Matador 2011).

30 For a detailed map of stores, see Yavuz Köse, *Konsum*, p. 459. See also Nur Akın, *19. Yüzyılın Ikinci Yarısında Galata ve Pera* (Istanbul: Literatür 1998), pp. 220–226 and p. 258. Though Istanbul

their advertising, such as Verdoux and Comendinger (mentioned in the next chapters), offered a wide array of luxurious commodities to customers who sought them and were able to afford them, as one can see in the various commercial guides (mostly in French) which catered to the upper classes.[31] A few had been granted the authorization to carry the title of purveyor to the court, which was a valuable publicity device. Toufoul Abou-Hodeib, writing mainly on Beirut, has indicated that some western commodities had already been offered in the *suqs* before the advent of department stores, and this may have been the case in more cities.[32] We suppose that the supply of certain new, imported articles escapes our attention owing to a one-sidedness of our sources. Altogether, research on consumerism in the Middle East still lags behind that of Europe and America.

Orosdi-Back: from a trading company to a chain of department stores

My study of the Orosdi-Back firm was triggered by research on the diffusion of sewing machines, a technological innovation, about which more below.[33] The store's name could occasionally be seen in advertisements but sounded puzzling for people unfamiliar with European family names; it was (and is) often misspelled in Turkish or Arabic.[34] For me it was a research challenge to investigate its history.

---

had a few Parisian-style *passages*, reminding us of Walter Benjamin's Aracades project, the outlay of the stores in most places was along main shopping arteries.

31 Usually called Guides Commerciales, Bulletins or Annuaires du Commerce, Renseignements, or Indicateurs and published in various bigger cities of the region by private parties or consular agencies. Logan Kleinwaks, Ottoman, Turkish Directories, https://groups.jewishgen.org/g/main/message/291598. The impressive supply of western luxury goods may also be seen from Orhan Koloğlu, *Reklamcılığımızın İlkyüzyılı 1840–1940* (Istanbul: Reklamcılar Derneği, n.d).

32 Abou-Hodeib, *A Taste for Home*, p. 149.

33 Chapter 2 is a reprint of my article on Orosdi-Back in Egypt. My book, however, deals with the company at large.

34 On the internet, which allows wider searches than I have undertaken, one may find various local corruptions of the name, e.g., Orozdibak (the firm itself sometimes transcribed the name in Ottoman-Turkish with a z), Orosdi Bach, Orozdy, but see also Orozdibek, Orosody Back, Uruzdy, Orisdi, Orizdi Oudodi, Oroze d'back, Orosdi Beick, Orosdi Pak, Pack or Bach, Oroz Dibak, and even the transcription Ursade Buck. Note that Bek sometimes led to the misunderstanding that it was an Ottoman form of the title *bey*, as spelled *bek*. Cf. also Işayağ, *Orosdi-Back Efsanesi (Horozdibeği)*, where the nickname *Horozdibek*, or Cock's Mortar, was a local pun derived from the store location's street name. The same applies to Omar Effendi, more formally correct 'Umar Afandi, but also found as Effindi etc.

The enterprise developed out of a ready-to-wear clothing firm set up in Galata, a district of Istanbul, in 1855 by a Hungarian refugee, Adolph Orosdi (magyarized from Schnabel). The two Jewish founding families Orosdi and Back were twice intermarried. Somewhat later, in the 1880s, a new business site called Ömer Efendi Han – for the Francophone readers advertised as Palais Omer Effendi –, became their main location in Istanbul (Fig. 2). It is noteworthy that in 1883, they still publicized themselves as a wholesale (*en gros*) firm, but soon combined this with a retailing business (*vente en gros et en détail*).[35] In 1888 they became a fully-fledged trading company, the *Établissements* Orosdi-Back, now headquartered in Paris.[36] Not only did the head office undergo a relocation from Central-Europe to France, its founders themselves settled in France and became part of the local upper bourgeoisie. Strikingly, the company remained in business for more than a century if we take the 1960s in Iraq and Morocco as their demise from the Middle Eastern scene. From the late nineteenth century, the multifaceted trading company gradually developed into a leading chain of department stores with outlets in eleven countries (according to present borders).[37]

Especially, their stores in Istanbul, Cairo, Beirut, and Tunis, could compare to European models. Though the Orosdi and Back family names have disappeared from the business scene, they live on in memories, nowadays prominently on the internet. The chain Omar Effendi itself survived as a brand in Egypt.[38]

---

[35] We have not been able to compare their French-language advertising strategy with the Ottoman-Turkish press which may contain more clues. This is equally true for the other media (Greek, Ladino, English, Italian, Yiddish etc.). Mayer and others advertised in German too, in the late Ottoman *Osmanischer Lloyd*. For an inspiring comparable study, see Sarah Abrevaya Stein, *Making Jews Modern, the Yiddish and Ladino Press in the Russian and Ottoman Empires* (Bloomington, Indiana UP 2004).

[36] The Statutes of 1895 defined their activities in the widest possible way: "L'exploitation et la création, en tous pays, de comptoirs et agences pour l'achat, la vente, l'importation et l'exportation de toutes marchandises et produits; la fabrication, le dépôt et le commerce à commission de tous articles ; les opérations de banque ; les entreprises immobilières et de travaux publics. Les transports par terre et par mer et en général toutes opérations financières, industrielles et commerciales". Paris, at the heart of their commercial strategy, gave their firm a French touch though they were of Eastern European descent, and their ready-to-wear clothing had initially come from there.

[37] In fact, Orosdi-Back still survived until as a shell company quoted on the Paris Bourse until 2014. Since the publication of my *The Orosdi-Back Saga* in 2007, former administrators have placed a series of financial reports (1895–1991) on the internet: www.entreprises-coloniales.fr/empire/Baratoux_Jules+Marcel.pdf. A commercial off-shoot seems still active in Beirut.

[38] Their reputation remained strong enough to be taken up in several places: e.g., by a fashionable bridal salon Orosdi Beck in Southfield MI, of which the owners answered my inquiry saying that they are Chaldeans; by an active book publisher in Sweden who (for unexplained reasons) not only used the name but even the historical logo with the elephant on a tricycle, but the firm was in

Fig. 2: A poster from app. 1928, printed by the Camis publicity firm in Paris, showing the Orosdi-Back store in Istanbul ("The Largest Stores of the Entire East"), partly in French, but in Ottoman-Turkish specifying the range of articles sold – whole-sale and retail –, from cloths, silk, flannel, lace, shirts, neckties, shoes, veils, ribbons, rubber band, handkerchiefs, perfumes, lavender, and small fashion goods for ladies, to purses, canes and parasols, and to glassware, cutlery, ceramics, iron bedsteads, watches and clocks. Note the Ottoman crescent, the crowd in western clothing, the delivery car (similar to the ones operated by Au Bon Marché in Paris), and the airplane (Leon Orosdi's son-in-law Emile Dubonnet was indeed a renowned aviator) (Source: Bibliothèque Nationale de France, by permission).

The expansion of the Orosdi-Back enterprise in the 19th century resembled a sort of spiderweb with its Parisian *siège social* in the center. (Fig. 3) This regional sprawl resembled smaller chains of foreign department stores, as well as stronger contemporary banks such as the Ottoman Bank.[39] In 1913, the firm proudly advertised with a list of purchasing offices in prominent manufacturing centers, each of which at the time could be associated with specific products. Places which had become sorts of reputed brands: Vienna, Manchester, Bradford, Birmingham, Lyon, Roubaix, Caudry, Milano, Barcelona, Chemnitz, Barman (Wuppertal), Gablonz, and La Chaux-de-Fonds.[40] The list of outlets comprises so-called port-cities, while their presence in inland cities indicates important railway nodes or strategic marketing points.[41] There were so-called *succursales* or *comptoirs* in Istanbul, Salonika, Izmir, Samsun, Adana, Beirut, Aleppo, Dayr al-Zur, Alexandria, Cairo, Tanta, Zaqaziq, Port Said, Tunis, Bizerte and Sfax. Earlier business in Plovdiv, Rusçuk, and Varna, and also in Bucharest, had proven a failure.[42] Even Yokohama and Kobe figure here, though their own Orba brand of Swiss-made watches in La Chaux-de-Fonds failed in Japan, allegedly owing to protective import duties.[43] Undeterred,

---

2016 taken over by Bonniers Förlag; by a restaurant Orozdi Beik in Beirut; and by others in the Middle East. A new E-commerce business in Iraq by the name Orisdi, which offers a broad range of products, admits that it is inspired by the once famous and popular department store, wanting to "reconnect with customers' memories of this well-known store", *Kapita News*, 18 July 2022.
39 See figure 3. On the whole, trade was one-directional, from Europe to the Middle East, though advertisements of the late 1910s and 1920s in the French and English language press of Turkey also publicize the export of Turkish and Perian rugs to Europe and America.
40 Vienna, for instance, could be associated with garments or furniture (possibly also Thonet chairs produced in what is today Czechia), Manchester with cotton goods, Bradford with woolen products, Birmingham with metalwork, Lyon with silk, Roubaix with textiles, Caudry with lace, Milan with textiles, Barcelona with calicoes, Chemnitz with knitwear, Barman (Wuppertal) with textiles, and Gablonz with artistic glass and crystal, and La Chaux-de-Fonds with watches.
41 Compare André Hauteman, *The Imperial Ottoman Bank* (Istanbul: OBARC, transl. 2002), p. 280 which had a similar but much larger network.
42 The terms appear to be interchangeable. *Succursale* means a permanent branch, and *comptoir* (counter in English) too denoted more than a trading post or a local agency. Most were real stores though we know little about some of the smaller ones. Interestingly, a historian of Mersin, H. Şinasi Develi. mentions a branch of "the famous Orozdibak" on the local Gümrük Meydanı (Customs Square) which we have not found in the official reports, https://www.yumuktepe.com/gumruk-meydani-h-sinasi-develi/.
43 Orosdi-Back closed their watch factory in 1909, and their agency in Yokohama even earlier in 1904. Whether they had expected to import Japanese products to the Middle East remains a matter of speculation, but Leon Orosdi's capitalist approach over the years was marked by a succession of closures and new adventures, including investments in other commercial branches such as the perfume industry, celluloid film, artificial silk, milk powder, banking, and the hotel business, see

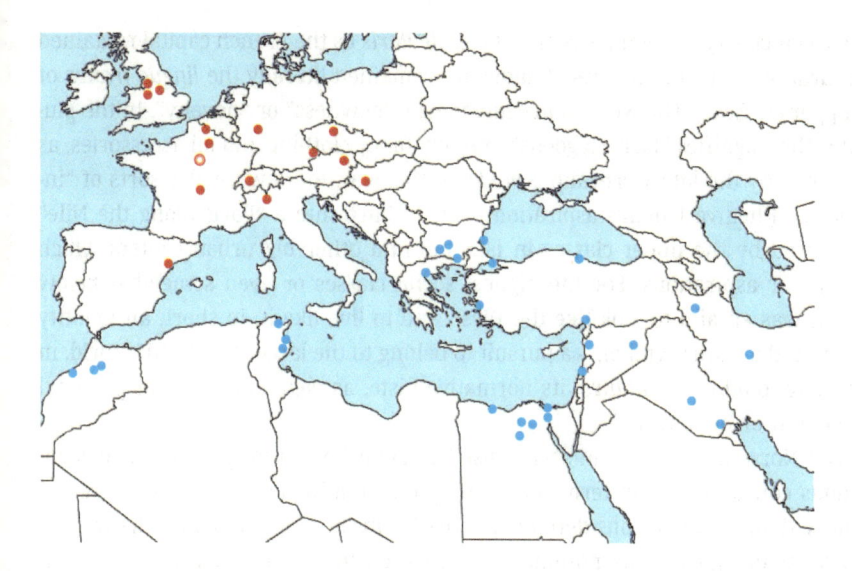

Fig. 3: Map showing Orosdi-Back's evolving commercial network between approximately the 1880s and the 1960s, with Paris as its headquarters, and the various purchasing offices (*maisons d'achat*) marked in red, and the outlets (*comptoirs*) in blue. Yokohama and Kobe in Japan fall outside this frame. The division clearly indicates the one-directional nature of the firm's international commerce (by author).

in one of their last stages, they tried to set up stores in Tabriz and Tehran, and as late as the 1950s, in Meknes, Fez, and Casablanca but these also did not survive. Their success fluctuated over the years because of economic setbacks, changing agricultural circumstances, hostilities, and fires, and ultimately nationalizations, but for a long time the quest for new opportunities went on.

So far, I have not been able to discover catalogues issued by Orosdi-Back, only advertisements, though we have a few indications that such publicity material may have existed, for instance, in Beirut in 1919.[44] If we had a series of catalogues, such as the big stores in France used to distribute, we might be able to follow the goods which were sold. After all, their merchandise changed and developed over time.

---

www.entreprises.coloniales.fr/empire/orosdi-back-18491895.pdf. Leon Orosdi's counterfeiting of cigarette paper (1898), and twice of perfume brands (1905 and 1922) led to law suits in which the company was convicted, "Leon Orosdi, Récidiviste en Contrefaçon", www.entreprises-coloniales.fr.

44 At least, the Parisian firm Au Printemps sometimes advertised (on several illustrated pages) in journals to encourage orders by post. In Istanbul, the G & A Baker store on Rue de Pera produced illustrated catalogues in different languages, http://www.levantineheritage.com/baker-catalogue.html (accessed 27 September 1922).

But we can clarify the emphasis on *articles de Paris* as the French capital remained a cultural beacon, and the French language remained literally the *lingua franca* of the upper classes.[45] The word *nouveauté* means "newness" or "novelty". In the plural it rather signifies "fashion goods", initially fancy clothing and all accessories, as prescribed by the latest prevailing *mode*. In this way, these were also sorts of "inventions." Khedive Ismail's aspiration to turn Cairo into a "Paris along the Nile" was shared by the upper classes in Istanbul and other big urban centers which had similar aspirations. For the highest social classes or even somewhat below them, it was an aim to look like the West, and to live like it, in short, an identity marker and a status symbol, – a pursuit to belong to the larger "civilized" world, in particular to adopt its *dhawq*, its normative taste, as well as a sense of personal satisfaction or hedonism.[46]

The Norwegian-American sociologist Thorstein Veblen long ago defined "conspicuous consumption" in terms of "using goods of a higher quality or in greater quantity than might be considered necessary".[47] In the Ottoman lands, the wish to acquire luxury goods was nothing new, since exclusive items such as Venetian glassware, Indian or French fabrics, or even Chinese porcelain, had for centuries been part of bilateral trade relations, and had entered the homes of the rich. Trends changed somewhat with newly imported commodities, and with the aspiration to compete with Europe and America

What had started in the mid-19th century as a business in ready-to-wear clothing gradually became a network of full-fledged department stores in the larger cities. This gradual transition corresponded with a growing awareness of novelties and increased demand: In addition, from about the 1880s, columns in cultural magazines on housekeeping and domesticity, on a modern outlay of upper- and middle-class homes with designated rooms, and on interior furnishing and decoration, as well as more frequent advertising (often with images) by department stores and luxury stores, had their impact on developing consumerism. The following decades

---

45 *Articles de Paris* are difficult to capture in one definition, see Todd, p. 131 ("marquetry and ornamental knickknacks"), or p. 136 ("endless incidentals and *et ceteras* of ornamentation and civilization, which can hardly be said to satisfy a previously-felt want, but create a desire for themselves..." as quoted from *The Economist* of 1886).

46 The notion *dhawq* for taste is pivotal in Abou-Hodeib's work see *A Taste for Home*, p. 126 *et passim*. Indeed, Todd, pp. 137–138, 151–156, 161, and 174, refers to the leading French role in a global commodification of "superior taste" and fashion.

47 Thorstein Veblen, *The Theory of the Leisure Class, an Economic Study of Institutions* (New York: Modern Library 1899).

also saw the appearance of didactic books on housekeeping and etiquette, as well as some related adaptations of school curricula.[48]

Ready-to-wear clothing, initially western-style dresses for women and costumes for men, soon developed into a wide array of accessories, from shoes, socks, neckties, hats and *bonneterie*, to imitation *bijouterie*s, and from umbrellas (assembled in Istanbul from imported parts) to travel goods.[49] The company perceived, for instance, of a commercial window to become – for some time – the largest of several suppliers of tarbushes (fezzes), the headdress which was a social marker of elite status for men. Oddly, but conforming to the industrial world of the time, over seventy per cent of the fezzes sold in the Ottoman empire, Egypt included, were manufactured in the Hapsburg-Austrian lands, later in particular by a syndicate in Straconitz, today Czechia. From 1898 Orosdi-Back held ts monopoly on sales in the Ottoman market.[50] For a firm with a French atmosphere, perfumes and cosmetics must have been an especially lucrative article. One such brand was sold under names such as Ramses and Ambre de Nubie, allegedly extracted from Egyptian aromatic herbs (Fig. 4).[51]

In addition to imported fashionable furniture (*mubiliyat*, from the French *meubles*) such as bentwood chairs and iron bedsteads, Orosdi-Back sold earthenware

---

[48] The term *tadbir al-manzal* (management of the house) became more applicable to the woman's tasks at home, see Mona Russell, "Modernity, National Identity, and Consumerism: Visions of the Egyptian Home,1805–1922", in Relli Shechter (ed.) *Transitions in Domestic Consumption and Family Life in the Modern Middle East* (New York: Palgrave Macmillan 2003), pp. 37–62; Fruma Zachs, "Cross-Glocalization: Syrian Women Immigrants and the Founding of Women's Magazines in Egypt", *Middle Eastern Studies*, vol. 50/3 (2014), and her "Gendered Reorganization in Late Ottoman Beirut: The Reciprocal Influence of the Domesticity Discourse and the Urban Space", in Yuval Ben-Bassat and Johann Buessow (eds.*) From the Household to the Wider World: Local Perspectives on Urban Institutions in Late Ottoman Bilad a-Sham* (Tübingen; Tübingen University Press 2022), pp. 151–155. For Istanbul, see Ersin Altin, pp. 42, 48, 111–112, and 117–118

[49] Şemsiyes (literally against the sun (*şems*), parasols too were a typical requirement for upper-class women. Orosdi-Back was not the only firm manufacturing them, e. g., Mayer sold them as well.

[50] Markus Purkhart, *Die Österreichische Fezindustrie* (PhD Thesis, Univ of Vienna 2006), pp. 165–166, 191–192 and *passim*. The researcher extensively describes the fez industry and refers also to the limited local production of fezzes in Turkey, Egypt, and elsewhere (with thanks to the author for putting the thesis at my disposal).

[51] Orosdi-Back did not shy away from marketing cheaper brands, which were in fact counterfeited. Leon Orosdi had financially supported a business founded in Paris in 1917 by a certain Ernest Coty and a Madame de Bertalot. This led to a court case with the much better-known perfumer François Coty, but the former's marketing continued till the mid-1930s. This was not the first time, because in 1905, Orosdi-Back had been convicted earlier for marketing a counterfeit perfume under the name "Mikado". See note 43 above.

vessels and porcelain services, silver cutlery, enameled objects, *quincaillerie* (a general term for ironmongery, metal utensils, and metal trays), and other European household items, travel goods, and children's toys, and, as we have seen, pocket watches with their own brand name.[52] As far as we could discern from a surviving

Fig. 4: Certain perfumes by the brand name Ramses with all its Egyptian connotations (Ambre de Nubie, and imagined Pharaonic workers) were marketed by Orosdi-Back from 1916 till about 1940 notwithstanding legal disputes, and even exported to the USA. The Ramses company was based at the Orosdi-Back headquarters in Paris (permission by Mrs. Grace Hummel).

---

52 Some articles began to be locally manufactured, e. g., European-like furniture, and also fezzes had been locally made.

Fig. 5: Interior of the large Orosdi-Back Istanbul store in Istanbul, probably in the 1910s, then still mainly a wholesale place, from an album made by the prominent Sebah photographers, kept at the University of Istanbul Library (Orosdi Back Şirketi Mağazaları).

series of interior photographs around 1910, clocks were an article permanently in stock, and there were also gramophones for sale (about which more below). In Tunis, Orosdi-Back issued photographic post cards of local tourist spots. Today, some of these occasionally surface as auction items on the internet.[53]

For the trade firm, which Orosdi-Back was, "stocks" (or *dépôts* in French) was a critical notion, as can be understood from their annual reports to the sharehold-

---

[53] Sometimes unexpectedly we can re-construct the popularity of material culture objects made and also sold by Orosdi-Back: Lorans Tanatar Baruh," Tracing the Painted-Tray Dealers in Istanbul, a Commercial and Spatial Reading" in Flavia Nessi and Myrto Hatzaki (eds), *Rituals of Hospitality, Ornamented Trays of the 19th Century in Greece and Turkey* (Athens: Melissa 2013), pp. 149–163. Such trays, no longer used as tables but to serve refreshments, were ornamented according to local taste.

ers.⁵⁴ This was an interface between imports and bulk sales to big merchants in the hinterland.⁵⁵ The above-mentioned series of photos from Istanbul still suggests a wholesale environment, rather than a department store (fig. 5). In any case, no women are seen on the photographs. The evolution of Orosdi-Back into the European sort of department stores which had then just begun, occurred over time.⁵⁶ In Istanbul, their location at Bahçe Kapı in an older merchant environment, that was neither in the grand covered bazaar, nor at the sumptuous Grande Rue de Pera where most of the other larger department stores were found, was a conscious choice. In Cairo, they settled for the Muski district (sometimes called Rue des Francs), and only later built their sumptuous flagship store, not far from the busy ʿAtaba square. Cultural geographers might still wish to say more about each of their localities anywhere.

The above "stocks" were a vital commercial concept for Orosdi-Back. In several places, the Orosdi-Back company acted as depots for certain brands.⁵⁷ This was the case for Singer sewing machines which subsequently led me to research this particular company. I note, in passing, that while the focus of my research had been on the personal rather than the industrial use of sewing machines, I did realize that Orosdi-Back, as well as similar firms, owed their commercial success to the mass manufacturing of ready-to-wear clothing for which the then innovative sewing machines had been essential. As clothing entrepreneurs they might even have felt a special affinity to them, but they anyway undoubtedly seized all businness opportunites. Though recognizing Singer's activist, if not exclusive, marketing strategies, we have no particulars on how that system of commercial representation actually worked. However, it made me aware that mechanical goods or consumption durables, or "small technologies" as we will call them, were no less at-

---

54 İşiyağ's book on Adana, p. 36, carries an interesting Orosdi-Back advertisement from the First World War by which the "Anglo-French Department Store" (sic) offers "military articles of all kinds". The advertisement is in English and may have been published elsewhere: Adana was then under French occupation.
55 Gad G. Gilbar has emphasized the role of big local Muslim merchants over minorities and foreigners as figuring in long accepted but biased sources, but this does not necessarily contradict. It seems that all sorts of commodities, were bought from Orosdi-Back storehouses as an interface by inland merchants.
56 Kupferschmidt, *Saga*, pp. 61–66. The series is kept by Istanbul University. A rare image on the internet from Baghdad clearly suggests an interior outlay closer to conventional department stores. E.g., see Fig. 6. Often called the "Harrod's of the East" they catered to British taste.
57 Whereas Orosdi-Back acted as an agent for Singer from the 1880s, the latter opened its own store in Istanbul in 1904, and soon elsewhere as well.

Fig. 6: Photo of the interior of the Orosdi-Back store in Baghdad, possibly from the 1940s, with the modern outlay of a conventional department store (courtesy: Dr. Amer Hanna Fatuhi, USA).

tractive novelties and status symbols of a would-be western society than fashionable clothing or household goods.[58]

---

58 From time to time, antique dealers offer for sale Singer sewing machines with the painted label "Orosdi-Back & Cie. Paris (Marque Deposé)" which must date from the 1880s or later. A metal vignette with an anchor, nailed on it, refers to the Anchor Mills in Paisley, Scotland whose yarns and threads were an important accessory also sold by Orosdi-Back. Seidel & Naumann machines were marketed by competing firms such as Em. J. Mertzanoff at Bahçe Kapı, Old Istanbul, and Demetrius Persides at Yüksek Kaldırım, Galata (see also Yağız and Ağır, note 60 below, pp. 45–46).

Fig. 7: The Orosdi-Back department of musical instruments in Istanbul, showing a clarinet, harmonium, accordion, guitar, and gramophone (University of Istanbul Library collection, Orosdi Back Şirketi Mağazaları).

A photograph of Orosdi-Back's music department in Istanbul shows mostly western instruments such as a harmonium and a guitar, but also a gramophone, which at the time was also considered as a musical instrument (Fig. 7). Remarkably, the Egyptian branch issued self-commissioned recordings of Arab music, e.g., of the famous Egyptian singer Yusuf al-Minyalawi, which can be considered as an adaptation to local taste. This is interesting because it indicates that not all local customers were geared to Beethoven or Wagner and other western music. In this respect, clocks and the company's own brand of pocket watches are no less fascinating because by the end of the 19th century, personal watches were still set to prayer time and not the other way round. (Fig. 8) Turkey before 1910 still maintained a double standard of time measuring but his did not mean that watches were useless.

Fig. 8: The Orosdi-Back department of clocks and watches in Istanbul (University of Istanbul Library collection, Orosdi Back Şirketi Mağazaları).

Orosdi-Back apparently also sold *vélocipèdes*, forerunners of modern bicycles, from the late 1880s onwards.[59] At some stores, doorbells, binoculars, small electrical appliances, and later, in Iraq, even cameras were offered. Such mechanical goods are less volatile and less subject to fashion than clothing and may be easier to define as necessities. This does not mean that they are easier to track because the firm not often advertised with images.[60] But the latter objects, with their im-

---

[59] Yavuz Köse, "Bicycling into Modernity in the Late Ottoman Empire: Ahmed Tevfik and his Bicycle Travelogue" in Ebru Boyar and Kate Fleet (eds.), *Entertainment Among the Ottomans* (Leiden: Brill 2019), p. 186 quoting Gökhan Akçura's book *Evvel Zaman Bisiklet*. Interestingly, the Orosdi-Back logo was an elephant on a tricycle (Fig. 9).
[60] Other firms did advertise with images. Leon Hazarossian at Rue de Pera who, among other articles also sold gramophones, and lighting fixtures, advertised its special bicycle department with illustrations, see Burcu Yağız and Aygül Ağır, "XIX. Yüzyıl Sonu İstanbulu'nda Batılı Tüketici Ürünlerinin Dolaşıma Girdikleri Kanallar ve Yarattıkları Hareketlenmeler: Şark Ticaret Yıllıkları Üzerinden Bir Araştırma", *Aralık* 2017, p. 43.

pact on society, as distinct from clothing and home furnishings, shifted my research into the direction of new mechanical goods, some American produced, in short, technologies.

Technologies as consumer goods, big and small

Whereas the word "technology" has a variety of meanings, among them techniques in the sense of skillfully performing a task, we mean here just tools or appliances to do something. Simply defined as such, technologies are of all times and all sizes, – even a pencil for writing or a needle for hand-sewing is a technological tool. But a technology is more than a mere object. We tried to place their use, or rather their usefulness, and their impact, in a broader human and cultural context.[61] Chapter Two was published in 2015 as an exploratory essay into the acquisition of new mass-manufactured mechanical devices and their use.[62]

Fig. 9: The somewhat enigmatic but eye-catching trade mark of Orosdi-Back: an elephant on a tricycle. Likely, they did sell tricycles for children, but as a real circus act it was probably known only from magazines or books. In advertisements it was often accompanied with the slogan "The Ultimate Luxury" (permission: ProMark, Paris).

---

61 My idea on biographies of things was to some extent inspired by Igor Kopytoff, "The Cultural Biography of Things: Commoditization as Process" in Arjun Appadurai (ed.), *The Social Life of Things, Commodities in Cultural Perspective* (Cambridge: Cambridge University Press 1986), pp. 66–68.

It is true that light bulbs, electric household implements, or automobiles could not function without a network or system which was usually supplied by the government or the rulers, in other words these were infrastructural "big technologies". But, at the same time, most of the small new technologies such as gramophones or cameras were not imposed by governments or rulers. They represented individual wants or desires, and ultimately "needs", and hence were a matter of consumer trade. Whether we call them "small technologies", "foreign objects", "new commodities" or "consumer durables", they were adopted, acquired, or appropriated not merely by the upper classes, including foreign residential elites, but gradually also by the upper middle strata. This happened more or less voluntarily, by the "agency" of those who appropriated them, and of course to the extent that they possessed or developed the financial means to do so.

We have suggested that some of the new technologies were not objectively needed for the subsistence of those who acquired them, but for reasons of imitation or prestige. If there was nevertheless an element of coercion to buy these goods, it was a need motivated by social competition (in their imagination, with the west, or with other locals), thus it became a matter of modern identity and status, "to be like them".

While most American innovative durable consumer goods of the 19th and 20th centuries (some of which were later manufactured in Europe) evolved in a sequence of trial and error on the market, the Middle East gradually imported most of them, one by one, albeit not all of them to the same extent, as we will see.

As a working assumption, we have devised a distinction between "big" and "small" technologies. By our definition, it is not the physical size of a new technological device that matters, but the relative ease by which a local person could purchase and install it, mostly at home for his or her own use and benefit. As examples of "big" technologies, one may count railway systems, telegraph networks, full-scale electrical grids, tramways, large hydraulic agricultural projects, modern urban planning and construction sites, and the like, which were usually imported and operated by colonial regimes, foreign entrepreneurs, and investors, or by dependent local groups. Often, they arrived with a relatively short time lag after they had been installed in Europe or America, and "turn-key", or in a few cases, accompanied by (invited) foreign experts or technicians. As a rule, such "big" technologies could not be acquired by local people privately.

True, over time, the "big" could become "small" in one particular sense, as their use became affordable to wider population strata, but this did not change their ownership and exploitation. For instance, by the turn of the nineteenth cen-

---

62 Some imperfections have been corrected or updated in our later chapters.

tury, "common people", – admittedly a problematic term –, could use tramways, and travel on (mostly third class) trains, or dispatch messages by telegraph to relatives, trade partners, or even to the authorities. But this did not turn the tramways or telegraph into small technologies.

This ambiguity equally applied to the technology of printing in the Middle East, a topic not dealt with here in depth because it had arrived earlier and had established itself in stages. Even though Jewish printers had been active from the 15th century in Istanbul, and certain Christian churches had installed presses from the 16th century onward, followed by a Hungarian convert, the entrepreneur Ibrahim Müteferrika in 1723.

Notwithstanding the printing projects of Napoleon and subsequently of Muhammad ʿAli in Egypt at the beginning of the 19th century, we consider printing primarily as a "small" technology. Private entrepreneurs and other parties who used printing presses, mostly acquired them to further their own particular interests.[63] The causes for this delayed diffusion, as compared to Europe, are still debated, as they seem rooted in a variety of cultural reasons. Even when the press went through a steep general process of acceptance in the 19th century, it could still be called a "small" technology.[64] Nile Green has described the purchase of second-hand Stanhope hand presses at the beginning of the 19th century by Levantine individuals. At that time, their price was advantageous due to the fact that in Britain, steam-powered presses, as a newer technology, had come into use around that time, and their earlier prototypes were then disposed of.[65]

An ambiguity in the reverse applies to military armaments which were essential to Imperialism and conquests. In the case of the Ottomans, they acquired equipment from the West and even imitated it. A range of such big military tech-

---

[63] Printing by and for Muslims in the region also evolved slowly as compared to Europe (and China), at least until the 18th century, but in the following century it picked up rather quickly.
[64] Ami Ayalon, *The Arabic Print Revolution, Cultural Production and Mass Readership* (Cambridge: Cambridge University Press 2016), esp. pp. 1–17. This applies also to the Ottoman lands.
[65] Nile Green, "Journeymen, Middlemen, Travel, Transculture, and Technology: The origins of Muslim Printing", *International Journal of Middle Eastern Studies*, vol. 41/2 (2009), p. 212. See also his *The Love of Strangers, What Six Muslim Students Learned in Jane Austen's London* (Princeton, Princeton University Press 2016), p. 301. However, in Iran, Fourdrinier paper making machines which had begun to spread in America and in Britain in the same period, remained a "big" technology for a long term. Moreover, in Iran this innovative machine was delayed by inadequate industrial know-how as well as by the lack of waterpower. See his "Paper Modernity? Notes on an Industrial Tour, 1818", *Iran*, vol. 46 (2008), pp. 277–284. In the center of the Ottoman Empire, however, this had been different due to a good ecological reason, the abundance of water: Paper suitable for mass printing had been produced from the 15th century onwards, and even improved from Mahmud II's time onwards for printing.

nologies were adopted and locally manufactured.[66] On the other hand, many small arms, from rifles to revolvers, were acquired among various populations, including rural and nomadic people, as symbols of power, privilege, or prestige.

One additional elucidation on "small" technologies may be relevant here. Though their actual size may differ, most of them are relatively easy to move and to handle, even if they need some basic familiarization and training.[67] The consumer can use a sophisticated mechanical instrument without requiring a profound knowledge of its inner working, as Bruno Latour has proposed with his term "black-boxing".[68]

Thus, "small" technologies stand in contrast to "big", a term which I first suggested in my study of the sewing machine in 2004 (here Chapter Three) to reflect the "agency" of voluntary acquisition and personal use.[69] My first inspiration to investigate such small technologies came from the Sreberny's book *Small Media, Big Revolution*, which harked back to Wilbur Schramm's similar distinction between small media (radio) and big media (television). These were small or horizontal in the sense of not being controlled by the State from above. Their study showed that during the critical episode of the Islamic Revolution in Iran, simple cassette recorders and fax machines in the hands of the people had been crucial notably in ousting the Shah's regime's apparatus.[70] Since then, newer small technologies have deeply impacted our societies, but our aim is to understand examples from the more distant past.

However, most of these small technologies, personal goods and household appliances – from sewing machines and typewriters to radio receivers, electrical appliances, photo cameras or automobiles –, had neither been primarily aimed at en-

---

[66] See for instance V.J. Parry and M.E. Yapp (eds.), *War, Technology and Society in the Middle East* (Oxford: Oxford University Press 1975).
[67] Sewing machine manufacturers (e.g., Singer) did provide training for customers. Various typing courses were given all over the region; and the use of musical instruments also needed lessons.
[68] Bruno Latour, *Pandora's Hope: Essays on the Reality of Science Studies* (Cambridge, Mass., Harvard University Press 1999), p. 304. Black-box, in the sense of a camera obscura, is meant metaphorically for a complex modern camera, or another technology of which the intricate workings do not need to be known to the user.
[69] In this way, small technologies could be clothing, buttons, or forks, in medieval times, or paper clips, pins and zippers in the 19th and 20th centuries, see for instance Chiara Frugoni, *Le Moyen Âge sur le Bout du Nez, Lunettes, Boutons et Autres Inventions Médiévales* (Paris: Les Belles Lettres 2011, transl. from Italian ed. 2001), pp. 133–134, and 137; Henry Petroski, *The Evolution of Useful Things* (New York: Vintage Books 1994). Also, the technologies discussed in Chapter One, including automobiles which are larger in weight or dimension.
[70] Annabelle and Ali Mohammadi *Small Media, Big Revolution: Communication, Culture, and the Iranian Revolution* (University of Minnesota Press, Minneapolis 1994).

hancing Imperialist or Colonial power or economic exploitation, nor – to the contrary – been meant to fight it. More often, I would say that as symbolic and social markers, they indicated compliance with contemporary western technological supremacy.

Nevertheless, my main contention is that these small technologies which came from Europe or America were not all accepted and integrated with the same enthusiasm. The processes of adoption and diffusion differed from country to country in extent and pace, in their cultural settings and occasionally in their adaptations, or in purchasing power and preferences of one population segment to another. Indeed, new technologies could be speedily adopted at one place, while other or parallel innovations were delayed, or in exceptional cases, even rejected. The growing scholarly literature on the diffusion of various modernizing technologies proves that their adoption and impact varied according to country, class, gender, or occupational sector.

Similar perceptions of such small technologies have been suggested by researchers who dealt with other parts of the world, where such new personally acquired material goods equally affected sizeable sections of the population. Frank Dikötter in his inspiring *Things Modern* (2007) has used terms such as "creative appropriation", "inculturation" or "indigenation of modernity" in order to describe the adoption of foreign commodities by individuals in China (2007). The Chinese even began to manufacture some of these appliances locally. In many cases, this was not only a striving of the upper classes for foreign modernity and prestige, but also a more general dynamic process of diffusion of new consumer goods.[71] Andrew Gordon published his book on the impact of sewing machines on manners of dress, and society in general, in Japan, in 2012.[72] Jean Gelman Taylor worked on similar lines for the Dutch East Indies (Indonesia), as did Arnold J. Bauer for Latin America (2001). [73] David Arnold's perceptive *Everyday Technology: Machines and the Making of India's Modernity*, in particular, analyzed how imported new technologies, namely sewing machines, typewriters, bicycles, and rice mills became inte-

---

[71] Frank Dikötter, *Things Modern, Material Culture and Everyday Life in China*, (London: Hurst 2007).

[72] Andrew Gordon, *Fabricating Consumers: The Sewing Machine in Modern Japan* (Los Angeles: University of California Press 2012). The sewing machine there was called *mishin*.

[73] Jean Gelman Taylor in her "The Sewing-Machine in Colonial-Era Photographs: A record from Dutch Indonesia", *Modern Asian Studies*, vol. 46/1 (2012), pp. 71–95; Arnold J. Bauer, *Goods, Power, History: Latin America's Material Culture* (Cambridge: Cambridge University Press 2001).

grated in daily life, and how each generated changes in the complex Indian society.[74]

Invention and in-use

The notions of invention, and innovation, or novelty, in my view, do not always go together.[75] For many of the small technologies, the term "invention" is less appropriate because they were merely a critical breakthrough in a chain of development, by trial and error, and sometimes occurred by chance. It reminds us that we cannot pinpoint the origin of the reading glasses on one "inventor" in Pisa in 1287. And with due respect, even Gutenberg's movable type press in 1453 was preceded by several experiments and is therefore also not undisputed.[76] Bartolomeo Cristofori was once considered as the eminent inventor of the piano-forte in 1700, but recent research has shown that it had a long genealogy of harpsicords and clavichords. In the nineteenth century, Joseph Nicéphore Niépce's camera (1816), Elias Howe's lockstitch sewing machine (1846), Christopher Sholes' typewriter (1868), Alexander Bell's telephone (1876) and Thomas Edison's cylinder phonograph (1877) each had its own forerunners. Formal dates of registration as patents, moreover, often serve as formal landmarks of invention, but diffusion is a different matter and less easily followed. [77]

---

[74] David Arnold, *Everyday Technology: Machines and the Making of India's Modernity* (Chicago, Chicago UP 2013)). Arnold has also used the term *indigeneity* which, indeed, might suit colonial India as a country which itself later manufactured sewing machines, and typewriters, see his "Global Goods and Local Usages: the Small World of the Indian Sewing Machine, 1875–1952", *Journal of Global History*, Vol. 6/3 (2011), pp. 407–429.

[75] The conventional use of the term "invention" may date back to the 16th century and denotes the proof of feasibility of something totally new, mostly a mechanical device. Innovation implies production, use, and practice. Novelty rather refers to quality, or visible and technical specifications.

[76] In Haarlem, the Netherlands, it is assumed that Laurens Janszoon Coster preceded Gutenberg by a decade or two. New insights on an earlier advent of blockprinting to Western-Europe have been put forward by Kristina Richardson, *Roma in the Medieval Islamic World, Literacy, Culture and Migration* (London: Taurus, 2022), pp. 103–126.

[77] The international registration of modern patents (in Berne) reflects an inherent, even biased, inequality between the West and outlying countries. According to my computation for 1893–2005: in the USA 6.687.653 were registered, in Japan 3.302.801, in France 2.269.242, in the UK 2.180.412, in Israel, a late starter, 80.426, and then in Turkey 35.771, in Iran 22.438, in Egypt 18.125, in Lebanon 4030, and in Saudi Arabia a mere 69! This remained true for the year 1975: in Egypt 396 new letters of patent were registered (of which only 18 were by Egyptians), while the total figure was 36.992 for Japan, and 46.603 for the US, see Dorothee Greans El Sirgany, *Les Brevets d'Invention en Egypte* (Cairo 1978).

The French historian, Jacques Le Goff, rightly observes in his introduction to Frugoni's book on *medieval* inventions that "Les inventions que présente Chiara Frugoni non seulement ne sont pas datées exactement pour la plupart mais le temps de leur diffusion qui est le vrai temps de leur naissance dure plus ou moins longtemps."[78] Similarly, David Edgerton proposed not to give undue emphasis to invention but to look at the first utilization of a particular technology.[79]

For our discussion of diffusion, we accordingly suggest a difference between two markers, namely of **in**vention of an object or a technology, and its becoming **in-use**, in other words, becoming a consumer item. We have not been able to survey the host of 19th and early 20th century magazines in Turkish, Arabic, and other languages which perceptively used to notice new inventions or patents, relying on European magazines, which announced their advent and printed commercial advertisements. With more digitization it will be feasible to do this in the future. Between the two markers, in the modern era, time was significantly shortened due to faster communication and transport, as well as by commercial interests.

There remains the question of personal use as a marker of diffusion.[80] What interested me was this process, in particular of some of the new "small" technologies of the late 19th and 20th centuries in the countries of the Middle East.[81] They were no less *nouveautés*, and not merely in the sense of fashionable clothing or household goods. Maybe consumer durables which are less frivolous are also easier to keep track of. True, there were vast differences in the need, usefulness, and affordability between people who wanted sewing machines, typewriters, cameras, pianos, gramophones, incandescent light bulbs, electrical household appliances, and automobiles, more or less in the same era.

---

78 Frugoni, p. xii
79 David Edgerton, *The Shock of the Old, Technology and Global History Since 1900* (London: Profile 2006), pp. 31–32.
80 While photography became known in Europe from the 1830s, the first foreign traveling and local studio photographers were already active in the Middle East since the 1850s, and as Kodak revolutionized photography with its hand-held cameras in 1888, accelerating amateur photography, we still do not have any systematic survey of the actual possession (and use) of cameras by non-professional persons in the region.
81 Reading another inspiring study, namely Carolyn Marvin's *When Old Technologies Were New* (Oxford, Oxford University Press 1990) on the impact of newly introduced devices of the late 19th century such as the telephone, the gramophone, electric light, and the cinema, it became clear to me how new technologies affected millions of lives within a relatively short time.

Diffusion: the Everett Rogers model

Arguably, the best-known theoretical model of diffusion had been proposed in the 1960s by the American sociologist, Everett Rogers. His book, *Diffusion of Innovations* has since gone through no less than five updated editions (5th ed. 2003). We do not intend to go into the shortcomings and biases which others have found in his model but consider it worthwhile to examine its applicability to the non-American world, and the Middle East in particular.

Rogers' deterministic timeline, starting from 2.5% innovators, and 13.5% early adopters, then gradually building up to a 34% early majority, and a 34% late majority, leaves us with a 16% remainder of laggards. His proposition of an S-curve consisting of four distinct stages called "Knowledge", "Persuasion", "Decision", and at last "Confirmation" (or "Implementation") is somewhat more appealing.[82] But none of these can be substantiated by hard data from our region for any of the objects we discuss. As so often, models which are based on the American experience of production management, marketing strategies, commercial publicity, and inherent systematic statistics, cannot be simply copied to suit the countries of the Middle East.

The conventional way would be to have a look at production statistics (state initiated, or published by individual manufacturers), at export and import reports, customs data, or at household surveys. However, what is relatively easy in the USA or Europe, is a problem in the Middle East because most of such data are deficient or lacking, at least for the era we are covering. Let me give one example. The President of the French Chamber of Commerce in Istanbul estimated in 1910 that the total imports of typewriters had recently gone up from 180 to 400 per year. But even relevant trade statistics are confusing as they may reflect random monetary values, prices, and tariffs, and if at all specified in this case, typewriters are given only in overall weight and value of entire batches. We may guess that a machine weighing about 25 kilograms or based on a price around $100–125 in the USA, was sold for around 24–28 Turkish Liras, or approximately 23 French Francs, in Istanbul.[83] We frequently lack truly national data, which began only in the 1880s, or we know only the port through which they entered, or at the most its immediate hinterland. It does not help us assess demand and actual use, so we do not know who and where the buyers and users may have been. For most common consumer goods, we lack household surveys from the era we are dealing with. For 1900,

---

[82] Everett M. Rogers, *Diffusion of Innovation*, one of his first examples pertains to typewriters (NY, 5th Ed. 2003) pp. 8–11.
[83] Chambre de Commerce Française, *Revue Commerciale du Levant* (1910), p. 169. Exchange rates are not really helpful if we have no data on average income or purchasing power.

we have some limited data of global exports or production of some small technologies, which can give us at least a relative idea of the backlogs (or "chasms") regarding their diffusion in the Middle East.

"Differential diffusion"

The idea that there were chasms in the process of diffusion is not new. [84] In the more distant past, inventions in Europe did not immediately become known elsewhere. Fatma Müge Göçek, in one of her books, *East Encounters West*, (1987) observed that for the Ottoman Empire there was a differential process of adoption. Firstly, mechanical clocks had little use in societies which lived by a different temporal culture based on sunset and sundown time, and European clocks could therefore be only exhibited for their decorative value.[85] Printing, her second example, was slowed down for over three centuries owing to various inhibitions, as we have mentioned before. Her third example shows how superior European weaving techniques were for some time barred by the Ottoman authorities, while they tried to improve their own.

Another critical point concerns the diffusion within the diverse regions of the Middle East. The strong bias in favor of a few major cities in the Middle East can be illustrated, for instance, by Nancy Micklewright's article on the adoption of European fashions in Istanbul and Cairo. [86] Hazel Hahn, too, who has examined consumer culture in various continents and cultural regions, widely defined geographically, has highlighted only three specific major urban centers for the Middle East: Istanbul, Cairo, and Beirut, while she has discussed other regions in the world as a whole.[87]

In the past decades, more detailed research on smaller centers has enhanced our knowledge of cosmopolitan port cities such as Salonika, Izmir, or Alexandria, as well as the Eastern Mediterranean and also Aleppo (though technically speaking

---

[84] Jared Diamond in his *Guns, Germs, and Steel* (NY: Norton 1997) pp. 247–249, has argued that not all societies are equally receptive to all innovations, ascribing this to relative economic advantage. social value and prestige, compatibility with vested interests (where he discusses the Qwerty keyboard), or ease of observation.

[85] Avner Wishnitzer, *Reading Clocks, Alla Turca: Time and Society in the Late Ottoman Empire* (Chicago: University of Chicago Press 2015), pp. 45–67 *et passim*.

[86] Nancy Micklewright, "London, Paris, Istanbul, and Cairo: Fashion and International Trade in the Nineteenth Century", New Perspectives on Turkey, vol. 7, Special Issue: The 1838 Convention and its Impact, Spring 1992, pp. 125–136.

[87] We are aware that our cursory remarks on Iran do not do justice to Iran with its fairly large cities such as Isfahan, Tehran, and Tabriz.

not a port city). However, the fact remains that most of our historical and literary sources, and not only the trade data, pertain mainly to the largest cities.[88] The gaps on the map of diffusion of new articles and technologies are not easy to fill in as far as smaller urban centers (capitals, port cities, railway nodes, and even the second or third range urban centers) and the rural peripheries are concerned. This applies all the more to the break-down into various urban classes, ethnic, confessional, professional, and other segments of the population.

The most wanting aspect of Everett Rogers' model for application to the Middle East is the gradual trickling down of new consumer goods from the highest classes to the lower echelons, whether we call these an aspiring or "probable" middle class, or a "middling group". Of course, class formation also differs from place to place.[89] The increasing impact of "common paraphernalia of daily life" as one historian has called it, is not so simple. Nor can we call it a "democratization of consumption" because not all the novel small technologies were necessities which had to be shared by all classes. We have only scant evidence of the considerations and motivations which advanced, moved, or delayed this process, as far as the "consuming knowledge" and persuasion of the consumers is concerned, not to speak of less exposed rural areas which lagged far behind.[90]

The stage of Persuasion mostly did not start from an immediate existential need but as a desire for social status and prestige. This is worth noting because diffusion connects with identity, as well as with an aspiration towards upward mobility.

Without the full awareness of the novelty consumer goods, and of the small new technologies in particular, these early "adopters" (Rogers' term) would hardly have had a chance. But among the literate public, knowledge of the physical arrival of new inventions often preceded their local availability.

We therefore ascribe a strong impact to the cultural and scientific periodicals in various languages as these gained a circulation of many hundreds and later thousands in the second half of the nineteenth century. These figures may not seem too impressive but must be multiplied as they were sometimes read out in cafes and other public places. Not a few were short-lived and rarely preserved, which makes systematic research more difficult. But some, such as *Servet-i*

---

88 Hahn's article "Consumer Culture" is an ambitious attempt to survey different geographical and political regions of the globe, but the Middle East comes under the joint sub-heading of the cities of Cairo, Istanbul, and Beirut, pp. 402–403.
89 Abou-Hodeib has connected new preferences and tastes to the emerging middle class as a trendsetter in Beirut, *A Taste for Home*, p. 38 et passim.
90 Heather J. Sharkey, *SCTIW Review* of Liat Kozma et al., *A Global Middle East, Review* in the *Journal of the Society for Contemporary Thought and the Islamicate World*, 28 April 2016, p. 4.

٥٠٢

الآلة الكاتبة العربية

باب السؤال والاقتراح

الآلة الكاتبة العربية

﷽ مونتريل كندا ﷽ جرجس افندي جرجوره حنا
وعدتم في الهلال السادس عشر من السنة التاسعة انكم ستنشرون صورة الآلة
الكاتبة العربية وتبينون مزاياها فهل لكم ان تنجزوا الوعد

صورة الآلة الكاتبة

﷽ الهلال ﷽ الآلة الكاتبة ويسميها الافرنج Typewriter من اختراع

Fig. 10: The magazine *Al-Hilal* (vol. 13, 15 May 1904) published for the first time a four page description of the Arabic typewriter as invented by Idris and Haddad, and manufactured by the American Writing Machine Co. in New York, a Remington competitor who later merged with it.

Fünun, Mecmua-i Ebüziyya, Tercüman-ı Ahval, or *İkdam*, published in Istanbul, or *al-Muqtataf, al-Hilal*, and *al-Lata'if al-Musawwara* in Cairo, were amply illustrated with engravings and later with photographs. Discussing a variety of social issues and conditions in the West, editors and contributors strove to be in the forefront of reporting on new inventions, and thereby aroused curiosity, or a desire to acquire the new technologies.[91] Novels and travelogues, too, added awareness and curiosity. There were also youth magazines which propagated the sale of new technologies, with photographs of children inspecting typewriters, automobiles, cameras, and the like.[92]

Advertising in the press by importers and retailers was, by definition, a complement of consumerism. But it is impossible to gauge the exact impact of advertisements as a stage of "Persuasion" (modern polls and surveys came much later).[93] At a certain point, sewing machines and much later washing machines were explicitly advertised as a need, but we do not know how much response this generated.[94] Other forms of commercial publicity were billboards or wall advertisements, showrooms, and the windows of commercial establishments in major shopping streets such as La Grande Rue de Pera in Istanbul, or 'Imad al-Din Street and even the Muski area in Cairo. This was the case also in similar shopping areas in other large Ottoman cities.[95] Obviously, the press reached only a segment of the total population, notably the literates, while lower classes with a "threshold fear" did not even venture to come there. But, no doubt, the consuming public was expanding.

Some of the commodities I discuss can be graded: This is particularly true for our specific case of eyeglasses. We found that cheaper types were sold by peddlers and hawkers, and more expensive brands in fashionable stores. Peddling and second-hand trade (of clothing, for instance) must have been quite common, but our

---

91 See a special 280-page presentation volume of 1930 by Hana Khabbaz (ed.), *Mukhtar al-Muqtataf*, reprinting older articles on science and technology.
92 Heidi Morrison, "Nation Building and Childhood in Early Twentieth Century Egypt" in Benjamin Fortna (ed.), *Childhood in the Late Ottoman Empire and After* (Leiden: Brill 2016), p. 78.
93 A standard work is Orhan Koloğlu, *Reklamcılığımızın İlkyüzyılı 1840–1940* (Istanbul: Reklamcılar Derneği 1999). See further studies by Relli Shechter, Fruma Zachs, Martin Strohmeier, Yavuz Köse, as well as by Roni Zirinski for a later period.
94 Abou-Hodeib, *A Taste for Home* pp. 154–155; Ersin Altin, p. 137, quotes Singer advertisements in Istanbul from 1898 to the effect that a sewing machine was a "necessity and requirement for all ladies".
95 Orosdi-Back and competing department stores were frequent advertisers in all larger cities with a local press. On the periodical *Ahenk* ("harmony") in Izmir see Elif. C. Nelson, "Advertisement at Izmir Press During the Early 20th Century", *Tarih İncelemeleri Dergisi*, vol. 34/1 (2019), pp. 168.

knowledge on it is scanty. Imported household items appeared in a middle range of marketing but pianos were, of course, in the highest echelon.

Delays and resistance

While the diffusion of fashionable *nouveautés* in certain classes and environments sometimes met cultural objections, some specific new small technologies could hardly be blocked. However, their adoption could be delayed for one reason or another. I have pointed out (in Chapter Two) that the introduction of motorcars was initially dependent on proper roads and services. Bicycles too may not have been suitable in all countries, owing to natural conditions such as mountainous or rocky terrains, but they certainly belonged to small technologies as Yavuz Köse remarked in his fascinating article on late Ottoman recreative cycling, and I should, indeed, have included them.[96] Around 1907 the Egyptian satirical writer Muhammad al-Muwaylihi was still poking fun at a judge riding a bicycle, but the fact is that they had meanwhile become popular in Egypt too, as one could gather from images of young men (by exception also women) making long cycling trips as reported by the illustrated journal *al-Lata'if al Musawwara*.[97] Likewise, it can be argued that electrical devices, especially washing machines, were held up by a lack of vital infrastructures.[98] But some of these hurdles existed also in other parts of the world. What interests me here are variants between the introduction of these small technologies in the West and in the Middle East (broadly speaking), in particular, any reservations, hesitations, and objections which differed from the so-called chasms which have been identified for the American commercial market.[99]

---

[96] Köse, "Bicycling", p. 190 n.41, as well as Ersin Altin, 173–177. Note the difference with India, which had a fairly good road system; there, bicycles were not only imported for recreation but also for the police and the postal services, Arnold pp. 51–52

[97] Mona L. Russell, *Creating the New Egyptian Woman, Consumerism, Education and National Identity, 1863–1922* (New York: Palgrave 2004), p. 43. *Al-Lata'if al-Musawwara*, 13 November 1922, 21 May 1923, 18 June 1923, 6 August 1923, 12 April 1926, 14 June 1926, or a racing team, 17 May 1926, and on 17 November an advertisement for Dunlop bicycle tires.

[98] Washing machines, for instance, are dependent on electricity, a running water supply, plumbing, and suitable detergents,

[99] For instance, Geoffrey A. Moore in his *Crossing the Chasms* (New York: Harper Collins 1991) noted a gap between the Early Adopters (visionaries in his words who were willing to pay well to be the first to have the new technologies) and the Early Majority who were more pragmatic and risk averse. In the Middle East, however, we work with different factors such as need, use, and affordability.

It may be said that very few of these new small technologies were barred by the ruling power, and then only temporarily as a matter of control. In the end they could not be withstood. The well-known example is of the often-maligned Sultan Abdülhamid (1876–1908) whose ban on typewriters (and other supposedly menacing appliances) around 1900 figures in Chapter Four. The Sultan's feeling of being threatened by publicists in opposition to his rule was not unique in history. The story of prohibiting electricity has already been revised by historians, and photography was even used by the Sultan for propagandistic purposes. Similarly, this applies to armed cars, and heavier armaments in general, including submarines, which he acquired, and which clearly belong to our category of "big" technologies. More recently, exhibitions and catalogues of his possessions have tried to rehabilitate him in this respect.[100] Whether the Sultanic Court in Istanbul, or others in Cairo and Teheran, themselves acted as role models, or how the upper classes in general, influenced the lower strata, still needs deeper investigation.[101]

To consider the question of why certain technologies or goods were adopted in one or more Middle Eastern environments, while others were delayed, or in exceptional cases even rejected, we have suggested (in Chapter Two) a common-sense mnemonic framework of four C's, (Control, Capital, Competence and Culture) as four sweeping parameters.[102] The above case of the typewriters, one of Control, was rare.[103]

Under Capital, here meaning affordability, we consider a relatively high price and a consumer's low purchasing power, which delayed the acquisition of many small technologies. Had Singer not devised its financial purchasing scheme (an efficient hire-purchase schedule, or sometimes called a "canvasser-collector system"), sewing machines might have remained unaffordable for many, and for a long time. It was followed by its competitors, and possible also in other trades (jewelry, pianos, automobiles) which remain to be investigated.[104] At least in the case of eyeglasses, this may have been different due to a diversified supply in the market.

---

[100] *Sanayi Devrimi Yıllarında Osmanlı Saraylarında Sanayi ve Teknoloji Araçları* (Istanbul: Yapı ve Kredi Bank 2004).
[101] See the predicate Purveyor to the Court, mentioned before.
[102] There exist somewhat different versions of a Four C's schedule – examining e.g., Culture, Competence, Control and Capital. These could be broadened from new communication technologies, which lately seem to occupy most researchers. The idea is derived from a Rand Co. discussion paper entitled "Assessing the Impact of Science and Technology Drivers in Regions" (2003).
[103] In Chapter Two I have discussed them in a different order.
[104] In the Ottoman Empire some taxes could be paid in installments. Advertisements by the music firm Comendinger in Istanbul indicated that goods could be bought "à termes", and pianos at a one year credit.

Competence could be mastered in specialized schools and courses, mostly commercial training institutes, but it is noteworthy that instruction and guidance was sometimes provided by selling dealers as well; one instance was offered by the Singer concern, which was, of course, an interested party. Obviously, typewriters were different, requiring prior literacy. The use of typewriters in offices, or playing the piano, must be considered in the light of the dominant factor, Culture; however difficult this is to define (see Chapters Four and Six).

One could have thought unjustifiably that 'ulama (Islamic scholars) objected to new technologies as being *bid'a*, meaning forbidden or sinful innovations.[105] This was from time to time the case with regard to certain new practices or beliefs on which believers wished to hear the advice of their spiritual leaders, but 'ulama hardly ever rejected new technologies as such. Their fatwas mostly came too late to keep the innovations out. Leor Halevi, who has systematically studied the relevant fatwas published by the eminent Rashid Rida in his journal *al-Manar*, has aptly coined the term "laissez-faire Salafism" to describe their method.[106] Rashid Rida himself, for instance, had reservations with regard to piano lessons for women by strangers, outside the household, but like most 'ulama he supposedly did not consider the piano itself as objectionable – rather, it seems that it was irrelevant. The gender issue was much more important.[107] Anyway, very conservative Islamic scholars frowned upon music in general.

Establishment 'ulama often took a pragmatic stand on big new technologies, convinced that these could advance religious or national causes. Even radical anti-western 'ulama such al Jamal-Din al-Afghani, favored the acquisition of big technologies, such as telegraphs and railways, to be able to combat the West with its own means. An additional explanation may be the fact that these were not part of a perceived cumulative chain of hostile inventions by "Infidels", or

---

[105] Stearns, *Consumerism*, though himself impartial, refers in several places to criticism and opposition stemming from religion, nationalism, socialism, frugality, or environmental activism. The Veblen, Adorno and Horkheimer (Frankfurt School), and Galbraith models are critically discussed by Juliet B. Schor, "In Defense of Consumer Critique: Revisiting the Consumption Debates of the Twentieth Century", *Annals AAPS*, vol. 611 (May 2007), pp. 16–30. Jean Baudrillard, *La Société de Consommation* (Paris: Folio 1970) is also critical, even denying consumer needs. However, none of them has engaged in a study of the Middle Eastern countries.
[106] Leor Halevi, *Modern Things on Trial: Islam's Global and Material Reformation in the Age of Rida, 1865–1935* (New York: Columbia University Press 2021). At that time Salafism was an Islamic movement of renewal seeking a balance between tradition and modernization. Halevy, p. 178, even ascribes to Rida a certain "bias in favor of technologies".
[107] Halevi, p. 155, discusses Rida's fatwa of 1929 on piano playing. See also his extensive chapter on the gramophone (esp. pp. 131–149) and his discussion of Rida's views on photography (pp. 178–184).

even by the Devil, but came alternately, not as part of one conceived ideological assault.[108]

However, in certain critical novels, or in published opinion pieces, and especially in the satirical press, one does encounter aversion to or derision of innovative technologies, in particular against the latest means of transport such as steam ferries, tramcars, motorcycles, and automobiles, which were considered as a physical or moral threat. [109] Indeed, there were certain misgivings and moral objections to imported western objects, fashions, or foreign aesthetics and colors in general. These applied to clothing, as well as to the acquisition of material objects or home furnishings, albeit more as criticism against ostentatious or "conspicuous" luxury (à la Thorstein Veblen). In a well-known study, the Turkish sociologist, Şerif Mardin, has analyzed the phenomenon of "super-Westernization", a tendency which was criticized by the novelist Ahmed Midhat and others.[110] In fact, during the second half of the nineteenth and the beginning of the twentieth century, nationalist and protectionist tendencies gained momentum, and would come to expression more clearly after the Young Turk revolution in 1908.[111] In Egypt, this also came to the fore. [112] We have quoted the Egyptian writer, Mahmud Taymur, who criticized the "market of worthless objects".[113] No doubt, similar criticism could be found elsewhere as well.

From the end of the nineteenth century, the Middle East experienced a series of occasional trade boycotts against foreign commodities, a topic in itself (and of course, the much earlier Boston Tea Party also comes to mind).[114] One may also think of the Tobacco Revolt in Iran but, in our framework, the well-known boycott in 1908 of Austrian department stores and demonstrations against other businesses, especially against Orosdi-Back is relevant, albeit an exception. It did affect the

---

108 For an exception, see Ruud Peters, "Religious Attitudes Towards Modernization in the Ottoman Empire: A Nineteenth Century Pious Text on Steamships, Factories, and the Telegraph", *Die Welt des Islams*, 26/1–4 (1986), pp. 76–105. Possibly, similar fatwas against new western technologies have been forgotten as they were, so to say, overruled by actual practice.
109 Palmira Brummett, *Image & Imperialism in the Ottoman Revolutionary Press, 1908–1911* (Albany, State University of New York Press 2000); see also Russell, pp. 38–47; and On Barak, *On Time, Technology and Temporarily in Modern Egypt* (Berkeley: University of California Press 2013). Basically, all three describe a process of criticism which ultimately turned into compliance.
110 Şerif Mardin, *Super Westernization in Urban Life in the Ottoman Empire in the Last Quarter of the Nineteenth Century* (Leiden: Brill 1974).
111 Deniz Kılınçoğlu, *Economics and Capitalism in the Ottoman Empire* (London: Routledge 2015).
112 Russell, pp. 38–47.
113 See p. 45 and p. 158.
114 Reynolds, *A City Consumed*, pp. 78–113.

sale of fezzes, but the so-called Hat Law of 1925 had a greater impact on sales.[115] But here too, on the whole, it seems, that the diffusion of novel "small technologies" could not be delayed. While local imitations were part of the diffusion in China, Japan, and India, in the Middle East hardly any substitutes could be manufactured locally, at least not until the mid-twentieth century.[116]

One consequence of this emerging nationalist awareness – and ultimately protectionism was an upshot of it – was the wish to replace imported goods by local ones. Part of it was also the introduction of the neologism *"istihlak"* for "consumption", both in Turkish and in Arabic, initially in the negative sense of waste or squander like the original English and French terms. Women, in particular, were encouraged to buy local wares, and in fact also played a role in this struggle.[117] Several researchers have elaborated on the foundation of competing local department stores and other new commercial businesses, as well as on new local industries.[118] Foreign-owned department stores such as Orosdi-Back would gradually be taken over by the new statist and nationalist regimes in nearly all countries in the region.[119] However, this only partly applied to small technologies, as they continued to be imported as long as the local products were not available.

Leapfrogging

Let us go back to our small technologies. The narrative of the adaptation of typewriter to vernacular scripts, and its relative delay in integration in offices and private homes, as discussed in Chapter Four, had an unforeseen sequel. The advent

---

115 Purkhart, p. 177–178. One of the directors, Joseph Back, was involved in negotiations with the Young Turk government toward a compromise in 1909. Ottoman as well as Egyptian rulers had set up factories to increase local production. In the early 1930s, Fathi Radwan's "Piastre Plan" was also directed against the import of foreign fezzes.
116 Cf. Empire Ottoman, *Coup d'Oeil Général sur l'Exposition Nationale Constantinople* (Istanbul 1863), which reflects the strenuous Ottoman efforts to compete with the western industrial and scientific achievements shown at the World Fair of the time. With regard to our discussion, locally hand-made pianos were no match, and sewing machines, typewriters, and optical products came much later.
117 Russell, pp. 49–96. See also Nicole A.N.M. van Os, "From Conspicuous to Conscious Consumers: Ottoman Muslim Women, the Mamulat-i Dahiliya Istihlaki Kadınlar Cemeyet-i Hayriyesi, and the National Economy", *Journal of Ottoman and Turkish Studies*, vol. 6/2 (2019), pp. 113–130. One activity could be seen in sewing workshops where – supposedly – foreign sewing machines were used.
118 Reynolds, pp. 92–98, 104–113, et passim.
119 The large store in Istanbul was taken over in 1932 by Sümerbank to become a department store for domestic products, and the others branches soon followed; Tunis in 1955; Omar Effendi in Egypt in 1958; and the stores in Iraq in the same year, 1958.

and fast diffusion of personal computers from the 1980s generated a case of "leapfrogging" in the Middle East, in the sense of bypassing one or more stages of technological development to catch up with its successors: True, there was a short intermediate phase in the development and marketing of IBM's electrical typewriters with modernized Arabic fonts but, all the same, personal computers made them obsolete. What remained after this transition was the keyboard, and part of the technical and practical terminology. This fascinating phenomenon of leapfrogging is not unique, as developing nations may more easily seize the opportunity to skip earlier phases of diffusion. Examples are the transistor radio which among many populations skipped over the immobile home radio, and the cellular telephone over the terrestrial telephone, and maybe credit cards.[120] Here belong, in a way, the less successful attempts to adapt the piano to being able to play quartertones, but the leapfrog which produced the synthesizer and other electronic keyboards has remained an incomplete or unsatisfactory solution.

Four case studies

My choice of four case studies (sewing machines, typewriters, eyeglasses, and pianos) may appear puzzling. All four formed new research challenges for me. In terms of need and use they fall into quite different categories; their consumers belonged to different sections of the population. Of the basic schools of consumerism, and their critics, Thorstein Veblen's model can be conveniently applicable to the piano as an example of conspicuous consumption, but sewing machines hardly qualify, mainly because of the installment plan which enabled popular home uses. Sewing machines began their triumphal course from below. It makes any sweeping Marxian interpretation (e.g., the Frankfurt school model) more difficult. Very rarely does one encounter misgivings regarding the sewing machine.[121] Typewriters, though advertised in the general press, and therefore seemingly qualifying as consumer products, remained, on the whole, in the business and administrative sphere, and are therefore not on a par with the other small technologies, at least not until our times of home computers and mobile telephones.[122] In the beginning,

---

120 Jeffrey James, "Leapfrogging in Mobile Telephony: A Measure for Comparing Country Performance", *Technology Forecasting & Social Change*, vol 26 (2009), pp. 991–998.
121 One case is quoted by Mona Russell, p. 45, from *al-Muqtataf* of 1884, where a conservative Lebanese reader defends hand-stitching. But as in India, in the Middle East, there was hardly any cultural resistance to the sewing machines, see Arnold, p. 49.
122 In contrast to the Arab countries, in India, typewriters were introduced much earlier by the government, to make the administration and courts more efficient. Between 1910 and 1950 about

typewriters were seldom acquired by private persons, which makes this advertising all the more curious. Eyeglasses were not ostentatious consumables in the era we are discussing (fashionable as they are today) and have therefore escaped discussion in the scholarly literature.[123] In my view, these four cases serve the central argument of this volume that among the small technologies in the Middle East each has its own diffusion narrative.

Sewing machines were among the earliest small technologies to be mass produced and acquired worldwide and, as I mentioned above, they launched my explorations in this field. From their ascent in the 1850s they proved eminently useful in private homes, notwithstanding Karl Marx's fears of their exploitative side in industrial enterprises, for which they had been destined. Sewing machines became a success story due to their convincing advertising, their proven efficiency without a need for adaptation, their relatively easy operation in homes and, in addition, certain appropriate instruction, and servicing and, we should not forget their purchasing schemes. Toward 1900, Singer was said to have already produced at least 13.500.000 machines, half in their factories outside the USA.[124]

Andrew Godley, however, who has thoroughly examined the company's worldwide sales from his view as a business historian, has calculated that while in 1900 a total of 1.080.000 machines were sold, his conclusion was that only 20.500 had gone to the Ottoman and Balkan countries. This may seem not very impressive but if we note that their sales in the region began with only 70 machines in 1880 (of 538.609 sold world-wide), and that this number rose steeply to 66.757 from 1910 (again temporarily declining owing to World War I), then the Middle East market was clearly growing.[125] Moreover, Singer's success is proven by the fact that other manufacturers (e.g., Pfaff in Germany) soon marketed competing models. By World War I, sewing machines had become widespread, even "domesticated", as Abou-Hodeib remarks for Beirut, which must have been the case elsewhere too. Often, in marriages they were part of the dower.[126] Especially with regard to traditional Middle Eastern gender roles, sewing machines had many

---

half a million were imported, on a population of app. 250–360 m. (see David Arnold, p.58 et passim).
[123] Jean Baudrillard's perceptions can be relevant to the study of department stores (and later shopping malls), see Mona Abaza, *Changing Consumer Cultures of Modern Egypt, Cairo's Urban Reshaping* (Leiden, Brill 2006), pp. 13, 43–44, 156 and 268.
[124] Edward W. Byrn, *The Progress of Invention in the Nineteenth Century* (New York: Munn & Co. 1900), p. 194.
[125] Andrew Godley, "Selling the Sewing Machine Around the World: Singer's International Marketing Strategies, 1850–1920, *Enterprise & Society,* vol. 7 / 2 (2006), pp. 266–314.
[126] Abou-Hodeib, *A Taste for Home* pp. 109–112.

benefits, as they empowered women within their homes, not being forced to go out to sweatshops or factories, or mixed work floors and public space, as was the case in the USA.

Almost everybody remembers that her or his mother or grandmother owned "a Singer". As small technologies in ordinary households they became iconic, and it is not surprising that researchers found every justification to investigate the local acceptance and use. Ever since, the scholarly literature has widened to include particular narratives of sewing machines in Mexico, Argentina, Spain, Greece, and Africa at large, which proves my point of divergent diffusion.[127]

Typewriters, in Chapter Two, were considered a "younger sister" of the sewing machines, though their commercial take-off in the USA ("invention" is problematic as we have remarked) came in 1874. This means a gap of twenty-three years, enough for a mother and daughter relationship. But the leading manufacturers of both devices, Singer, and Remington, remaining far ahead of their competitors in global marketing, made it logical to compare their diffusion in the Middle East. And yet, in the region, they were a pair apart, because typewriter manufacturers had to overcome more obstacles than their sewing machine counterparts.[128]

Typewriters, initially, could appeal only to modern business sectors. For that reason, the potential consumer market in the countries of the Middle East was extremely limited. It has been estimated that by 1896, the USA had produced 450.000 typewriters of which 150.000 were in use there, but only a very small percentage had reached the Middle East.[129]

The mechanical adaptation to the Arabic vernacular scripts needed time. In our understanding, however, this was not the major hurdle. There were much more complex non-western scripts, in particular Thai, Chinese and Japanese,

---

[127] To the above-mentioned studies by Frank Dikötter on China (2007), Andrew Gordon on Japan (2012), Jean Gelman Taylor on Indonesia (2012), David Arnold on India (2013), we can add Paula A. de Cruz Fernandez on Mexico and Spain (2013), in a way preceded by Ruben Gallo, *Mexican Modernity: The Avant-Garde and the Technological Revolution* (Boston: MIT Press 2010); Gabriele Mitidieri, "'Un Autómata de Fierro': Máquinas de |Coser, Ropa Hecha y Experiencas de Trabajo en la Ciudad de Buenos Aires en la Segunda Mitad de Siglo XIX", *Historia Crítica*, vol. 85 (1 July 2022); Argyrios Sakorafas, "The "Great Civilizer": The Global Diffusion of the Sewing Machine and its Impact on Greece during the Late 19th-early 20th Century", *Global History Blog*, and E. McKinley on *Africa* and others. See also Leda Papastefanaki, "Sewing at Home in Greece, 1870s to 1930s" in Malin Nilsson et al. (eds), *Home-Based Work and Home-Based Workers (1800–2021)*, Leiden: Brill 2021), pp. 74–95.
[128] George Nichols Engler, *The Typewriter Industry: The Impact of a Significant Technological Innovation* (PhD Thesis, UCLA 1969), pp. 10–25 *et passim*.
[129] Byrn, p. 182.

which in the end proved to be surmountable, albeit with considerable delays as compared to Roman, Greek, Cyrillic scripts.[130]

Secondly, in our view, there was a wide gap between the Middle East and the United States with its large, capitalist corporations and their typically capitalist maxim of "Time is Money".[131] The Middle East, by the late nineteenth century had very few such large organizations which could see the immediate advantage of the new appliance. Neither the state apparatus nor its agencies (customs, postal services, railway companies, and telegraph managements), nor indigenous private enterprises, maintained modern offices in the American style. Indeed, most of the latter were still in foreign hands (banks, department stores, large importers, etc.). Even in countries which were not under direct colonial rule, most large trade relations with western countries were handled by foreigners, and in European languages. Private users were for a long time very few in number. In that sense, the typewriter could maybe still be called an imperialist device.

Thirdly, as it would take decades before typewriters became integrated in an office culture of the scope known in the west, it did not generate a large clerical working force of female shorthand-typists (then often called "steno-typists"). Turkey evolved somewhat more quickly in that direction. Indeed, surprisingly little has until today been written on this transition in the Arab countries and Iran, except – to some extent – on Turkey after its adoption of the Roman alphabet in 1928. This transition involves not only offices, but also the availability of word processing and use of other computer programs in domestic languages, beyond an alleged "digital divide" which has often been discussed. In fact, the arrival of personal computers, together with more women entering the clerical work force, led to a certain breakthrough, maybe even a little-noticed revolution. This was the leapfrog which we have mentioned.

Eyeglasses are a completely different consumer article but nonetheless an important one, albeit neglected in the literature. We have no grip on the statistics of imports and diffusion. In the Middle East, it took a long time before all, or most, people who in fact needed them, became aware of their utility, or were able to obtain or afford them. But we might call it the most "democratic" consumer article of the four.

---

130 Thomas S. Mullaney, *The Chinese Typewriter: A History* (Boston: The MIT Press 2008), also Arnold's extensive discussion of India.
131 As suggested by Benjamin Franklin, but such capitalist attitudes were not rooted in the Middle East, at least not in the same way. Neither was Ernest Giraud, President of the French Chamber of Commerce impressed by the 214 words p.m. which an American typist allegedly could reach: "Il est vrai que nos tapeurs [sic] à Constantinople sont extrèmement éloignés de ce chiffre", *Revue Commerciale du Levant*, vol. (1910), p. 161

Eyeglasses are a very different small technology as their diffusion is not limited to one particular segment of society. Having become "in-use" from the 14th century onward in Western Europe, eyeglasses had been widely marketed since the Renaissance to the point that some researchers have connected it to the subsequent Readers' Revolution. However, as to the Middle East, we have tried to investigate the reasons for their slow diffusion. Ultimately, popular demand, beyond privileged persons, rose in the nineteenth century, as increasing literacy and reading habits, as well as medicalization, gave diffusion a major impulse. As in other places, the late 18th century pince-nez, lorgnette, and monocle, disappeared, and eyeglasses gave way to more global fashions, including sunglasses.

Pianos, in our last chapter, belong to yet another class of use and need altogether, as they began as a luxury object owned by certain upper classes. In Ersin Altin's words: "Objects like [the] piano that were located in domestic space functioned primarily as tools that symbolized class differences, social status, and intellectual hierarchies; on the other hand, sewing machines became a shared value between women from different social strata…".[132]

From 1700 onwards, pianos had become ever more sophisticated, and were therefore highly technological instruments. Spread over many parts of the world, many millions have since been sold. In the year 1890 alone, the world output from Germany, the USA, Britain, and France was estimated at 232.000, and this would allegedly rise to 600.000 in 1920.[133] This to the extent that Peter Stearns even associated pianos with household furnishings: "The addition of the piano to the must-buy items, for middle classes, showed how innovation could swell this product category".[134]

But nowhere in our region would even the (somewhat) less expensive upright pianos conquer a significant segment of a middle-class market as was the case in European and American milieus. Our Chapter Six illustrates, more than the others, the predominance of the Cultural factor, custom, and taste. Music, like food, and often clothing, remained to a large extent entrenched in tradition, or rather traditions in the plural, – including imagined traditions. In this respect, the diffusion of pianos hit a limit of relative uselessness, and thus became a relative failure. While other small technologies, to one extent or another, trickled down from the upper classes, the piano never gained a stable foothold on the popular musical scene.

---

132 Ersin Altin, p. 134.
133 Mario Igrec, *Pianos Inside Out*, partly summarized as *"Marketing History of the Piano"* http://www.cantos.org/Piano/History/marketing.html (accessed 27 December 2011). See further Sonja Petersen, "Piano Manufacturing Between Craft and Industry; Advertising, and Image Cultivation of Early 20th Century German Piano Production", *Icon*. Vol 17 (2011), pp. 12–30.
134 Stearns, *Consumerism*, p. 52.

True, in the 19th century, pianos could be seen and heard in Middle Eastern courts and salons, in many bourgeois circles, as well as in missionary schools and institutions. But in the end, Middle Eastern musical traditions with their *maqam*s and quarter tones, did not suit the piano.

It seemed technically possible to redesign and superficially "re-tune" a piano, as had been done with the typewriter, but this never struck roots. I have tried to follow the strenuous attempts to adapt the piano to the modes of Middle Eastern music. After all, Western type violins could be tuned accordingly and became accepted instruments. But while Remington and other manufacturers of typewriters had seen a commercial purpose in adapting them to other languages, no western piano builder aimed at the Middle Eastern market. Even if there were mutual influences, European or American designers of quartertone pianos worked mostly to serve composers who experimented with new modes of classical music. And those in Middle Eastern countries who entertained the ambition that such a piano would bring their music in line with western standards certainly did not take an interest in advanced western classical music.

Similar to the arrival of the personal computer succeeding the typewriter, there appeared a new technological challenge in the form of electronic keyboards and synthesizers, but the final stages of "Persuasion" and "Implementation" still remains to be seen. It does not seem to be another successful leapfrog.

In conclusion

Though eyeglasses had been developed as small technologies over preceding centuries, the four consumer items here discussed in detail became western industrial mass products only in the nineteenth century. Their use had to be proven in their countries of invention before they could become "Persuasive" as an indispensable need elsewhere.

With the piano as a relative exception, most small (western) technologies were accepted and integrated in the countries of the Middle East. It is noteworthy that they originated in countries which could be identified as antagonistic or hostile by nationalist segments who were nevertheless local users. It turns out that small technologies were not primarily associated with Imperialist superiority, not to speak of an Infidel intrusion, as their utility was beyond doubt. In many respects, it was an irresistible, self-regulating process in which awareness, availability, affordability, want, need, and use went together.

The present collection of articles and chapters does not exhaust this field of research. Neither have all small technologies in the Middle East been neglected. But the social impact of many small technologies such as primus gas burners, thermos flasks, agricultural pumps, small firearms, bicycles, medical instruments, fans, subsequent generations of telephones, and air conditioners, though not ignored, still calls for a closer investigation.[135] In short, it is an area which begs for more research.

---

[135] Of course, any small technology might be re-invented, as cell phones for instance, which have today more varied uses and social effects than what was conceived a century and a half ago, see Burçe Çelik, "Cellular Telephony in Turkey: A Technology of Self-produced Modernity", *European Journal of Cultural Studies*, vol 14/2 (2011), pp.147–161.

1 Who needed department stores in Egypt?

1.1 From Orosdi-Back to Omar Effendi

Although the ascendancy of the Free Officers regime in Egypt in 1952 and the ensuing Revolution are no longer understood as an absolute watershed, major changes in Egyptian society and economy took place within a decade. Most foreign interests were nationalized or liquidated, and the daily life of the foreign, as well as the indigenous elites, was drastically changed. But seen in that light, while a new economic regime was to begin on both the production and consumption sides, the existing department stores paradoxically were not liquidated, but nationalized.

This is remarkable, because department stores were commonly identified as typical foreign capitalist interests, and undesirable symbols of foreign and bourgeois conspicuous consumption. This notion, however, needs qualification. True, apart from intermittent political boycotts against imported goods, social criticism had also occasionally been voiced against such consumption. A playwright such as Muhammad Taymur had satirized the "worthless stuff, nauseating objects, ornaments in red and white which we do not need", as sold by the famous Cicurel store.[1] More emblematic was the fact that some of these well-known department stores, such as Cicurel, Benzion, Gattegno and Hannaux were the objects of arson in 1948 and again 1952.[2] This may show that at least part of the Egyptian people, such as supporters of Muslim Brotherhood, identified them as unwanted foreign interests. The specific case of Cicurel may, on the other hand, prove the complexity of such prejudices. Its patriotic Egyptian record during the nationalist revolt of

---

1 A. al-Din Wahid, *Masrah Muhammad Taymur* (Cairo: Hayat al-Kitab, 1975), pp. 126–127 (from his play 'al-Hawiya'). References to the history or significance of department stores in Egypt, or to consumption in general, were until recently rare, see as a notable exception J. Berque, *Egypt, Imperialism and Revolution* (London: Faber & Faber, transl., J. Stewart, 1972), pp. 349 and 466–483. The contemporary Cairene journalist-historian Samir Raafat has made a major contribution in reviving interest in the stores and their owners, see S. Raafat, *Cairo, the Glory Years* (Alexandria: Harpocrates, 2003), and his earlier articles. On conspicuous consumption see M. Barakat, *The Egyptian Upper Class between Revolutions, 1919–1952* (Reading: Ithaca Press, 1998). See also, more recently, a novel by R. Ashur, *Qit'a min Uruba, Riwaya* (Cairo: Dar al-Shuruq, 2003), p. 74.
2 On 19 July 1948, Cicurel and Oreco were partly destroyed by bomb explosions. During the following weeks Benzion and Gattegno were attacked. In the Great Fire of 26 January 1952, five large department stores were badly damaged, namely Cicurel (which completely burnt out), Chemla, Ades, and Orosdi-Back. For photographs see '50 Years – the Great Cairo Fire', *Misr al-Mahrusa*, vol. 16 (Jan. 2002).

1919 had earned it the predicate of a politically "approved shop". After the devastating fire of 1952 this was even translated into government support for its restoration. But it could hardly change the existing image.[3]

On the part of the former foreign residential elite, the dark memory of the events of January 1952 – even today – also symbolizes nostalgia.[4] For many of those who had to leave Egypt, it evokes their erstwhile sumptuous lifestyle.[5]

More than anywhere else, it would seem, department stores in Cairo and Alexandria were a Jewish branch. The tendency to consider modern department stores as a typically Jewish niche of trade is widespread, particularly in Western Europe (especially pre-Second World War Germany) and Northern America. There, however, it may be arguable in view of the many non-Jewish entrepreneurs who entered this field at an earlier stage. The preponderance of Jews founding *grands magasins* in Cairo certainly is a statistical exception. It might have had a lot to do with their mobility and international business connections.[6]

The events of 1952 must be put in the right perspective: many of the businesses attacked, although Jewish property, represented foreign rather than specifically Jewish interests. Some objects targeted symbolized Western lifestyles or foreign consumer goods.[7]

Department stores with their fixed prices, their large stocks, their quick turnover, their merchandise on display, their salesmen and saleswomen and their customers, have in recent years attracted much attention and academic research. France, the United States, Britain and Germany vie with each other to claim its "invention". This is now also commonly reflected in the recent scholarly literature on the topic – but we would like to give the credit to Au Bon Marché in Paris.[8] As with

---

3 On personal memories of some of the department stores attacked see 'Memories of Egypt', http://www.rootsweb.com/-nafrica/EGYPT/Burining/Cairo.html; on Cicurel see especially J. Beinin, *The Dispersion of Egyptian Jewry, Culture, Politics, and the Formation of a Modern Diaspora* (Berkeley, CA: University of California Press, 1998), pp. 21–22. Cicurel also participated in the nationalist Banque Misr project (1920).
4 The aspect of nostalgia will be elaborated on in a forthcoming article.
5 See, for instance, A. Farhi, 'Back to the Nile', http://www.farhi.org/Documents/backtothenile.htm; V.D. Sanua, 'A Return to the Vanished World of Egyptian Jewry', http://www.sefarad.org/diaspora/egypt/vie/001/0.html, and V.D. Sanua, 'The Sepharadim in Egypt', http://www.sefarad.org/publication/lm/033/6.html; G. Mizrahi, 'Ma vie au 19ème siècle', http://www.monimiz.com/giacomo.html.
6 See 'Department Stores', *Encyclopedia Judaica*, CD-Rom edition (Jerusalem, 1997).
7 One could argue that in 1948 the Muslim Brothers sought out Jewish, or alleged Zionist, objects.
8 Classical studies by P. Goehre, *Das Warenhaus* (Frankfurt aM: Ruetten & Loening, 1907); H. Pasdermadjian, *The Department Store* (London: Newman, 1954); M.B. Miller, *The Bon Marche, Bourgeois Culture and the Department Store, 1869–1921.*) (Princeton, NJ: Princeton University Press, 1981); W. Lancaster, *The Department Store, A Social History* (London: Leicester University Press, 1995); an important recent addition is G. Crossick and S. Jaumain (eds.), *Cathedrals of Consumption:*

so many features of West European and later North American life, department stores soon also came to the Ottoman Empire, including Egypt. The well-known reformist educator 'Ali Mubarak, in his lesser-known work *'Alam al-Din*, was probably one of the first to make Egyptian readers familiar with at least one exemplary (but unnamed) large Parisian store. He guided them alongside its crystal dome, its large windowpanes, its mirrors and lighting, and above all drew attention to the exemplary orderly display of luxurious commodities, the fixed prices, and the women shopping there.[9]

Ironically, in Paris itself, and elsewhere in Western Europe and North America, some of the new "cathedrals of consumption" were initially called bazaars (a term from Persian, while the word *magasin*, via Italian, is of Arabic origin); at the same time, the word adopted for department stores in Turkish was *bonmarşeler*.

In Cairo, the branches of business, which soon developed into department stores, were started by different categories of migrants, be they Levantine Syrian families, Sephardi-Mediterranean Jewish families, or Ashkenazi Jewish families, as well as a few others. Most were in the hands of families (fathers and sons, brothers, etc.), which was the common form of business: Cicurel (from Izmir),[10] Chemla (from Tunis),[11] Gattegno, Hannaux,[12] Ades, and Benzion,[13] to mention only the largest.[14] The two main exceptions were Sidnawi[15] and Davies Bryan,[16] but these were

---

The European Department Store 1850–1939 (Aldershot: Ashgate, 1998). See further R.D. Tamilia, 'The Wonderful World of the Department Store in Historical Perspective: A Comprehensive Bibliography Partially Annotated' (Working Paper, Centre de Recherche en Gestion, March 2002).

9 A. Mubarak, *'Alam al-Din*, Vol.3 (Alexandria: Matba'at Jaridat al-Mahrusa, 1882), pp. 816–823.

10 On the history of Moreno Cicurel (who settled as a tailor in Cairo in 1870) and his sons Salomon and Joseph, see: A. Wright and H.A. Cartwright, *Twentieth Century Impressions of Egypt* (London: Lloyds, 1909), p.377; N.A. al-H. Sid Ahmad, *al-Hayat al-Iqtisadiyya wal-Ijtimaiyya lil-Yahud fi Misr, 1947–1957* (Cairo: al-Hay'at al-Kitab, 1991), pp.38–40. Cicurel had branches in Alexandria, Ismailiya, etc. See also S. Raafat, 'The House of Cicurel', *al-Ahram Weekly*, 15 Dec. 1994, and his *Cairo*, pp. 258–61.

11 The Chemlas came to Cairo around 1905, see http://www.chemla.org/caire.html. See further, Sid Ahmad, *al-Hayat al-Iqtisadiyya wal-Ijtimaiyya lil-Yahud fi Misr*, pp. 40–41.

12 Hannaux started as Au Petit Bazaar in the Muski (1882), see Wright and Cartwright, *Twentieth Century Impressions of Egypt*, p.337.

13 Sid Ahmad, *al-Hayat al-Iqtisadiyya wal-Ijtimaiyya lil-Yahud fi Misr*, p.42. The founders of this firm, still popularly known as *Binzayun* was apparently also related to old-time Jewish families such as Suares.

14 Others were Carnaval de Venise, Chalons, Cohenca, Salon Vert, Pantremolli, Rivoli and Simon Artz (the latter in Port Said). There may be more, nowadays forgotten, e.g., Frances, see Wright and Cartwright, *Twentieth Century Impressions of Egypt*, pp. 376–367.

15 Also spelled Sednaoui, originally from the village Sidnaya in Syria, the two brothers, one trained as a European tailor, the other with commercial expertise, arrived in Cairo in the 1870s,

similarly founded by outsiders (Syrian Christians and Welshmen respectively). To these, after the First World War, branches of a few truly French *grands magasins* such as Au Bon Marché, Le Louvre, Au Printemps and Galeries Lafayette have to be added.[17] Most of these companies had branches in Alexandria, and sometimes in Port Said and some of the towns in the Delta (Tanta, Zaqaziq, Mansura) and even beyond (Asyut, Minya).

One conspicuous category was the "Austrians" who had their roots in the expanding confection business in Vienna, and who opened stores in Cairo and Alexandria (and in Istanbul) during or soon after the Crimean war. A relatively well-documented example of their beginnings is that of Alfred Mayer, whose brother Sigmund was a textile entrepreneur (born 1831 in Pressburg/Bratislava) and a local politician in Vienna.[18] The latter described how ready-to-wear clothing began to be exported to the Balkans and Istanbul, and how his brother Alfred set up shop in Cairo around 1866.[19]

To this category also belong other families such as Salomon Stein's (born 1844 in Jassy), who had opened a clothing store in Cairo in 1863 or 1865 at 'Ataba Square in Cairo, and in 1875 (or 1879) also one in Alexandria.[20] Or take Victor, Gustav and Konrad Tiring. After learning the trade as a "Tuerkenschneider" in Vienna, Victor, together with his brothers, set up clothing stores in Istanbul (1842), Cairo, Beirut,

---

setting up a clothing business in the Muski area, S. Raafat, *Cairo Times*, 29 May 1997, and *Cairo*, pp. 56–58. See also *al-Hilal*, 1 May 1908, pp. 471–478; D.M. Reid, 'Syrian Christians, the Rags to Riches Story, and Free Enterprise', *International Journal of Middle East Studies (IJMES)*, Vol.1 (1970), pp. 358–367.

16 S. Raafat, 'Davies Bryan & Co. of Emad el Din Street', *Egyptian Mail*, 27 May 1995, and *Cairo*, pp.38–41. See also Wright and Cartwright, *Twentieth Century Impressions of Egypt*, pp. 322, 376, 468. Davies Bryan's case is exceptional because he came to Egypt in 1886 for health reasons. In 1887 he was joined by his brother Joseph.

17 S. Saul, *La France et l'Egypte de 1882 a 1914, Intérêts économiques et implications politiques* (Paris: Ministère de l'Economie, 1997), pp.15, 517. See further N. Carnoy, 'La colonie française du Caire' (PhD Thesis, Paris: Presses Universitaires de France (PUF), 1928), pp. 50–58. Carnoy was a teacher at the Khedivial Law School in Cairo.

18 Alluded to as 'my maternal uncle' by E. Hobsbawm, *Age of Empire* (London: Weidenfeld & Nicolson, 1987), pp.1–2.

19 S. Mayer, *Ein Juedischer Kaufmann, 1831–1911* (Leipzig: Duncker & Humblot, 1911), ch.4; see further R. Agstner, *Die oesterreichisch-ungarische Kolonie in Kairo vor dem ersten Weltkrieg, das Matrikelbuch des k. und k. Konsulates Kairo 1908–1914* (Cairo: Austrian Cultural Institute, 1994).

20 Wright and Cartwright, *Twentieth Century Impressions of Egypt*, pp. 369–374, 470; with branches also in Istanbul, Salonika, Tanta and a factory in Vienna, and allied to the Raff and Blumberg families. See further R. Agstner, 'Das Wiener Kaufhausimperium "S. Stein" im osmanischen Reich', *Wiener Geschichtsblaetter*, Vol. 59 (2004), pp. 130–140.

and Salonika.[21] A similar story is that of the brothers Emanuel, Leopold and Gustav Stross, and the former's Alexandria-born sons Rudolph, Karl and Oscar who had started their business in Cairo in 1865. Emanuel Stross even became a member of the Alexandrian Municipal Council.[22] The Orosdi-Backs who are the object of this study will be discussed below. As a group they soon came to dominate the ready-to-wear branch in Egypt – the "Austrians" according to one source controlling more than half of it.[23]

The name Orosdi-Back, which to some may sound like Orosdi Bek (Bey), refers to two families.[24] The founding father of the firm was Adolf Orosdi, an officer in the army of Lajos Kossuth, the leader of the Hungarian revolt of 1848. Originally called Adolf Schnabel, he had magyarized his name to Orosdi, also spelled Orosdy.[25] With the defeat of the revolt in 1849, and Kossuth's taking refuge in the Ottoman Empire, Adolf apparently lived for a few years in Aleppo. In any case, his son Leon was born there in 1855. He may have belonged to the group of Hungarian officers who converted to Islam and went to Aleppo.[26] Another son, Philippe, was born in Istanbul in 1863. Indeed, we know that Adolf Orosdi set up a store in Galata in 1855, which is considered to be the founding date of the later firm. In the course of the next decade or two, with supposedly excellent connections in the Ottoman capital, Adolf acquired a plot in the Bahçe Kapi area (close to Eminönü). It is not entirely clear whether the Ömer Efendi Han, as the location was known, was a preexisting store or waqf. Not many Egyptians will realize that the name of the well-known Omar Effendi stores was originally the name of a (modern) *khan* in Istanbul.[27]

---

21 Victor and Konrad were born in Istanbul (1849 and 185?), Gustav in Livorno. Raafat, *Cairo*, pp. 51–53.
22 Wright and Cartwright, *Twentieth Century Impressions of Egypt*, pp. 322, 442, 459.
23 S. Raafat, 'From Mag-Arabs to al-Magary', *Egyptian Mail*, 13 April 1996.
24 Orozdibak in modern Turkish orthography.
25 For more on the family history (and extensively on the business history) see U.M. Kupferschmidt, *European Department Stores and Middle Eastern Consumers: The Orosdi-Back Saga* (Istanbul: Ottoman Bank Archives and Research Centre, 2007).
26 K. Karpat, 'Kossuth in Turkey. The Impact of Hungarian Refugees in the Ottoman Empire, 1949 – 1851', in K. Karpat (ed.), *Studies on Ottoman Social and Political History, Selected Articles and Essays* (Leiden: Brill, 2002), pp. 169–184. Adolf Orosdi even taught Kossuth Turkish; another source has it that Adolf changed his name to Ali (cf. the story of General Jozsef Bem who served as a short-lived governor of Aleppo under the name Murat Pasha, see Kupferschmidt, *European Department Stores and Middle Eastern Consumers*.
27 The modern building, which opened in 1907 and which was enlarged at least once, serves as a department store owned by a holding company of the Sümer Bank, which bought it from Orosdi-Back in 1943. The building was recently renovated.

We do not know exactly when and how the first linkage between the Orosdis and the Backs was established; the Backs may have had earlier roots in the Viennese *confection* business (*commissionairs* of ready-to-wear clothing). One of the four founders of the later firm, Hermann Back, was born in Galgocz or Hlohovec in 1848, and his brother Joseph was born in Czechia of today. In any case, the two families twice intermarried, Maurice Back with Antoinette Orosdi, and, in the following generation, Hermann Back with Mathilde Orosdi.[28] The registers of the Ashkenazi Jewish community in Istanbul, the so-called *Qehillat Tofrei Begadim* (the Tailors" Community) with the famous Schneidertempel, bear witness to the birth of their daughters in the 1880s. It is interesting that at the time they still identified, or had to identify, themselves as Jewish.

The Orosdi-Backs launched their business in Cairo in 1856, one year after Istanbul, if we may believe the date given for the celebration of their centenary in 1956 (which was the opportunity to open a new Omar Effendi branch on 'Adli street).[29] It started, not surprisingly, on the main street of the Muski quarter, or Rue des Francs, where the firm even today keeps a branch. Indeed, the area was the first commercial district outside the "bazaar" – if that term is right for Cairo – where European merchants and modern fashionable commodities made their appearance.[30]

By then, a process of change in consumption was already underway. One description of 1883 speaks of a complete transformation, rich *"magasins"*, replacing modest *"boutiques"*.[31] For about half a century, hardly any tourist guide missed a description of the area, contrasting the old "picturesque" bazaars with the "large shops, built on the French pattern, with plate-glass windows, gilded fascias, etc.".[32]

Due to the lack of company archives, the historian must reconstruct the narrative of the firm from a host of different sources. Annual reports show that its

---

28 Typically, we have only the French surnames.
29 At the time the store in Istanbul was owned by Adolf Orosdi. We do not know whether this business in Cairo was his too, or maybe Back's. The new store was opened on 3 March 1956, *l'Egypte Nouvelle*, 2 and 16 March 1956.
30 The Muski area had also been the starting point for Moreno Cicurel, Albert Mayer and Hannaux.
31 H. de Vaujany, *Le Caire et les environs* (Paris: Plon, 1883), p.124.
32 E.A. Wallis Budge, *Cook's Handbook for Egypt and the Egyptian Sudan* (London: Cook & Son, 1911), pp. 458–459; D. Sladen, *Oriental Cairo, The City of the 'Arabian Nights'* (London: Hurst & Blackett, 1911), pp. 68–69. In her PhD Thesis, 'Commodity Communities, Interweavings of Market Cultures, Consumption Practices and Social Power in Egypt, 1907–1961' (Stanford University, 2003), Nancy Reynolds has convincingly shown that the dichotomy between 'traditional' and 'modern' is misleading in an overlapping reality.

geographical expansion was quick.[33] Already in the 1880s, the Orosdi-Backs had established branches in Bucharest, Plovdiv, Salonika, Izmir, Aleppo, Beirut, Tunis, and purchasing offices in a host of industrial cities in Western and Central Europe, as well as in Japan. Thereafter came stores in Rusçuk (Rusu), Varna, Adana, Samsun, Beirut, Bizerte, Sfax, Baghdad, Basra, for a short period in the 1920s even Dayr al-Zur, Tabriz and Tehran, with a swan song in Casablanca, Fez and Meknes in the 1960s. Their business network in Egypt comprised not only Cairo, but also Alexandria, Port Said, and at some time Tanta and Zaqaziq.

The Orosdi-Backs opened branches mainly in port cities, and major railway nodes, with sizeable minority and foreign populations. It is a story which belongs to "another age of globalization", in Egypt facilitated by a capitulatory regime. Between the 1860s and the 1940s, some 250.000 foreigners settled in Egypt – among them at least 65.000 Jews – with large concentrations in Cairo and Alexandria, and considerable purchasing power. From trade *en gros* they gradually moved into trade *en détail*. Each branch or *succursale* could, of course, tell its own story – if only we had more records. As far as we know so far, the firm never engaged in export trade to Europe, for instance like the German department store Wertheim which had a representative in Istanbul to import Turkish rugs to Germany.[34] It was furthermore still unthinkable that international trading firms such as Orosdi-Back would themselves produce commodities in the Middle East for the European market. At the most, local ateliers or seamstresses finished clothing or textile items for customers, but this sort of sub-contracting generally remains hidden.[35] Only in Istanbul did Orosdi-Back maintain a workshop, which produced, or to be more precise assembled, umbrellas from imported parts.

There is no doubt that Orosdi-Back and its limited circle of shareholders conducted a profitable business for many decades, which, however, does not mean that they did not go through financial and other crises. We will not elaborate here on the effects of wars, major population movements and exchanges, or fires, which from time to time devastated one branch or another. From 1907 the consumer market in Egypt suffered from "marasmes", slumps, in the sense of temporary cotton crises. Cairo was perhaps never their most profitable branch. Samir Saul shows that for 1914, for instance, net profits for Istanbul, Izmir, Salonika, and

---

33 Kept at Roubaix, Centre des Archives du Monde du Travail (CAMT), series 185 AQ-344 and 205 AQ-170, see also 65 AQ-T136/137. A few of the branches were sold or closed again. The fully-fledged department stores were in Istanbul, Cairo, Beirut, Tunis and later in Baghdad. For more details see Kupferschmidt, *European Department Stores and Middle Eastern Consumers*.
34 Goehre, *Das Warenhaus*, p. 146.
35 E. Brakha, *Galei Hayyai* [Riding the Waves of My Life] (Tel Aviv: Teper, 2005), p. 20, remembers his mother sewing tags on sheets and tablecloths for Hannaux in Alexandria.

even Alexandria were higher, Cairo came in only ninth place.[36] The branches in Tanta and Zaqaziq had to be closed in 1911 and 1928, respectively; Port Said did not survive either. Altogether, however, it would seem that Orosdi-Back showed vitality and resilience, presumably through business acumen, for instance, by speculating with large stocks (e.g. during and after the First World War). The overall management of the chain store business basically remained a family affair until well into the 1920s, with few additional shareholders, which is the reason why annual reports to shareholder meeting yield few details.[37]

Philippe Back, Hermann's brother, served as the managing director of the Cairo branch of Orosdi-Back. His name even figures on one of the commemorative plaques of the Sha'ar haShamayim synagogue, which was built between 1899 and 1904.[38] This is interesting because of a more or less open identification as being Jewish (and rich). Around 1907 he became actively involved in Egyptian archaeology and contributed towards the excavations in Jamhud (or Gamhud, near Bani Suwayf).[39] In 1910, he left Cairo for Paris, where he started a perfume business.[40]

Meanwhile, with the development of the modern Isma'iliya quarter, a splendid new store was constructed between 1905 and 1909 on the corner of 'Abd al-'Aziz and Rushdi streets.[41] The location suggests a compromise between the Muski, which was never entirely abandoned, and the opulent 'Imad al-Din street, where most of the other department stores were moving.[42] It was not simply a matter of location and architecture, but one could say that the commodities themselves, and salespersons (and consumers) too, all gave flavour and colour to the aspira-

---

36 Saul, *La France et l'Egypte*, p.173 (table includes Paris in the first place).
37 Saul., pp. 170–174.
38 Raafat, *Cairo*, p. 46.
39 The excavations conducted by the Polish archaeologist Tadeusz Samuel Smolenki (Reichstein) and the transfer of ancient Coptic relics to Budapest are supposed to have given a boost to Hungarian interest in ancient Egypt. Back received a minor ennoblement for his involvement. Thanks to information given by Dr Peter Gaboda of the Museum of Fine Arts in Budapest, who has two recent (2002) publications on Back. P. Gaboda, 'Back, Fülöp', *Magyar Múzeumi Arcképcsarnok* (Budapest: Pulszky Társaság, 2002), pp. 29–30; 'Gamhudi Asásatás, 1907', *Múzsák Kertje, a Magyar Múseumok Születése* (Budapest: Pulszky Társaság, 2002), pp.46–47.
40 Information from Nicole Back. At the outbreak of the First World War he moved to Spain and died there in 1958. Interview, Paris, 12 Feb. 2004.
41 We have not yet been able to establish how the plot was acquired, see J.-L. Arnaud, *Le Caire, mise en place d'une ville moderne, 1867–1907* (Paris: Sindbad, 1998), who describes a slightly earlier phase of development.
42 We do not know whether the choice of the location was a strategic decision, but the Orosdi-Back store in Istanbul too was constructed near the central bazaar area, at Bahçe Kapı, and not in Pera where the other major department stores were.

tions of the Khedive Isma'il to build a "Paris along the Nile". One could speak of an epoch of "Parisization".

An important move, in this respect, was the choice of Paris as the company's *siège social* (registered head office). In 1888, Leon and Philippe Orosdi, and Hermann and Joseph Back registered their firm as Établissements Orosdi-Back; it became a limited stock company in 1895.

Indeed, at some stage around the turn of the century they all appear to have moved to Paris, where they began an upward social career. Hermann Back would add "de Surany" to his family name (after a small Czech town from which his family probably originated), and his daughters married into high French nobility. Before his death in 1925 he also served as Consul-General of Persia in Paris. Leon Orosdi too saw his descendants marry bourgeois French businessmen (and he also served for a short time as Consul-General of Costa Rica). He died in 1922, leaving a valuable art collection. It is noteworthy that both had converted to Catholicism, as would appear from the funeral services held for them in the St. Honoré d'Élau church in the aristocratic XVIth arrondissement.

We assume that the fact that Orosdi-Back survived in business in Egypt until 1958 is due to the timely transfer of their *siège social* to Paris. Be it by diplomatic insight and versatility, by business acumen and intelligence or by sheer luck, Orosdi-Back prospered. In the First World War, which posed challenges in terms of their Austrian origin (and trade interests), and their new French allegiance, they could claim to be on the Allied side.

During the First World War, their counterpart Austrian firms in Egypt, such as Stein, Stross, Mayer, Salamander and Tiring came under heavy British pressure. In 1916, on the orders of the newly promulgated British Protectorate, 100 mostly smaller firms, such as 24 shops and drapery stores, were liquidated as "enemy interests". Others obtained special licenses to continue their business, albeit not trade with Germany and Austria.[43]

Orosdi-Back too had felt the pressure from the beginning of the war, albeit rather from the French side; there appeared to be more than just strategic war interests. In December 1914 the French consul requested Paris to verify whether the Orosdis had true French sentiments: "La Maison Orosdi-Back, qui est sans doute constituée à l'aide de capitaux français, n'a jamais eu la réputation d'être "française" au Caire. Au mois de juin 1913 … elle ne comptait qu'un Français et sept ou huit Protégés Français dans son personnel, mais on l'accusait de vendre

---

[43] *Report on Policy Adopted in Restraint and Liquidation of Enemy Trade, June 1917* (Cairo: Government Press, 1917), pp. 2, 17–18, 45–46, Nantes, Centre des Archives Diplomatiques (CAD), le Caire 547. These measures related also to insurance and shipping companies, as well as banks.

principalement de la camelote [sic] allemande ou autrichienne."[44] Similarly, in 1915 the French consulate demanded the replacement of an Austrian store manager by the name of Siderer by "un vrai français". Orosdi himself refused to comply and sent a list of personnel emphasizing that some of them had even been mobilized. The French remained unhappy, to say the least, as Orosdi-Back "ingeniously" continued to trade in Austrian commodities, apparently via France itself. In order to stave off further criticisms, the department store started to fly a French flag on its building, and its owners made a generous donation to the local French and Allied support committee (Comité d'Assistance Français et du Vestiaire des Alliées).[45]

Most Austro-Hungarian firms gradually went out of business in the aftermath of the First World War because of sequestration of enemy property during the war, as well as lack of supply immediately after. Mayer was sold to locals, Tiring to an Italian, and Stein was to become Morum's, a British firm.[46]

As we saw, in 1914 a point of contention between the French consulate and Orosdi-Back had been the national composition of the latter's personnel. They had countered with a detailed list, which was meant to show its mixed composition, but which in fact disclosed a tilt towards foreign higher-level employees and Egyptian workers and servants.[47] A few personnel fiches of higher-level employees in Alexandria have miraculously survived, showing such names as Morgenstern, Harari, and Marinovitz. Indeed, almost everywhere we find local Jewish managers.[48]

While many of the floor managers were men, customers were mostly attended to by saleswomen (though this aspect, to the best of our knowledge, has never been locally accorded the literary status of Zola's *Au Bonheur des Dames* or some other similar novels). An interesting group photograph of part of Orosdi-Back's personnel in 1944 – supposedly with numerous Jews among its employees and cadres, however, shows less than a quarter women.[49]

---

44 In translation: 'The firm of Orosdi-Back, which has without doubt been constituted with the help of French capital, has never had the reputation in Cairo of being "French". In June 1913, it counted merely one French national and seven or eight French protégés among its personnel, but it was blamed for selling mainly German or Austrian junk', CAD, le Caire 546, Letter, 19 Dec. 1914 and further correspondence. On the national/ethnic composition, see also below.
45 Letter Consul Bonsery to Paris, 11 Aug. 1915, Ministry, CAD, Le Caire 546.
46 *Report ... June 1917*, p. 79.
47 CAD, le Caire 546, encl. in report, 4 Nov. 1914. See also Reynolds' analysis in 'Commodity Communities', pp. 206–207.
48 Collection of M. Paoli, last secretary of the Établissements Orosdi-Back (EOB) Board.
49 *Juifs d'Egypte, Images et Textes* (Paris: Éditions du Scribe, 1984), p. 208. Cf. Reynolds, 'Commodity Communities', p. 216 and note.

At least in the haberdashery department, one knowledgeable source, indeed, describes all saleswomen as Jewish and "de bonne famille".[50] Saleswomen had to speak several languages, or, as a prominent French diplomat wrote in his memoirs: "Actuellement, au Caire, les vendeuses des grands magasins, des maisons Chemla ou Sednaoui par example ... doivent parler de quatre à cinq langues; l'arabe, l'anglais, l'italien, parfois le grec et toujours le français."[51]

The nostalgia of the former residential elite, which we mentioned earlier, may, at least in part, be connected to the ostentatious visibility of the "multi-storied Parisian styled" department stores, "no less grand than Le Printemps, the Galeries Lafayette or Au Bon Marché".[52] Even today, the eye-catching cupola of the former Tiring building at 'Ataba Square, still symbolizes the epoch.[53] Although the opposite Stein's Oriental Stores has been largely demolished, the nearby Sidnawi (Sednaoui) building still graciously stands (and functions).[54] But the Orosdi-Back building which had opened in 1909 is perhaps the pearl in the crown of Cairo's remaining department stores.[55] The building was designed in an imitation Rococo style by architect Raoul Brandon (1908–9), a graduate of the École Nationale des Arts who at the time even operated an office in Cairo (Fig. 11 and Fig. 12).[56]

The pink-coloured "Palais Omer Effendi" has survived, but the French relief inscriptions on its outer walls, each announcing a specific commodity, such as "dentelles" (lace) or "soieries" (silk), have since been effaced. The original drawings show the carefully planned layout of its six floors, its large window displays, its elevators, and abundant electrical light. Within a few years, the two upper floors,

---

50 S.V., private correspondence, 9 March 2003.
51 In translation: 'At the moment, in Cairo, the saleswomen in the large stores, Chemla, Sednawi, for instance ... are required to speak four or five languages: Arabic, English, Italian, Greek, and always French': J. d'Aumale, *Voix de l'Orient, souvenirs d'un diplomate* (Montreal: Editions Varietes, 1945), p. 75. A requirement to speak Arabic would indicate also middle-class customers.
52 C. Myntti, *Paris along the Nile, Architecture in Cairo from the Belle Epoque* (Cairo: AUC Press, 1999), p. 12. See further M. Volait, *Le Caire-Alexandrie, architectures européennes, 1850–1950* (Cairo: IFAO, 2001), and Raafat, *Cairo*.
53 As in Europe and the US, famous architects were engaged to design the department stores (e.g. Gustave Eiffel designed the Au Bon Marché (ABM) store). Tiring's was designed by Oscar Horowitz (1912–13); in Galata, Istanbul they had a building with similar cupola; Stein's store was designed by Schoen.
54 Volait, *Le Caire-Alexandrie*, pp. 149–150.
55 A. Raymond et al., *Le Caire* (Paris: Citadelles, 2000), p. 402 and illustrations. See further Raafat, *Cairo*.
56 Brandon had also designed the new Bourse Khédiviale and Villa Hug in Zamalik, see Raafat, *Cairo*, p. 183. A collection of Brandon's drawings and photographs has recently been acquired by the Musée d'Orsay. For a description of his work see *La Revue du Musée d'Orsay*, Vol.16 (Spring 2003), pp. 46–49.

Fig. 11: The building in Cairo (opened in 1908), an iconic landmark even today, resembling Au Printemps in Paris, in advertisements often depicted as a lighthouse emitting beams to the surroundings. (photo by the author, 2010).

which had been reserved for wholesale business, were also turned into retail space.[57] Today, its interior has been changed, but it is still impressive.[58] Although all such buildings have lost much of their former glory, fortunately heritage consciousness in Egypt is growing. In 2006, the building was officially classified as an "Islamic" monument.[59]

---

57 EOB Annual report, 19 Nov. 1911.
58 In Cairo, it would seem that Stein was the first to have elevators. In Beirut, Orosdi-Back (built 1901) had for years a 'monopoly' on elevators.
59 *Egypt Today*, Vol.27, No.1 (Oct. 2006).

Fig. 12: A souvenir fan with an image of the store in Cairo, probably from the 1940s, which was recently auctioned in Paris and acquired by the Fan Museum in London (courtesy: Mrs. Hélène Alexander).

It would be too simplistic to maintain that "Ces magasins ne sont, à vrai dire, spécialisés dans aucun genre".[60] They rather offered one wide but specific genre of commodities which particularly appealed to a certain layer of consumers, namely "nouveautés" (novelties), or "articles de Paris". Indeed, according to one calculation, in the 1920s, five of the seven *grands magasins* in Cairo were French.[61] It was the experience of seeing, being surprised, and, ultimately, purchasing the latest fashionable articles at a distance from the cultural capital of the world. "Dernières nouveautés" [sic], "dernières créations", "recent arrivals" as they were advertised. Truly, almost everything sold was imported (and not exclusively from France). This passion of the Cairene upper classes to be in the forefront of modern global taste was shared with elite populations elsewhere, but it was accentuated by a colonial perspective.

The development of consumption, and of a new community of consumers, is, therefore, strongly connected to social class and to behavioral patterns, which connect, in the case of Egypt, with the tremendous influx of foreigners.[62] Yet it is not easy to follow changes in consumption of dress, furniture and household articles

---

[60] In translation: 'These stores, in actual fact, are not specialized in any category', Carnoy, 'La colonie française du Caire', p. 50.
[61] Our computation from Carnoy, p. 50.
[62] Reynolds, 'Commodity Communities', has elaborated on the term community of consumers.

by mere trade statistics, be they simple consular reports or more analytical descriptions.[63] Often, the general and the specific are confused. Clothing and "textiles", or beds and "ironware" are put into the same generalized categories. Clearly, by the turn of the century, Austria had a lead in textile goods, Britain in furniture and ironware, and France in fashion and luxury goods, but statistics reveal little on demand and competition in the consumer market. Austrian, British and French consular reports, on the other hand, are all biased towards the marketing of their own national manufactures.

We can, however, in a somewhat unsystematic way, follow advertisements in the press, in French (e.g. *La Bourse Égyptienne*) and English, and later Arabic as well.[64] The impression is that in Egypt, as elsewhere, department stores were among the largest advertisers.[65] Unfortunately, so far, we have seen no catalogues from Orosdi-Back or the parallel department stores operating in Egypt.[66] It is the sort of material printed for customers, not for historians, and thus generally thrown into the dustbin. They must have looked like the sort of catalogues sent out annually by Au Bon Marché. This firm, for instance, which had a sizeable mail order business, sent out 1.500.000 catalogues in 1894 alone, of which 260.000 went abroad.[67] In addition, we have conducted some selected interviews with former residents of Cairo and of other cities with Orosdi-Back stores.[68]

The Crimean War in the mid-nineteenth century is generally assumed to have been a watershed in local consumption, especially with regard to Istanbul. Imports started to rise steeply. Also, foreign visitors to the Middle East, who would formerly have adopted local costume, now began to travel around in European clothes. Austrian ready-to-wear clothing was the first to make headway, first for men, then for

---

63 More analytic, but also not satisfactory, is for instance E. Weakly, *Report on the Conditions and Prospects of British Trade in Syria* (London: Board of Trade, 1911).
64 Advertisements in Ottoman-Turkish mostly carried the firm's name Orosdi-Back (also) in Latin characters. For Baghdad – a different social setting – we found advertisements in the *Iraq Times* with a logo in Arabic. On this era's commercial publicity, see M. Russell, 'Creating al-Sayida al-Istikhlakiyya: Advertising in Turn-of-the Century Egypt', *Arab Studies Journal*, Vol.8/9 (2000/01), pp. 61–96 and her recent *Creating the New Egyptian Woman, Consumerism, Education, and National Identity, 1863–1922* (New York: MacMillan, 2004).
65 H. Hillel, *'Yisrael' beQahir, Iton Tzioni beMitzrayim haLe'umit 1920–1939* ['Israel' in Cairo, a Zionist Newspaper in Cairo] (Tel Aviv: Am Oved, 2004), pp. 186–187.
66 We assume that they existed, see, for instance, an advertisement in *Ruz al-Yusuf* ('ask for our catalogue'), 5 Nov. 1934. Also, one of the persons interviewed on Bizerte is certain that Orosdi-Back distributed catalogues of its own.
67 Miller, *The Bon Marche*, pp. 61–62, but without details on the Ottoman lands.
68 Some Baghdadi and Beiruti memoirs, mention Orosdi-Back as a landmark (physically or in terms of consumption).

children, and lastly for women. Clearly, the consumption of various Western commodities, ever more mass-produced, was on the increase.

Also, the new habit of "shopping" – as a non-specific consumer pursuit by *flaneurs* – emerged, though not, of course, for all classes alike. Indeed, even until the mid-twentieth century it was not common for middle- and higher-class women in the Middle East to stroll around in bazaar areas; it was the men who went out. In Cairo, the Muski area was probably the first where women went to shop freely. It is an aspect which still needs more research, and, in particular, the use of cash or credit.[69] However, in the inter-war period, in Cairo, and in other large cities with specific new business districts and suitable public transport, this had already become more common for the upper strata of the middle classes.

This brings us to the topic of fashion and fashion-consciousness, which was closely connected. One researcher, mainly focusing on Istanbul, has ascribed the change in taste, particularly in women's dress, to the "presence of European women" in Istanbul.[70] As to Cairo, we may argue that Istanbul had been a trend-setter, but this had probably changed by the turn of the century.[71]

For Orosdi-Back, ready-to-wear clothing for men, women and children remained throughout the leading category. It could be made to measure in ateliers on the premises. Only the local Au Bon Marché apparently had a fully-fledged local line of production.[72] Shirts were offered for sale too, and – as mentioned more than once by former clients – underwear as well.

Bonneterie or hosiery (that is products knitted as a hose, theoretically from one thread) was an important new category of European manufactures. A new article made possible by more advanced machinery, and one relatively inexpensive at that. By extension this included not only socks and stockings, but also vests,

---

[69] A. Blind, *L'Orient vu par un médecin, Egypte, Palestine, Syrie* (Paris: APM, 1913), pp. 93, 134–135. In Istanbul, the Pera area fulfilled the same role.

[70] N. Micklewright, 'London, Paris, Istanbul, and Cairo: Fashion and International Trade in the Nineteenth Century', *New Perspectives on Turkey*, Vol.7 (1992), pp. 125–136. The author ascribes the beginning of the phenomenon to the 1838 Convention and to contemporary changes in trade and trade. True, in the 1850s, some women of the former Egyptian Pasha's court allegedly became big spenders in Istanbul, see A. Cevdet, *Ma'azurat* (Istanbul: Çağrı Yayınları, 1980), pp. 7–8. For Cairo itself, however, it would seem that this development is somewhat slower and later.

[71] Cf. Sherif Mardin's term 'super-westernization' in an ideological rather than a material sense, and on changes in material culture. S. Mardin, 'Super-westernization in Urban Life in the Last Quarter of the Nineteenth Century', in P. Benedict *et al.* (eds.), *Turkey: Geographical and Social Perspectives* (Leiden: Brill, 1974), pp. 403–445.

[72] Carnoy, 'La colonie française du Caire', p. 51.

waistcoats, cardigans, shawls, baby clothes, and underwear.[73] Orosdi-Back was credited with introducing this commodity to Istanbul – we assume that this was the case in Cairo as well – but it was soon offered by all foreign department stores.[74] With modern shoes being more generally adopted in Egypt, a member of the Back family could imagine that the store was "to educate Egyptians to wear socks".[75] But before long local firms started to enter this field of production as well; Shurbaji, a firm owned by Syrians, did very well.[76]

In 1899 Orosdi-Back proudly announced that it had taken the initiative, after many earlier efforts, to implement a merger of all the tarbush (fez) manufacturers. Capable Bohemian manufacturers, which had the lion's share of the international market, felt threatened by price competition and increased output from France (which had previously cornered a large part of the market), Italy, Germany, Belgium, and, indeed, the Ottoman regions themselves.[77] With the Kredit-Anstalt in Vienna, a joint company was formed with a capital of Fr. 3.2 m. In return, Orosdi-Back acquired the exclusive right of its representation in the entire world (sic).[78] Indeed, the new syndicate of fez manufacturers in Strakonice (Bohemia) long remained the largest in the world. Orosdi-Back became "seule dépositaire [exclusively authorized agency] pour la Turquie". The 1908 boycott affected this typical Orosdi-Back branch of activity, though we do not know exactly to what degree this was also the case in Egypt. In Turkey in 1925, when the fez was abolished, it became a nationalist symbol in Egypt, particularly of what was called the "effendi class". When Ahmad Husayn's "Young Egypt" successfully staged his Piastre Plan in

---

73 It would take until the 1920s for local manufactures to compete for this market, see Reynolds, 'Commodity Communities'.
74 *Revue Commerciale du Levant*, No. 207 (June 1904), p. 925.
75 Nicole Back quoting her aunt (see note 40).
76 See Reynolds, note 88, pp. 337–344.
77 L. Tanatar-Baruh, 'A Study in Commercial Life and Practices in Istanbul at the Turn of the Century: The Textile Market' (MA thesis, Boğaziçi University, 1993), pp. 197–199. In Egypt Kaha also had a share of the market, while in Turkey the Karamürsel company tried to capture part of it. For a short history of Austrian tarbush manufacturing see: K. Batheit, 'Nachrichten über den Fezexport Oesterreichs nach dem Orient im 19. und beginnenden 20. Jahrhundert', *Vierteljahrschrift für Sozial- und Wirtschaftsgeschichte*, Vol.29 (1938), pp. 296–303 (with thanks to Manfred Sauer). The tarbush industry has recently been studied by Markus Purkhart, *Die Österreichische Fezindustrie* (Unpublished PhD Thesis, University of Vienna, 2006).
78 Rapport to AG of 12 July 1899. The Strakonice syndicate comprised the factories of Volpi, Loussi, Furth, Wolff & Co., Stein Str., J. Stein, Bondy, Klein, Munch, Zucker and Gulcher. This development was criticized by the Istanbul Chamber of Commerce as hindering free competition, Tanatar-Baruh, 'A Study in Commercial Life', p. 199. See also *Revue Commerciale du Levant*, No. 151, 31 Oct. 1899, p. 799.

1931–32, the money collected was invested in a tarbush factory. It was Cicurel, not Orosdi-Back, which propagated the locally manufactured fezzes.

Orosdi-Back not only kept large stocks of cloths, textiles and yarns (e.g. Clark & Co. machine cotton threads). Some went there to buy quality cloth – of European origin – or other material for dressmaking, or for school uniforms. No less important was the representation of the Singer sewing machine company. Orosdi-Back even served as its depot in, for instance, Beirut (rather exceptional as Singer always wished to hold sales, dealership and service in its own hands).[79]

One should remember that clothing is not the same as fashion. Department stores are almost by definition adaptable to market opportunities, they can seek out new niches, and can, in fact, generate new demands. This applied not only to hats, shoes, boots and gloves (for both men and women), and canes. Women could find perfumes and cosmetics in the store, but also paste jewellery (*bijouterie fausse*), a new trend first imported from Gablonc, Bohemia, in which Orosdi-Back proved strong. It was undoubtedly a novelty, which ran counter to the long-established tradition of using real gold for adornment.[80]

Umbrellas or parasols were also an important product, though we do not know whether those sold in Cairo were assembled from imported parts or manufactured by their own workshop in Istanbul.

The company had always sold watches. By 1898 it had established a watch factory of its own in La Chaux-de-Fonds. However strange this may sound in our era of established Japanese manufactures, it would seem that Orosdi-Back had mainly that market in mind. But Japan raised its import duties in 1903 and the venture failed.[81]

In 1908 Orosdi-Back tried to popularize another new article, gramophone records. Noteworthy here, because of the actual or potential audiences, and their tastes in music, is the case of the popular Egyptian singer Yusuf al-Minyalawi, who recorded some songs for them.[82] We know about a law suit between him

---

[79] P. Fesch, *Constantinople aux derniers jours d'Abdulhamid* (Paris: Plon, 1905), pp. 517, 604. See also U.M. Kupferschmidt, 'The Social History of the Sewing Machine in the Middle East', *Die Welt des Islams*, Vol. 44, No.2 (2004), p. 203, here Chapter Three.
[80] *Revue Commerciale* (March 1909), pp. 439–450. In the post-Second World War years, Orosdi-Back would maintain a stake in Burma Bijoux, which even today is famed at Bd. des Capucines in Paris. See www.bijouxburma.com.
[81] Orosdi-Back maintained establishments in Yokohama and Kobe until 1904, possibly also to purchase merchandise.
[82] Press cuttings, CAMT, 65 AQ T-136–7.

and Orosdi-Back with regard to exclusive recordings made for them in 1907–8.[83] The marketing of this sort of music is intriguing because such recordings catered to local taste – we assume, in any case, not primarily to the taste of most foreign residents. Also, those buying gramophones and records must, of course, have had the means for it. It seems, however, that this line was not pursued.[84]

From a clothing store, Orosdi-Back had gradually become a department store. It added leather ware (*maroquinerie*), household goods, glassware, travel goods, ironware, and some furniture to its range. Though some packed or canned groceries such as biscuits, chocolate and sardines were sold as well, a "service d'alimentation" (food department), maybe initially set up to cater for foreign troops during the First Wold War, was closed in 1927.[85]

The supply of ever changing *nouveautés* was, of course, only one side of the story. For our narrative it is important to establish how these commodities were "democratized" from the conspicuous consumption of the upper classes to at least the upper layers of the middle classes. What do we know about the clientele of the department stores? Samir Saul, referring to the beginning of the twentieth century, says: "des colonies étrangères et des Egyptiens dont les réserves sont d'un niveau moyen ou élevé".[86] A study from the 1920s, however, states: "Ces magasins, indépendamment de la clientèle européenne, s'adressent surtout à la clientèle arabe [sic] riche."[87] Diffusion, or "democratization" of Western commodities, for better or for worse, is a critical aspect.

It is probably correct to distinguish – as anywhere in Europe or America – between relatively stratified clienteles. Such a hierarchy may also have been changing over the years. Without doubt, Cicurel was the top trend-setting store in Cairo, probably followed by Sidnawi. Orosdi-Back, owing to its location closer to the more "popular" 'Ataba area – a major public transport nexus point in the city – probably served upper middle class rather than upper class customers.

---

83 See clipping from *Journal du Caire*, 30 Jan. 1908, CAMT, 65AQ T136–7. On al-Minyalawi see A.J. Racy, *Making Music in the Arab World* (Cambridge: Cambridge University Press, 2003), pp. 45–46, 69, 160, 170.
84 HMV opened its first store in Cairo in 1905, and soon thereafter branches in other cities and towns, Wright and Cartwright, *Twentieth Century Impressions of Egypt*, p. 472.
85 CAMT, 65 AQ T137, SEF 1928.
86 In translation: 'foreign colonies with average reserves or more', Saul, *La France et l'Egypte*, p. 517.
87 In translation: 'these stores, apart from their European clients, applied themselves above all to rich Arab customers', Carnoy, 'La colonie française du Caire', p. 50.

The story of changing consumption patterns has only recently begun to be written.[88] In Egypt, social advancement, political awareness and economic activism together clearly had an impact on the production side. Maybe the first harbinger of change was the 'Awf department store near al-Azhar (1914). Further indigenous initiatives and consumptive self-assertion in the 1930s led to the founding of several Egyptian national clothing retail stores, e. g. Misr lil-Muntajat al-Misriyya (owned by Banque Misr), Mahallat al-Muntajat al-Misriyya and Sharikat Bayt al-Masnu'at al-Misriyya. Although these never attained the prestige of the foreign stores, they clearly captured part of the growing consumer market. This was not typical of Egypt. In strongly etatist Turkey, however, with a campaign to buy local manufactures, Orosdi-Back was phased out of business, and finally had to sell its main store in Istanbul in 1943.

There is one additional business aspect which ought to be drawn into this discussion, even though our source material is, unfortunately, meagre. Consumer habits and ethics must have had an impact on the entire business operation. It was said that the Egyptian public bought much, but paid poorly (cash payments were rare). In the inter-war period a complaint could be heard against women purchasers, "spécialement les plus riches – rechignent à l'idée de solder un compte. Les factures sont indéfiniment renvoyées à la justice, elle se voit souvent contrainte de subir des amputations de facture".[89] While all department stores had to be careful with credit to customers, it would seem that Orosdi-Back was generous in this respect.[90]

All department stores advertised end-of-season sales, or sales towards Ramadan or Christian and Jewish festivals, back-to-school sales, as well as special exhibitions. Tombolas and lotteries were very popular at the time, and apparently served as an additional means to attract customers. For some time Cicurel publicized monthly lotteries, reimbursing the value of recent purchases. In 1923 one could even win a Berliet car.[91] In 1925, Orosdi-Back awarded all customers spending over 100 PT (*piastre*, one tenth of an Egyptian pound) the right to participate in

---

[88] N. Reynolds, 'Sharikat al-Bayt al-Misri: Domesticating Commerce in Egypt, 1931–1956', *Arab Studies Journal*, Vol.7/8 (1999–2000), pp. 75–107; M. Russell, 'Modernity, National Identity and Consumerism: Visions of the Egyptian Home, 1805–1922', in R. Shechter (ed.), *Transitions in Domestic Consumption and family Life in the Modern Middle East* (New York: Palgrave, 2003), pp. 37–62; Berque, *Egypt*, pp. 331–333.
[89] In translation: 'in particular the richest balked at the idea of settling an account. Bills are indefinitely sent to the Courts, which see themselves often forced to arrange for a reduction of the bill', Carnoy, 'La colonie française du Caire', pp. 64–5.
[90] Reynolds' PhD Thesis, 'Commodity Communities', p. 141.
[91] Carnoy, 'La colonie française du Caire', p. 52; *L'Egypte Nouvelle*, 7 Jan. 1923.

a monthly lottery whereby 26 premium shares of the Crédit Foncier were granted.[92]

Even before the Second World War, it could be seen that the golden years of French hegemony in fashion and taste were already in decline. The potential clientele, and especially rich women, could more easily and more freely travel to Europe and make their purchases over there. "Estivage" (summering) in Europe became more popular. There was even a strong element of snobbism attached to it.[93] Still, in the years immediately after the Second World War, Orosdi-Back resumed its purchases in Britain and in France, which shows that their business expectations had not declined significantly.[94]

But it could be sensed that times were changing. Even before the nationalizations and the accelerated exodus of the foreigners, it would seem that there was a changeover in personnel. The company law of 1947, which aimed at replacement of foreign nationals by Egyptians, had its effect. By that year, at Cicurel, Egyptians (and Syrians) already outnumbered foreigners.[95] In the words of one interviewee: "There were [at Orosdi-Back] more and more Arabs [sic] who were "nonchalant" and we noted how the whole atmosphere was different. I do not know if the Egyptian government imposed a quota of Arab employees, but I do remember that the Jews left there were a small handful."[96] Indeed, we know that Hannaux under the new company laws of the late 1940s complained bitterly about the difficulty of hiring suitable saleswomen.[97]

Moreover, one woman whom I interviewed was or claims to have been under the impression that after 1948 Jews were no longer allowed to live within a radius of 3 kilometers from 'Abdin Palace, which caused a lot of problems for businesses and residents, and which also meant more problematic access to the Orosdi-Back store.[98] Life had become more difficult for Jews and others who belonged to the foreign residential elite.

Our main point, however, is that the economic capacity to buy Western-type consumer goods was trickling down to the upper layers of the middle classes. Be-

---

92 *L'Egypte Nouvelle*, 31 Jan. 1925.
93 Carnoy, 'La colonie française du Caire', pp. 64–65; cf. Barakat, *The Egyptian Upper Class*.
94 In Dec. 1946, two 'acheteurs', Zaki Pandelis and Pandelis Demitriades, were in England and requested a visa to enter France, CAD, le Caire 249.
95 F. Karanasou, 'Egyptianization: The 1947 Company Law and the Foreign Communities in Egypt' (PhD Thesis, University of Oxford, 1992), pp. 272–274.
96 S.V., private correspondence, 9 March 2003.
97 MAE, Egypte, carton 83, K/82/8, 22 April 1948 [Egypte with E accent]. Cf. Karanasou, 'Egyptianization', p. 273.
98 J.S., private correspondence, 20 June 2003.

cause of the lack of consistent local sources, this complex process is difficult to reconstruct, but advertising in Arabic is one marker. The earliest advertisements in the Arabic press (which we have found so far) date from the 1930s.[99] Let us also consider the economist Galal Amin's recently published memoirs:

> When my father wanted to purchase some clothes for us he would take us to one of the emporia in downtown Cairo's Fouad Street (now 26 of July) or 'Ataba Square, most of which had foreign names like Avarino, Cicurel, Chemla, or Sidnawi. The garment industry was the preserve of non-Egyptians and Europeans, many of them Jews or Levantines, mainly Lebanese and Palestinians. There were a few exceptions to this rule ... And so the situation remained until the end of the 1950s, when the sweeping nationalizations of foreign property occurred as well as the wave of Egyptianization that overtook industry and commerce.[100]

One assumes that the new Officers' regime, in spite of indigenous roots and a different socioeconomic background, and no special affinity to the French taste or style of the upper classes, had also become accustomed to the use of department stores.

The Nasirist Revolution ousted the foreign residential elite, neutralized its domestic bourgeoisie, sequestered and nationalized its assets, but did not liquidate the department stores as erstwhile carriers of luxury and conspicuous consumption. "Concentration, Egyptianization, and nationalization" were aimed against the exclusive ownership – or "monopolies" in the ideological discourse of those years – of "a few large Levantine families. Some commercial societies were even sold "à l'amiable" to local Egyptian business families.[100] Though cutting down commodity imports drastically, the new regime apparently still did see a role for department stores as such. It did not, in any case, abolish them as redundant, but hoped to use them as outlets for its local industries. Even the Soviet regime had maintained the Gosudarstvemy Universalny Magazin (GUM) stores for its own purposes.[101]

The upshot of Omar Effendi's sequestration and nationalization appears to be an interesting story in itself. While negotiating in Switzerland with France about compensation for French companies, the Egyptians, in a show of force, or intimi-

---

99 E.g., *Ruz al-Yusuf*, 5 Nov. 1934; in *al-Ahram*, Orosdi-Back is seen advertising from the early 1950s.
100 H. Riad, *L'Egypte Nasserienne* (Paris: Minuit, 1964), p. 129 and note. Cf. *La voie égyptienne vers le socialisme* (Cairo: Dar al-Maaref, [1962?]). A critical observer attacked 'Abd al-Nasir's daughters for continual shopping expeditions to Paris, see I. Powell, *Disillusion by the Nile, What Nasser has Done to Egypt* (London: Solstice, 1967), p. 37, but there was definitely a big change in consumer habits even among the old upper echelons of society.
101 M.L. Hilton, 'Retailing the Revolution: The State Department Store (GUM) and Soviet Society in the 1920s', *Journal of Social History* (Summer 2004), pp. 939–964. Cf. also the case of Turkey.

dation, possibly bought off Orosdi-Back in an underhand deal, even before the final settlement of August 1958.[102]

For reasons that were not made explicit, the new management of Orosdi-Back chose to use the old name Omar Effendi (which had never been completely abandoned). It could hardly serve as an inversion of the former Ottoman title, because the ascribed status designation of "effendi" had also been made obsolete by the Revolution.

In the words of Samir Raafat, referring to Elie Sidnawi (Sednaoui), but probably true for all department stores: "An army officer had come in to replace him. The age of drab counters and khaki-colored cooperatives was about to begin."[103] Indeed, it has often been argued that nationalization led to Egypt's department stores declining into "gray bureaucratic mediocrity, selling undistinguished goods at inflated prices".[104] Another complaint was the large amount of unsold stock. However, this may be only partly true because there must have been something to sell. Since its nationalization, Omar Effendi has expanded from some 20 stores in 1961 to 83 outlets and 15 depots today all over the country. It thus became the largest chain in Egypt and attained a rather large turnover (£E 463 m in 2000). One store can even be found at Sayyida Zaynab square, a place often cited as typical of "average" popular Cairo.[105]

The *Infitah* has once more shuffled the cards for department stores. From 1996 onwards, the Egyptian government embarked on a privatization program in which, thus far, the sale of the Omar Effendi chain has played a salient role. Overly bureaucratic management, unsold stock, inflated employment rolls, as well as a con-

---

[102] Developments are not entirely clear from PRO (NA), FO 371/ 125577, 125467, 125467, 131353 and 131354. See R. Tignor, *Capitalism and Nationalism at the End of Empire* (Princeton, NJ: Princeton University Press, 1998), p. 145; Riad, *L'Egypte Nasserienne*, p. 129 n.100. See also Ministère des Affaires Étrangères (MAE), République Arabe Unie (RAU), April 1958–Dec. 1959, carton 1–2, on Egyptians, old and new businesspeople, aspiring to take over abandoned French interests. Actually, Orosdi-Back remained active in some Middle Eastern countries until the 1960s, after which the firm focused on metropolitan France. The company as such ceased operations when taken over by Robert Lascar, a French tycoon, in 1992, but remains registered on the Paris stock exchange.
[103] *Cairo Times*, 29 May 1997.
[104] *Egypt Almanac* (Wilmington: Egypto-file, 2003), p. 180.
[105] G.E. El-Din, 'Scramble Over Department Stores', *al-Ahram Weekly*, 23–29 Sept. 1999; G.E. El-Din, 'New Lease of Life for Flagship Stores', *al-Ahram Weekly*, 18–22 Dec. 1999; N. Ryan, 'Disposable Dinosaurs', *al-Ahram Weekly*, 7–13 Dec. 2000; *Egypt Almanac*, 2003, p. 180. Meanwhile Sidnawi also expanded to 74 stores and 51 entrepôts. See http://www.ajoe.org/grandmag/gm.htm. Cf. American Chamber of Commerce in Egypt, *Business Monthly* (April 1999).

dition on the part of the governmental Egyptian holding company to protect the rights of the employees, deterred potential buyers.[106]

Over the years, no less than four rounds of bidding, in which various Egyptian, Spanish, and Kuwaiti parties appeared to be interested, failed. The government then made an effort at reorganization, started a retirement plan for personnel, and managed to show a profit in 2005 for the first time in years. At last, in September 2006, the chain was sold to the only remaining bidder, namely the Saudi Arabian Anwal United Company, for £E 589.5 m, and a series of additional financial commitments on debt settlement and overdue upgrading of the stores.[107]

This all led to some public debate on privatization in general (that is on distrust of the government and on corruption), which was to some extent echoed by the opposition in parliament. Opponents of the sale stated that the price (ironically for what had been a symbol of bourgeois consumption and then one of nationalist assertiveness) was too low. The land alone, or even the trade name itself, would be worth more, some argued. The press also wrote about Omar Effendi as a "faded highstreet diva", or "the grande dame of Egyptian department stores", and a "national icon". Sometimes its Jewish roots were even mentioned.

It remains to be seen what the new Anwal era will bring. The company runs stores in a dozen Saudi Arabian cities, but this is apparently its first venture abroad. On the other hand, it holds franchises for selling a number of prestigious French and American brands, which parallels old times.[108]

Clearly, since the *infitah*, there has been a shift in upper- and middle-class consumer patterns in urban Egypt, in particular towards private boutiques and imported quality goods. This also explains why the department store chains faced further decline. But department stores everywhere may have passed their heyday, including in Egypt.[109] The more up-to-date version of shopping malls may offer an attractive and, even more, a culturally palatable alternative, a space more closely resembling the old bazaar.[110]

---

106 *Egypt Almanac 2003*, p.180; K. El-Sayed, 'Egypt's Last Effendi, on the Government Efforts to Privatize Department Stores', *The Chronicles* (AUC), Vol.1, No.4 (April–June 2006), pp. 22–24.
107 The Egyptian government keeps 10 per cent of the shares, as well as the land. The two 'historic' buildings in Cairo and Alexandria remain protected. See *al-Ahram Weekly*, 27 April-3 May 2006; *The Daily Star*, 10 July 2006; *al-Ahram Weekly*, 28 Sept.–4 Oct. 2006 and 5–11 Oct. 2006, and various articles on the internet.
108 E. g. Etam and Oshkosh.
109 P. Schemm, 'Cautious Consumerism', *Cairo Times*, 26 Dec. 2002–1 Jan. 2003. In addition, so far in Egypt large supermarkets have been unable gained a foothold – as was attested a few years ago by the failure of the British Sainsbury's chain.
110 M. Abaza, 'Shopping Malls, Consumer Culture and the Reshaping of Public Space in Egypt', *Theory, Culture and Society*, Vol. 18 (2001), pp. 97–122.

1.2 Postscript

Since the publication of this chapter as an article in March 2007, further vicissitudes in the broad public debate on nationalization versus privatization have taken place in Egypt. In fact, Omar Effendi was a critical part of it. Under the impact of accusations of mismanagement and other irregularities, tainted by government corruption, and in the wake of the Revolution of 2011 (the so-called "Arab Spring"), an Administrative Court annulled the sale of Omar Effendi to the Saudi Anwal company. Ownership of the chain reverted to Egyptian hands, and the company has meanwhile embarked on modernization measures, seeking an upgraded image in accordance with its pre-1952 reputation: higher quality products and foreign brands, since then some have even been offered on-line. This also includes a full renovation of the flagship store at 'Abd al-'Aziz Street in Cairo. But bringing in a new Egyptian investment group has so far apparently failed. The question is whether all this will be an adequate solution for the company's losses and debts; exploitation results remain unclear. Lately (2021) there has been talk of Carrefour, the French supermarket chain, setting up shop at fourteen Omar Effendi branches. It may not be the end of the story. Yet the department store chain nowadays proudly advertises itself as having been founded in 1856, and its over eighty branches continue to operate, which indicates that there is a contingent of consumers which needs and patronizes them.

2 On the diffusion of "Small" Western technologies and consumer goods in the Middle East

The first wave of globalization was characterized by the intensified and accelerated movement of ideas, commodities and technologies between Europe and America on the one hand and the countries of the Middle East on the other. The technological, scientific, and industrial revolutions, the ensuing mass production and mass marketing, steam navigation, new and faster communication networks, and the development of the private press, had, at least for the new era, made the transfers one-directional.[1]

It was the age of accelerated chains of new inventions – indeed few were the result of one brilliant *eureka!* – many of which soon became available in the sense that powers, rulers, or elites were able to acquire and use them, for better or for worse, and subsequently individuals could purchase them and benefit by them.

The region of the Middle East was not part of an organic chain of inventions, but discovered them one by one, sometimes through cultural journals and translated textbooks, by description rather than by commercial publicity, and more often than not – after their physical arrival.

"Technoscapes", which in principle flow in two ways, became, on the whole, one-directional in this period.[2] There could be occasional skepticism, satire, or

---

[1] The absoluteness of this direction, as reflected also in our secondary literature, needs some caution. One example which comes to mind is the adoption of local techniques of vaccination, propagated by Lady Montagu, and later scientifically developed in Britain by Edward Jenner (see Lady Mary Wortley Montagu, *The Complete Letters* (Aix, 1796), vol. I, pp. 76–105. The principle of the IUD, also, which had been practiced by Bedouins for their camels was taken up and refined first in in Europe and later in the USA. Even in the nineteenth century, there still were instances of local individuals who tried to compete with Europe. In 1855, a mechanical reaper, developed in Algeria, competed at an exhibition in Paris with other models from the United States and Britain, but it lost out: Edward W. Byrn, *The Progress of Invention in the Nineteenth Century* (New York, 1900), p. 201. The Armenian engineer Serkis Balian, of the well-known family, exhibited an agricultural machine at the National Ottoman Exhibition in Istanbul in 1863: Empire Ottoman, *Coup d'Oeil Général sur l'Exposition Nationale á Constantinople, le Octobre 1863* (Istanbul, 1863), pp. 166, 176. Similarly, a certain Iskandar Ilyas Nasru in Egypt was reported to have developed new agricultural mechanical devices, albeit with German help: *al-Muqtataf* 20 (1896), pp. 662–669.

[2] In Arjun Appadurai's spectrum of finanscapes, mediascapes, ethnoscapes, ideoscapes, and technoscapes, the latter refers to 'the global configuration, also ever fluid, of technology ... both high and low, both mechanical and informational, [which] now moves at high speeds across various kinds of previously impervious boundaries': see *Modernization at Large* (Minneapolis, 1996),

https://doi.org/10.1515/9783110777222-003

criticism, but the flow of new technologies could hardly be halted. Not even by 'ulama, supposedly a conservative element both in the establishment and in opposition to it (e.g., al-Afghani): most, as a matter of fact, took an instrumentalist stand, convinced that new technologies could advance their causes.[3] Rather, one may find reservations about Western innovations in the realm of material culture and luxury commodities.

Most studies have placed their emphasis on what we call "big technologies", those which were often connected with imperialism. Steam navigation, railways, telegraph networks and electric power grids have received most attention.

In this sense, "big" is defined as moved by regimes, rulers, foreign entrepreneurs or by dependent or related indigenous elites, and operated by them to their particular advantage.[4] Current literature tends to view such technologies critically, as embedded in colonial or imperialist power structures and working in their interest, which is the reason why some nationalists (Arab, usually not Turkish) entertain criticisms. Such technologies arrived often "turn-key", in certain cases even accompanied by (invited) foreign experts or technicians. By "big" we do not primarily refer to a physical dimension or a large capital investment, though they mostly were, but to an infrastructural system or grid, and centralized management of personnel. Think of railways and telegraph lines.

The new technologies excited certain segments of the indigenous populations, as evidenced by the discussion in cultural and scientific magazines which added didactic illustrations, and later also photographs. Over time, wider populations, short of owning them, could also take part in or even benefit by them. One case would be the telegraph which was introduced into the region by Britain for its imperial interests, and was adopted by the Ottoman government as a means of control, but which at a certain point began to serve merchants, bankers, journalists and others, followed by travelers, relatives and friends in different locations, or

---

p. 34. Regarding uni-directionality, consider also the international patent system, which started in 1883 and was from the beginning dominated by the USA, with few patents from European and smaller industrial countries. This continues to this day. While in 1975 in Egypt, for instance, 396 new letters of patent were registered (of which only 18 by Egyptians), this figure was 36.992 for Japan, and 46.603 for the US. See Dorothee Greans El Sirgany, *Les Brevets d'Invention en Égypte* (Cairo, 1978).

3 Cf. a rare opposition tract in Rudolph Peters, 'Religious attitudes towards modernization in the Ottoman Empire', *WI* 26 (1986), pp. 76–105. In fact, fatwas against new technologies which became accepted may have become redundant and were lost.

4 E.g., Daniel R. Headrick, *The Tentacles of Progress* (Oxford, 1988) and his other works, lately *Power Over People, Technology, Environments, and Western Imperialism, 1400 to the Present* (Princeton, 2010). Some of the systems could be locally adopted.

even petitioners and complainers.[5] Similarly, the "common man" could travel on (at least, third-class) trains, though railways had been constructed primarily for larger economic or commercial purposes. Thus, where governments initiated and dominated the "big" technologies, the individual could nevertheless have a "small" share in them.

2.1 "Big" and "small"

Yet we suggest making a schematic distinction between 'big" and "small" technologies. The idea occurred to me when reading Annabelle and Ali Sreberny-Mohammedi's book *Small Media, Big Revolution*.[6] Their study showed how crucial the spread of audio cassettes, a media both accessible and not controlled from above, among the population had been in the making of the 1979 Islamic Revolution in Iran against the background of the last Shah's grip on the "big" mass media.[7] Some of the recent new communication technologies play a similar role.

Many "small" technologies, depending on one's vantage point, may also be called consumer products, consumer durables, or dry goods. They were adopted, acquired, or appropriated as commodities on the market, initially by elites, but gradually trickled down to lower strata, according to their means, needs and desires.

This overlap between "technology" and "consumer goods" needs clarification. We are dealing here with different aspects of the same thing. "Technology" is usually understood to include not only materials and tools, but also know-how, especially "what for?" Thus, a "small" technology is more than merely an object. Think of automobiles, which need drivers who have undergone training (or even a specialist chauffeur), suitable roads, gasoline and supply stations, tires and spare parts, garages, as well as a servicing environment with experienced personnel.

I argue that many of the objects which form the practical application of technological innovations are more than simple "black boxes" in the user-friendly

---

[5] Among others, Yuval Ben-Bassat, *Petitioning the Sultan, Protests and Justice in Late Ottoman Palestine* (London 2013), pp. 34–35, 42, 57 *et passim*. In Iran, there were local protest actions in which telegraph stations turned into *bast* (refuges).
[6] Annabelle and Ali Sreberny-Mohammedi, *Small Media, Big Revolution* (Minneapolis, 1994).
[7] The authors also used the term 'small' 'to denote 'horizontal', for people-to-people communication. One hardly needs to elaborate on a similar dynamic process with regard to the use of new mass media evident in the latest developments in the Middle East.

sense of Bruno Latour.[8] At least at the beginning, they require the user's familiarization and sometimes guidance, e.g., sewing machines, cars, electrical devices. Thereafter, more significantly, they generate new social relations and habits, or create new professions.

Local individuals or groups with the capacity or ability to acquire the tools or appliances as commodities at their initiative, and at an affordable or equitable price, used them to their own benefit. Rather than Michel Foucault's heavy emphasis on power structures, or older theories on the compulsive consumption of luxury goods put forward by Thorstein Veblen or the Frankfurt School, we discover consumers possessing a measure of agency with regard to free acquisition. Undeniably, some wished to appear "modern" (or "civilized"), at par with their European counterparts.

Elsewhere we have described the advent of a host of such new commodities in the countries of the Middle East. Department stores and specialized agencies and dealers offered *nouveautés* or *articles de Paris*, from clothing and *bonneterie* to fezes, and from costume jewelry to umbrellas, watches and children's toys, but also sewing machines and musical instruments, which are the more typical "small" technologies.[9]

Many, maybe most, of the technological innovations appeared on the market as mass-produced consumer goods and their impact soon began to affect millions of lives.[10] In the imagination of a world market, it would at first glance seem that the USA took a leading position with regard to mass products, and often it was one firm which dominated: McCormick reapers (before 1858 when they registered a crucial patent), Singer sewing machines (1853), Remington typewriters (1873, following successful arms manufacturing), General Electric incandescent light bulbs (1879), Kodak roll-films and cameras (1888), Gillette safety razors (1902), Ford Model T cars (1908), Waterman and Schaefer fountain pens (1912)[11] and Frigidaire refrigerators (1918), to mention but a few. The fact is that all of these had

---

[8] Bruno Latour, *Pandora's Hope* (Cambridge MA, 1999), p. 304 and passim, uses the verb blackboxing for ' ... when a machine runs efficiently ... one needs focus only on its inputs and outputs, and not on its internal complexity'. The term probably derives from the idea of a camera obscura.
[9] Uri M. Kupferschmidt, *European Department Stores and Middle Eastern Consumers: The Orosdi-Back Saga* (Istanbul, 2007), pp. 37–40.
[10] Europe may be said to have already had mass manufacturing, e.g. of print, for a long time, but now the Taylorist-Fordist mode was adopted. See Carolyn Marvin *When Old Technologies Were New* (New York, 1990) to sense the fast impact of newly introduced devices of the late nineteenth century such as the telephone, the gramophone, electric light and the cinema.
[11] Like other 'inventions', it is claimed that the idea of fountain pens originated in the Muslim world (tenth century) but it did not lead to diffusion. Cf. C.E. Bosworth, 'A mediaeval prototype of the fountain pen?', *Journal of Semitic Studies* 26 (1981), pp. 229–234.

competitors, and cheaper European brands often competed successfully for the Middle East, as can also be seen from advertisements.[12]

2.2 Diffusion and the process of acquisition

There is good reason to make a distinction between dates of invention or first patenting of the technologies we are talking about, and their coming into common use. The time lag between their invention abroad and their first demonstration in the Middle East, however, was continuously diminishing, and could ultimately be a matter of a few years. That process, however, was uneven for different technologies and goods.[13]

Everett Rogers posited a determinist S-curve that passes through the phases of "knowledge", "persuasion", "decision", "implementation", and at last, "confirmation".[14] The American model is not wholly applicable to the countries of the Middle East. One reason is that the said technologies were not invented or developed within the region but were for the first time noticed or described when they had already gained mass acceptance in North America or Western Europe and were for that reason adopted after some more time had passed. Within the region, as it was around 1900, we have to consider the role of the colonial and the so-called foreign residential elites as models. Secondly, it cannot be ignored that many other layers of the indigenous populations were too poor to afford them.

It is difficult to trace exact import trade statistics. They are often too general ("iron ware", "instruments", etc.), or confound weight, volume, value and quantities. Reports by foreign chambers of commerce are often biased. Only few large industries have kept business records relevant to the Middle East. Besides, commercial trade catalogues (where they were published at all – there were debates on their utility) have survived only in small numbers, and most were probably thrown away. As a consequence, insight in the diffusion of "small" technological consumer goods suffers from generalizations.

---

[12] This was not only a question of relative poverty, geographical distance and imperial leverage, but, judging from US trade reports, also because of the American unwillingness to supply credit terms as European manufacturers and agents did (Singer is an exception, see below). Certainly, American goods were less triumphant in the Middle East than in Europe, cf. Victoria de Grazia, *Irresistible Empire: America's Advance through Twentieth-Century Europe* (Cambridge MA, 2005).
[13] Cf. David Edgerton, *The Shock of the Old* (London, 2006), pp. 31–33. The 'indigenation of modernity' still calls for a deeper investigation. See Frank Dikötter, *Things Modern* (London, 2007), on the diffusion and adoption of various consumer goods in China, as a case in itself.
[14] Everett Rogers, *The Diffusion of Innovations* 4th ed. (New York, 1985), pp. 10–11 *et passim*.

On the other hand, the discourse in the journals of the time, and all concomitant publicity, e.g., advertisements, are a relatively useful window on the process of diffusion, at least insofar as powerful manufacturers and agents persisted, e.g. Singer, Kodak and the major car manufacturers, though one has to take into account that even so they cannot serve as proof of wide diffusion.[15]

2.3 About use and need

Not everything could be sold, or mass-marketed, or diffused at the same speed: let us begin illustrating this with two older examples, namely reading glasses and mechanical clocks.

Eyeglasses were developed in thirteenth-century Italy and, since the Renaissance, widely marketed by colportage all over Europe.[16] Some researchers have connected the post-Renaissance "reading revolution" not only to education but also to the use of reading glasses. It would seem that the process in the Middle East was quite different. Remarkably, the Middle East had seen a few outstanding scientists learned in optics (e.g. Ibn Firnas in the ninth century and above all Ibn al-Haytham in the eleventh century), as well as great clinical ophthalmologists, and the region also possessed a renowned glass-making tradition. But this never led to a local craft of manufacturing reading glasses, or even to imitating known types from Italy.

According to one source, thousands of pairs of eyeglasses were exported from Italy to the Ottoman and Moghul Empires.[17] We have a few rare references to reading glasses in medieval Muslim historiographical and poetic sources, as well as an occasional image.[18] We also know that in 1532 the Ottoman *defterdar* requested the

---

15 There is also good reason to search personal memoirs and literary works for reminiscences of newly acquired technologies and consumer items. What used to be very time-consuming is coming within reach in our era of digitization.
16 Cf. my 'From Ibn Haytham to Abu Nazara', [Here Chapter Five]. Cf., on China, Dikötter, *Things Modern*, p. 64 *et passim*. See also idem, 'Objects and agency, material culture and modernity in China', in Karen Harvey (ed), *History and Material Culture: A Student's Guide to Approaching Alternative Sources* (London, 2009), pp. 158–172. In China, spectacles were often acquired to display wealth and education rather than to help vision.
17 Toby E. Huff, *Intellectual Curiosity and the Scientific Revolution* (Cambridge, 2011), pp. 120–122, 130, 206; Vincent Ilardi, *Renaissance Vision from Spectacles to Telescopes* (Philadelphia, 2007), pp. 117–125.
18 Chiara Frugoni, *Le Moyen Âge sur le Bout du Nez* (Paris, 2011). Later than paintings of persons – often monks – wearing glasses in Europe (e.g. Jan van Eyck in the fifteenth century), we have a painting of Ridā al-'Abbāsī (ca. 1635) and a few others. Also there is a cryptic poem on glasses

Venetian *bailo* to procure him a set of spectacles, and we have seen at least one later case of reading glasses in an inheritance list of an Ottoman scribe (1755).[19]

Only by the end of the nineteenth century do we find local professional directories, of the sort that mostly catered to foreign and other elites, which began to list opticians and carried their advertisements. One French manufacturer, Lamy at Morez in the French Jura, where close-by watch production was another specialization, multiplied its trading partners in the Ottoman lands from 21 in 1887 to 243 in 1914.[20] It would seem that there is no evidence of them having spread massively till the beginning of the twentieth century, or even later. We cannot completely exclude the possibility that the incidence of presbyopia was lower than in Europe, but the most plausible explanation is that the low rate of literacy retarded the demand for eyeglasses, the reverse of Western Europe where their diffusion enhanced reading habits.[21]

The other instance are pocket watches (after World War I: wrist watches), the sort of personal timekeepers one carries.[22] Our understanding of their diffusion is still impressionistic. Evidence for the sale of watches by means of advertisements is there, but no reliable statistics.[23] Lately, several excellent studies on temporal culture and on the long surviving *ghurubi* (sunset) time tradition have been done.

We know that by the end of the nineteenth century personal watches were worn in Mecca, but still set to the *azan* of prayer time.[24] Cesar Zivy, a Jewish dealer in Cairo, regularly advertised watches with a strong recommendation by the *mu-*

---

by a certain Ahmad al-'Attār al-Masrī and a reference by al-Sakhāwī to the fifteenth-century calligrapher Sharaf Ibn Amir al-Mardanī wearing glasses. See Huff, *Curiosity*, see above. A. Mazor and K. Abbou Hershkovits, "Spectacles in the Muslim World: New Evidence from the Mid-Fourteenth Century", *Early Science and Medicine* 18/3 (2013), pp. 291–305.

19 Bernard Lewis, *The Middle East, 2000 Years of History* (London 1995), p. 232; idem, *What Went Wrong?* (Oxford, 2002), p.127; see further Joel Schinder, 'Mustafa Efendi: Scribe, gentleman, pawnbroker', *IJMES* 10:3 (1979), pp. 415–420.

20 Of the latter, 66 were in Beirut, 50 in Alexandria and 27 in Istanbul. See 'Des Lunetiers Moreziens', p. 10, in www.hal.archives-ouvertes.fr/.../Des_lunetiers_a_l_echelle_du_monde.pdf (accessed 1 July 2013).

21 The stereotype al-Misri Effendi (see Chapter Five) always wore glasses, supposedly to emphasize his pseudo-intellectual status.

22 The Islamic world had sundials and water-clocks.

23 Orosdi-Back had its own factory in La Chaux-de-Fonds that was primarily geared to the Japanese market but was closed in 1903. Kupferschmidt, *Orosdi-Back Saga*, p. 40. Occasionally, such preowned watches still come up for auction in Japan.

24 Christian Snouck Hurgronje, *Mekka in the Latter Part of the Nineteenth Century* (Leiden, 1931; repr. 1970), pp.71–2. Wearing a watch could be a 'symbol of modernity' among dignitaries, also in Iran. See photograph (c. 1890) discussed by J. J. Witkam in *Journal of the International Qajar Studies Association* 1 (2001), p. 54.

waqqits (timekeepers) of the al-Azhar and Husayni mosques.[25] One Western observer remarked on the Islamic aspect, "it is to know the exact time for prayer and just as an ordinary compass is known by the name of "Mecca pointer" (Kibla) ... At the beginning of Ramadhan, for example, there is often a brisk and increasing trade in timepieces of every description ...".[26] Possessing a modern watch thus preceded *alla franca* time-keeping, with its secular social rhythms and obligations. We don't know to what extent the transition to modern time-keeping around 1910 stimulated the purchase of personal watches.[27]

Indeed, mechanical clocks are one of three examples adduced by Fatma Müge Göçek, in her *East Encounters West*, by which she illustrates delays in the pace of diffusion and adoption of certain Western technologies in the Ottoman Empire.[28] In fact, she cites mechanical clocks as the first case. Such clocks had allegedly been exhibited for their decorative value, as she concludes from the many timepieces in the Topkapi collection.[29] Printing, her second case, was held up for over three centuries owing to various inhibitions (which are still under academic debate in our literature). At the beginning of the nineteenth century – we may add – the declining costs of acquiring second-hand, compact, iron Columbian and Stanhope presses gave a boost to private printing, and thus turned printing into a "small" technology.[30] A third case shows how certain superior European weaving techniques could, for some time at least, be barred by the authorities, while they were trying to improve their own.

While her account may need some refining, we do take up her point of diversity in the pace of diffusion. Not all devices could be marketed successfully or instantly in the region, even though a US trade consul in Beirut in 1913 proudly reported on a Syrian visitor, "who wore American shoes, American rubbers over them, carried a Chicago fountain pen in his pocket, had shaved with a New York safety razor and had just dropped in ... to listen to some Arab songs of the late lamented Cairene singer, Sheikh Yusuf [al-Minyalawi] on an Edison phono-

---

25 *Al-Muqtataf* 26 (1902), following p. 576.
26 S. M. Zwemer, 'The clock, the calendar and the Koran', *The Moslem World* 3 (1913), p. 273.
27 Cf. Roni Zirinski, *Ad Hoc Arabism. Advertising, Culture, and Technology in Saudi Arabia* (New York, 2005), pp. 23–45.
28 Fatma Müge Göçek, *East Encounters West* (New York, 1987), pp. 103–115.
29 There was a community of Western clock makers in Istanbul. Typically, Jean-Jacques Rousseau's father, upon presenting a special clock to the Sultan, was asked to maintain the timepieces in the palace.
30 Nile Green, 'Journeymen, middlemen: Travel, trans-culture and technology in the origins of Muslim printing', *IJMES* 41:2 (2009), p. 211; idem, 'Persian print and the Stanhope revolution: industrialization, evangelicalism, and the birth of printing in early Qajar Iran', *Comparative Studies of South Asia, Africa and the Middle East* 30:3 (2010), pp. 473–490.

graph. Probably this is quite as convincing as long statistics would be!". However, further reading in US trade reports shows that this was not at all a foregone conclusion. Apropos, Western shoes, while competing with local footwear, at the time sold mostly to certain westernized segments of the population.[31] We have no specific data on fountain pens but may assume that they were mostly in demand as prestige objects by westernized elites. Safety razor blades, which recently had come into widespread use elsewhere, were allegedly delayed, as many Middle Eastern men were in the habit of shaving their heads (often still with sharpened irons on a wooden handle or straight blades) but not their beards. Anyway, Gillette had to compete with European brands and even with counterfeit imitations.[32] Consider also doorbells and door locks where urban stores were closed with shutters and padlocks, and houses had doormen guarding the gate. An equally telling example is the (somewhat debatable) comment of a certain Istanbul merchant who annually imported 60.000 dozen spoons, but 40.000 dozen forks, "because the Turk of the interior often uses the former only and dispenses entirely with the latter, using his fingers instead".[33]

But let us give a few more specific examples of larger items to support our main argument on the differential pace of diffusion.

2.4 The sewing machine

Of all the "small" technologies, it seems that the sewing machine has received a place of honor in our secondary literature. [34] Despite a pedigree of unsuccessful earlier machines, it was the Howe-Singer model, as well as later "Singers",

---

[31] Nancy Reynolds, *A City Consumed, Urban Commerce, the Cairo Fire, and the Politics of Decolonization in Egypt* (Stanford, 2012) has a fascinating chapter (pp. 114–144) on shoes and socks. It is not surprising that an American trade report mentions that sock-suspenders were no article for the region.

[32] Sidney E.P. Nowill, *Constantinople and Istanbul, 72 Years of Life in Turkey* (Kibworth, 2011), pp. 104, 108, 119–36, 166; Gordon McKibben, *Cutting Edge: Gillette's Journey to Global Leadership* (Boston, 1998), pp. 359–60. Typically, this semi-official Gillette history mentions India and China more elaborately. We asked the firm for past sales figures in such different societies as Turkey, Lebanon, and Egypt, but have not yet received them.

[33] G. Bie Ravndal, *Turkish Markets for American Hardware*, Special Consular Reports, no. 77 (Washington, 1917), pp. 22–23. Numbers may be multiplied by five for the Istanbul district at large.

[34] For a more extensive discussion, see Uri M. Kupferschmidt, 'The social history of the sewing machine in the Middle East', *WI* 44:2 (2004), pp. 1–19, here Chapter Three.

which led the conquest of world markets from the 1860s onwards.[35] It was followed by competing brands from the USA and Europe. Mention of its invention can be found in *al-Muqtataf from* 1883 onwards. Business historian Andrew Godley has calculated that by 1920, 12 per cent of Ottoman households had acquired a sewing machine.[36]

Karl Marx had been suspicious of the new invention, as he foresaw its use in factories leading to more exploitation of women and children. Surely, the sewing machine did enter Middle Eastern workshops and factories, and it may even have had positive significance for the faltering textile and garment industry, but this is not our concern here.[37] Marx, however, failed to see its overall benefits in domestic environments; in fact the sewing machine "came home". There, it proved its potential to add income and savings, and very clearly could empower women. Many women could now make a living, working from their homes, or save money by making clothing for their own children.

The rapid spread of sewing machines, especially the Singer manufactures, can therefore be called a success story in Middle Eastern societies (as it was in the USA and numerous other countries too). This was also due to innovative hire-purchase schemes which overcame purchasing thresholds. The operation of the machine itself was easy to learn, and it could be carried around. Besides, Singer maintained an efficient service and guidance system.

Sewing machines thus became eminently useful in a region where cultural factors often inhibited women from working in factories. It would take a long time before (some) Middle Eastern countries would witness the mass employment of seamstresses as had developed in the United States and in some Western European countries by the end of the nineteenth century. Nor did this happen in our region with regard to typical "pink collar" occupations such as typists, stenographers and secretaries in general, at least not until recently.[38]

---

35 The Singer factory at Clydebank, Glasgow, manufactured a total of 36 million sewing machines between 1884 and 1943, and at its high point it employed 11.000 workers. We may also think of Enfield-Martini rifles (1895) and Mauser revolvers (1870), Philips and other light bulbs (1892).
36 Andrew Godley, 'Selling the sewing machine around the world: Singer's international marketing strategies, 1850–1920', *Enterprise and Society* 7:2 (2006), pp. 266–314. This estimate is lower than the quarter-to-third for European households, but higher than in India and Japan.
37 Donald Quataert, *Manufacturing and Technology Transfer in the Ottoman Empire* (Istanbul and Strasbourg, 1992), pp. 22–25.
38 By the 1880s in the USA women typists had already overtaken men.

2.5 The typewriter

One may call the typewriter the younger sister of the sewing machine.[39] Here as well, we discover the initial dominance (not monopoly) of one American firm, Remington, an erstwhile arms manufacturer, which was able to make the essential commercial breakthrough in a long series of more or less experimental machines (1868).[40] In the USA, half a million typewriters had already been sold by 1900. In a society which believed in the motto "Time is Money", they became the preeminent office machines. But writers, too, such as Mark Twain, were already known to use them.[41]

We find early descriptions of typewriters and of their advantages in *al-Muqtataf* from 1884 onwards, as well as advertisements, albeit mostly in the local foreign-language press. But it is clear that in comparison to the USA or Europe, the introduction of the typewriter was delayed in the Ottoman lands and in Iran.[42]

One might ascribe this delay to the differences between Roman and Ottoman-Turkish, Arabic and Persian scripts. But this could only be a minor factor. Obviously, the adaptation of the keyboard was a technical hurdle to be overcome. It had to be re-aligned from right to left and from 23 Roman characters to 28 characters for Arabic and a few more for Ottoman script. More complicated were the necessary initial and final forms and the different width of the ligatures in cursive script, which added up to over 70 characters.[43] This in itself was a fascinating challenge. None other than Theodor Herzl may have been a catalyst. As a journalist, knowing the advantage of this innovative tool, he decided to present Sultan Abdülhamid with an Arabic-Ottoman typewriter, in the hope that this miracle of technology would facilitate the obtaining of a Zionist charter for Palestine. While no such machine existed at the time, Herzl asked Remington to design one, and even enlisted

---

39 For details see Uri M. Kupferschmidt, 'Leap-frogging from typewriter to personal computer in the Middle East', paper presented to the Eighth Mediterranean Social and Political Research Meeting, Montecatini, Italy, 2007 [Here Chapter Four].
40 In fact, 1873 is the initial date of the Sholes-Glidden model with qwerty keyboard; typewriters became more or less standardized in 1910.
41 Friedrich Nietzsche, too, is among the many users always mentioned, and world literature abounds with references to typewriters. Arab writers, however, emphatically continued to write in longhand (e.g. Naguib Mahfouz, Yusuf Idris). Apparently, this began to change only in the 1960s, e.g. the Iraqi writer Najm al-Wali who acquired a typewriter in exile. (Private ownership there had been prohibited for some time.)
42 This is also true for Thai, Japanese and Chinese typewriters, which had to overcome linguistic barriers as well.
43 The QWERTY keyboard which made the Remington-Sholes machine superior, remains controversial. Note, however, that presently 'Arabic' numerals have replaced traditional 'Indian' ones.

the help of the orientalist Richard Gottheil in New York. He had, however, not reckoned with the Hamidian prohibition on importing typewriters, which remained in force till 1908.[44] This Ottoman Remington machine (model 7) somehow disappeared till it was rediscovered during a thorough search of the Sultan's belongings at the Yildiz palace following his dethronement.[45]

Remington had been very proud of its new product; in fact the very first issue of its journal *Remington Notes* in 1907 opened with a description of it.[46] But theirs was not the only attempt to produce a typewriter for the Middle Eastern market. Somewhat earlier, around 1898, Philippe Wakid and Salim Haddad (two Lebanese living in Cairo, the latter a painter of some fame too) had started working on an Arabic typewriter, which was duly patented in the USA in 1901. It would seem that it came into production in 1904 (by Smith apparently). Also, no less than Edward Said claims that his father later designed the Arabic keyboard for the Royal firm.[47]

Still, the typewriter made no headway. Even before World War I, while the new implement came increasingly into use by government offices and commercial firms in America and Europe, it would seem that in the countries of the Middle East it was still a rarity. At the Sublime Porte there was maybe one in use, and a few more at official agencies around the empire, such as the customs in Alexandria. The Khedive was advertised as having one, along with other royal and noble figures, though owning a typewriter and actually using it might be two different things. On the whole, in the region the typewriter remained restricted to Western-oriented businesses. And almost all of these used French and English.

---

44 Heavily criticized by *The New York Times*, 19 and 20 May 1901.
45 *Remington Notes* 2:5 (1908), p. 14: 'The gift of this Remington – which was certainly unusual, since the Arabic Remington was then a novelty of novelties – was duly accepted by the Sultan, but it is not on record that it was effective in helping the Zionists' cause. Indeed, the former Sultan is reputed to have been unfriendly to the typewriter, even as he was notoriously hostile to the printing press.'
46 *Remington Notes* 1:1 (1907), p. 1. The Persian Ambassador in Washington too had presented the Shah with a model. The former Shah had used an earlier model, allegedly, not before first cleansing it in boiled water to drive out evil spirits. Another prototype in simplified Japanese Katakana script was donated by the Remington Company itself to the Mikado. Soon, Remington regularly advertised that it supplied machines in over thirty different languages.
47 *The Phonographic Magazine* 17:3 (February 1903); *The New York Times*, 21 August 1904. Descriptions of western typewriters can be found in *al-Hilāl* from 1897. Haddad's invention was first noticed in the issue of 15 May 1901 (vol. 9, p. 470), but only on 15 May 1904 (vol. 12, pp. 502–505) did the magazine offer a detailed description, and advertising apparently started in 1908 Cf. Edward Said, *Out of Place* (New York, 1999), p. 94; his father owned well-known office supplies stores in Jerusalem and Cairo.

Explanations are not difficult to find: The (male) scribal occupational conventions of the time, the limited usefulness in small businesses, let alone in private homes, and restricted literacy in general. For decades to come, no Arab or Turkish writer of fame can be mentioned who prided himself on using a typewriter. Also, only few local commercial schools taught typewriting and stenography. No office culture, in the American sense, existed. Besides, as a technology derived from the typewriter keyboard, linotype machines were not turning up in large numbers either.

In Turkey, typewriters became more utilized with the transition to the Latin alphabet in 1928. Turkey even designed a "national" keyboard of its own, recalculated after the qwerty pattern.[48] But in spite of all the official encouragement, it took a relatively long time for typewriters to conquer a significant place in Turkish society or business, and even longer in the Arab countries.[49] Egypt enacted legislation in 1942 to make Arabic compulsory in commercial dealings.[50] The region of the Middle East remained, by and large, an unmentioned or minor consumer in the world trade of typewriters.[51]

While this falls outside the scope of this first period of globalization, we have here an interesting example of "leapfrogging". From the 1990s, the PC jumped, as it were, over the typewriter, though the keyboard in essence persisted. Remarkably little has been written on the transition from typewriters to personal computers (and word processing) in general, let alone in Middle Eastern countries. The process now became swift and logical.[52] Nowadays, typewriters have become museum objects.[53]

---

48 Turkey reverted back to QWERTY in the 1980s under the impact of Microsoft.
49 From 1956 a specialist magazine, *Sekreter Daktilograf*, briefly appeared in Istanbul. (The terms *daktilo* and *sekreter* were, of course, taken from the French.) There were also frequent speed-typing competitions.
50 Floresca Karanasou, 'Egyptianization, the 1947 Company Law and the Foreign Communities in Egypt' (PhD Thesis, Oxford, 1992), pp. 53–54.
51 US Department of Commerce, *World Trade in Typewriters 1948–1958* (Washington, 1959), pp. 20–22; George N. Engler, 'The Typewriter Industry: The Impact of a Significant Technological Innovation' (PhD Thesis, University of California, 1969), who emphasizes the declining US-American share of world production.
52 The phenomenon of leapfrogging, that is, skipping a 'natural' phase of development or drastically shortening it, applies in parts of the Middle East also to the transistor radio jumping over the immobile home radio and the cellular telephone over the terrestrial telephone. A social history of the telephone in the Middle East is long overdue, particularly with the fast diffusion of smartphones and their impact. We find some similarity to the Dutch historian Jan Romein's perceptions on 'the dialectics of progress' (his so-called Law of the Handicap of a Head Start). Recently, the *International Herald Tribune* (29 December 2011) remarked on the lack of a researched history of word processing.

2.6 Photographic cameras

Photography qualifies to our mind as a "small" technology, too, but the phases in the introduction of the various categories of cameras to the Middle East still need to be unraveled.[54] Photography as such has enjoyed increasing interest from researchers of the Ottoman and Iranian lands, and a considerable literature on pioneers of landscape, urban scenes, and object documentation, as well as on studio photography, has sprung up.

From the fact that, initially, mostly Christians engaged in this new field, and hardly any Muslims (and few Jews), it has sometimes been concluded that there were religious inhibitions, and even today there are communities and individuals who refuse to be photographed.[55] Indeed, there had apparently been fatwas against photography, e.g. by Rashīd Ridā and the Şeyhülislam, but they were eclipsed by religious endorsements, and as probably happened so often, by the irresistible success of the technology. The Damascene Salafi 'ālim Jamal- al-Din al-Qāsimī, though a conservative in spirit, mentions the professional photographer positively among the very few modern occupations in his compendium.[56] Around the date of his writing, the first photographs must have appeared in the vernacular press, and supposedly added to a greater awareness of the technology, as did postcards and visiting cards (which used to be photographs).[57] Landmarks of actual use in each of the various places would be the discourse in the indigenous press and the first advertisements there.[58]

---

53 The Koç museums in Istanbul and Ankara, and also the Safir Office Machine Museum in Tehran show old typewriters as if they were intensively used there (with thanks for the Tehran reference to Willem Floor). A last typewriter factory that had survived in India closed in 1991.
54 Walter Benjamin opens his famous essay on the history of photography (1931) with the phrase that it is clouded in comparison with the history of printing, but he does not distinguish between technological invention (known dates) and artistic or literary application ('in-use').
55 On opposition in Egypt, see Maria Golia, *Photography and Egypt* (London, 2010), pp. 15 and 55. Muhammad 'Ali was not the only one to call it 'the work of the Devil'. There must have been 'ulamā', too. The question is, however, whether it was simply greater familiarity, which gradually achieved the breakthrough, or certain populations becoming convinced of photography's benefit.
56 *Dictionnaire des Métiers Damascains* (Paris, 1960), vol. I, p. 445.
57 Before, engravings had been drawn after photographs. On the beginnings of press photography, which does not fall into this framework, more ought to be said, but see for Egypt, Golia, *Photography*, p. 74.
58 Mona Russell, *Creating the New Egyptian Woman: Consumerism, Education, and National Identity, 1863–1922* (New York, 2004), pp. 68–69, with a deep analysis of an adapted Kodak ad from the early 1920s. See also Beth Baron, *Egypt as a Woman, Nationalism, Gender, and Politics* (Berkeley 2005), pp. 82–101, especially on the popularization of photography.

The aspect that concerns us here is therefore not the actual arrival of photography – the first photographs of this or that place – or studio photography, but the wider familiarization with and affordability of the technology, in short: amateur photography as it diffused from the elites to middle class youth and beyond. This is a process that evolved over several decades, and not only in the Middle East. It had everything to do with technological advances. In the USA, amateur photography can be said to have begun with roll films and hand cameras, which hit the market in 1888. It was suddenly not too difficult or too expensive to purchase a one-dollar Brownie box, and as Kodak advertised: "You push the button, and we do the rest". Individual foreign travelers, "Kodakers", now appeared in our region too. Kodak opened branches in Istanbul, Cairo and elsewhere, and by the first decade of the twentieth century, amateur pictures could be developed and printed in many a city and tourist resort. There was an aspiration to modernity, which translated into a strong connection with photography, as Stephen Sheehi has shown, albeit without clearly distinguishing between studio and amateur photographs.[59]

To the extent that they could (begin to) afford a camera, Middle Eastern populations, like others elsewhere, could bring it into their homes, into family celebrations, or weddings.[60] Even though amateur photography remained long only in the upper bourgeois sphere, a breakthrough in its spread to a somewhat wider public occurred around World War II.[61] We have a few family albums, which give us more insight into that development.[62] Fortunately, some institutions in the region, such as the Arab Image Foundation in Beirut, have begun to collect the evidence.[63]

---

[59] Stephen Sheehi, 'A social history of early Arab photography or a prolegomenon to an archaeology of the Lebanese imago', *IJMES* 39 (2007) pp. 177–208.

[60] Roger Allen (ed.), *Muwaylihi, A Period of Time. Hadith 'Isa ibn Hisham* (Oxford, 1992), pp. 281–282, whose satire makes fun of cameras brought into wedding parties and photographs subsequently shown and sold to outsiders.

[61] As anywhere else, Kodak was an active advertiser, in Egypt mainly aiming at the higher classes. "Go immediately, before the moment passes, and get a 'Kodak' from the closest photographic dealer". Mothers were depicted saving a "Kodak moment", photographing her children leaving the house on their way to school (*al-Lata'if al-Musawwara*, 6 dec. 1926). At first drawing on bricolages from American sources, Kodak later switched to local imagery (Russell, pp. 68–69), as was the case also in India (cf. Jennifer Orpana's 2018 article on Kodak advertisng in *The Times of India* 2018). While the new 35 mm cameras such as Leica (1925), Contax (1935) and the like, were affordable only for the rich, we have to look at the sale of cheaper roll-film (120–127 types), box cameras (Kodak, Brownie, etc.), and similar less expensive cameras.

[62] Rifat Chadirji, *The Photography of Kamil Chadirji* (Surbiton, 1991).

[63] Akram Zaatari, *The Vehicle. Picturing Moments of Transition in a Modernizing Society* (Beirut, 1999), published by the Arab Image Foundation. Lucie Ryzova has described private albums with photos by young Egyptian effendis which she found at flea markets. Idem, 'My notepad is my friend. Effendis and the act of writing in modern Egypt', *Maghreb Review* 32 (2007), pp. 323–

2.7 The piano

Another fascinating case, to our mind, concerns the piano, an instrument that may not immediately be identified as being on par with the previous "small technologies", but we see it as such. Though "invented" around 1700 by Bartolomeo Cristofori in Northern Italy, it became a highly sophisticated instrument, due to a series of improvements, during the nineteenth century (frame, hammers, strings all evolved alongside technological advances in production methods and materials). By the end of the century, pianos consisted of some 2.500 different parts.

From an artisan craft, piano making became a fully-fledged mass-producing industry.[64] Pianos were much in demand and were manufactured in Austria, Germany, France, and Britain, and also in the United States.[65] The output worldwide was impressive. It is estimated that it increased from 50.000 a year in the 1860s to 650.000 in 1910, grands as well as uprights, and that production remained rising till about the 1920s.[66]

Unlike the sewing machine and the typewriter, pianos did not contribute to labor productivity or to the acceleration of life rhythms, but their mechanical development can be said to have influenced the refinement of European music. This included (imagined) *alla turca* modes by eminent composers; hence, for some time, a so-called Janissary stop was installed, a pedal to sound drums or bells.[67] However, this short-lived addition had never been intended to conquer a Middle Eastern audience or clientele.

It has been said that the first piano arriving in the Middle East was one donated by Napoleon to the Shah of Persia. Such royal (in fact national) gifts were often pretentious exercises in superiority. The said instrument remained an orphan, as nobody at that time was capable of playing it. This would take time. A tile at the Golistan Palace shows Shah Nasr al-Dīn (1848–96) Shah listening to a piano performance.[68] Meanwhile, at the Ottoman court, a few more imported first-class grand

---

348. See also idem, 'I have the picture: Egypt's photographic heritage between digital reproduction and neoliberalism', part 1 and 2, in www.photography.jadaliyya.com/pages/index/8297 and 9725 (accessed 1 July 2013).
64 Sonja Petersen, 'Piano manufacturing between craft and industry: Advertising and image cultivation of early 20th [century] German piano production', *ICON, Journal of the International Committee for the History of Technology* 17 (2011), pp. 12–30.
65 Steinway started in Germany but became a leading manufacturer in the USA after 1860.
66 Cyril Ehrlich, *The Piano: A History* (Oxford, rev ed. 1990), p. 108 *et passim*.
67 Edwin M. Good, *Giraffes, Black Dragons and Other Pianos* (Stanford, 2nd ed. 2001), pp. 136–137, 182, 267.
68 Jennifer Scarce, *Domestic Culture in the Middle East* (Edinburgh, 1996), p. 54. The said tile was made from a photograph and installed on a frieze in 1887.

pianos had appeared, and remarkably, all latter-day sultans were able to play them, and they even composed music for the piano.[69] By 1901, one could find at least two Italian piano makers and five tuners in the Ottoman capital (1901).[70] In Egypt, ever since the inauguration of Khedive Isma'il's opera house, the Khedivial, and later royal, household imported several pianos from Germany.[71] Even the Hashimite court seems to have owned at least one piano.[72] Royal courts thus appear to have played a role with regard to further diffusion.[73]

Pianos gradually made their appearance in literary salons of the late nineteenth century and in bourgeois circles of the Middle East.[74] Manuals on the ideal (upper or upper-middle class) household recommended them.[75] Piano lessons became a must, especially for girls, exactly like in bourgeois circles of the West, and we hear not a few of them complain about the discipline it required.[76] In

---

69 Ayşe Osmanoğlu, *Babam Sultan Abdülhamid (Hatıralarım)*, (Istanbul, 1984), pp. 119–120; Leyla (Saz) Hanimefendi, *The Imperial Harem of the Sultans* (Istanbul, 1955), passim; Vedat Kosal, *Western Classical Music in the Ottoman Empire* (Istanbul, 1999), passim. (Thanks to the Istanbul Stock Exchange for providing me with the book.) Pianist Vedat Kosal has recorded composition by Sultans Abdülaziz and Murad, along with Donizetti and Guatelli Pashas: idem, *Ottoman Court Music*, 2003.
70 Daniel J. Grange, *L'Italie et la Méditerranée (1896–1911)* (Rome, 1994), p.486. Local piano manufacturing is also mentioned in the *Exposition Nationale de Constantinople*, p. 121.
71 For a Weissbrod piano imported by the Hugo Hach firm for the court around 1909, see http:// egyptianroyalfurniture.blogspot.fr/ (accessed 6 July 2013).
72 Thanks to a former student, C.G., who related that Emir Abdallah donated a Schimmel baby grand to his chief veterinarian Haim Appelboim, upon his retirement from the British army (it is not clear whether it had become redundant at the Palace); it later ended up with relatives of hers in Binyamina. See also Cyrus Schayegh, 'Connecting the Dots', paper submitted to the conference 'On the Move' (Jerusalem, January 2012), pp. 1–2, on the second hand frigidaire that Cecil Hourani in Lebanon obtained from Haifa
73 Missionaries also brought pianos along, e.g. to Lebanon, and sometimes over long distances, and not without some damage. An example is the Labaree family to Urumia in 1894 (thanks to Robert Labaree for the information). It is difficult to assess their impact.
74 E.g. the salons of Maryan Marrāsh (Aleppo), Mary 'Ajamī (Damascus), Hudā al-Sha'rāwī Cairo) and the princess-novelist Nazli Fādil. See 'Women's Literary Salons and Societies in the Arab World', in www.en.wikipedia.org/wiki/Women%27s_literary_salons_and_societies_in_the_Arab_world (accessed 7 July 2013), and Joseph T. Zeydan, *Arab Women Novelists, the Formative Years and Beyond* (Albany, 1995), p.51, quoting Sufuri (a pseudonym) from *al-Muqtataf* 72:6 (1928), p. 686. In Cairo, the Ramatan Museum exhibits Taha Husayn's reception hall as it was with a piano, as well as a gramophone with western records. See also the images of home interiors on the website of the Arab Image Foundation.
75 Mona Russell, 'Modernity, national identity, and consumerism: Visions of the Egyptian home, 1805–1922', in Relli Shechter (ed), *Transitions in Domestic Consumption and Family Life in the Modern Middle East* (New York, 2003), p. 48, quoting from Francis Mikha'il's advice manuals.
76 Mona N. Mikhail, *Seen and Heard: A Century of Arab Women in Literature and Culture* (Northampton, 2004), pp. 7–9. On Randi Abou-Bakr in Egypt, see Elizabeth Warnock Fernea (ed), *Remem-*

the main urban centers, we find specialized stores advertising the most notable manufactures.[77] Little systematic data on sales to the countries of the Middle East is available, but we know that in Egypt, for instance, in 1923 over 3.000 pianos were sold.[78]

Europe saw a few attempts to construct quartertone pianos. They ensued from avant-garde experimentation with new musical modes, notably by the Russian-French musicologist Ivan Wyschnegradsky and the Czech composer Alois Hába (Fig. 13). The innovative instrument was never meant to bridge the divide between the musical traditions of Europe and the Arab world. Actually, the building and playing of intricate quartertone pianos was never successful anywhere.[79]

Pianos could produce quartertones, somehow characteristic of Middle Eastern music, but merely in a mechanical way, not in an endearing musical or melodious mode. Still, the development of an "Oriental" (quartertone) piano was sometimes discussed also in the Middle East.[80] Four or five models were demonstrated at the 1932 Arab Music Congress in Cairo, a major event in the history of oriental music, attended by outstanding regional musicians as well as European composers such as Béla Bartók, Paul Hindemith and Alois Hába himself.[81] But even the young Umm Kulthum failed to harmonize her singing with the innovative instrument. Decisions of that congress also left the adoption of the conventional piano in sus-

---

bering Childhood in the Middle East (Austin, 2002), pp. 350–354; Frederic Lagrange, 'Musiciens et Poètes en Egypte au Temps de la Nahda' (PhD Thesis, Paris, 1994), p. 96. Boys also played. See Said, *Out of Place*, pp. 96–7, who received lessons from the famous Ignaz Tiegerman, and became – less remembered – a music critic in his own right. Prince Faruk also took piano lessons from him.
77 E.g., A. Comendinger (Istanbul) who also published note music, and J. Calderon (Cairo).
78 Jacques Berque, *Egypt, Imperialism and Revolution* (London, transl. 1972), pp. 332–333, is one of the few historians to have mentioned pianos. Some models, by the way, had decorations to suit local taste.
79 Wyschnegradsky ordered a quartertone piano from the famous French manufacturer Pleyel, which disappointed him. Later, the Czech-German firm Förster built a few quartertone pianos in cooperation with Hába (my thanks to that firm for information). The Pleyel piano was noted in the Egyptian press and appeared with a photograph in *al-Latā'if al-musawwara*, 11 September 1922.
80 The musicologist Alberto Hemsi in Egypt referred in 1928 to various local experiments with the 'oriental piano' but concluded that it would remain without a repertoire, and that Eastern and Western music had to be taught in parallel. See idem, *La Musique Orientale en Egypte* (Alexandrie, 1930), pp. 11–14. A few other contemporary Western composers – e.g. Béla Bartók in a few pieces and the American Charles Ives – used quartertones. (Bartok may have been inspired by Arab music.)
81 Linda Fathalla, 'Instruments à Cordes et à Claviers dans les recommendations du Congrès du Caire', in collective volume of CEDEJ, *Musique Arabe. Le Congrès du Caire de 1932* (Cairo, 1992), pp. 99–103 with relevant figures.

Fig. 13: The Czech composer Alois Habá at Förster's quatertone piano, from a brochure of the manufacturer (1920s) (courtesy; Wolfgang Förster).

pense.[82] In the next decade, Muhammad 'Abd al-Wahhāb, who had participated in the congress, would include the piano, along with some other Western instruments, in a few of his cinema productions but this remained a limited experiment.[83] The *qānūn*, the *santūr*, the *'ūd* and other traditional instruments continued to reign supreme.

We conclude that the piano, though not completely absent from the regional musical culture, remained limited to the higher classes. It never gained a popular foothold in the region nor became integrated into local music.[84] However, modern

---

[82] *Al-Latā'if al-Musawwara*, 4 April 1932, carries a story on the piano 'invented' by George Shaman. Experiments with regard to producing a quartertone instrument continued in the 1940s. See Abdallah Chahine, whose family still owns a large music store in Beirut. Occasionally new types emerge in the West, but not for Eastern music.

[83] During the phase of the silent cinema, pianos accompanied movies. Alter 1917 Istanbul saw the influx of a number of refugee pianists from the Soviet Union. See Jak Deleon, *The White Russians in Istanbul* (Istanbul, 1995), pp. 63–66 (Thanks to Canan Balan for the reference). See further Nabil Salim Azzam, 'Muhammad 'Abd al-Wahhab in Modern Egyptian Music' (PhD Thesis, University of California, 1990), pp. 137–138, 260. He employed a Greek pianist. In Lebanon, the Rahbani brothers used the piano in some of their arrangements.

[84] Local conservatories do teach the piano. But as a whole, the region, with the relative exception of Turkey, has few performing artists, especially in comparison with the Far East. Cf. Richard Kurt Kraus, *Pianos and Politics in China* (New York, 1989). China is a major piano producer today.

synthesizers, which can produce quartertones, created new challenges, and may therefore be seen as another technological leapfrog.

We may raise two comparisons here. The Western-type violin, in its ultimate form made primarily by Italian craftsmanship that could hardly be improved by new technologies, gradually began to appear or replace, at least in part, the traditional Middle Eastern *kemence* or *kaman*. Strangely, little has been written on this acculturation, but it is clear that violin strings could be tuned to Arab music in a fairly simple way. It suffices to have a look at Umm Kulthum's usual *takht* to grasp the Italian-type violin's success. The transition is sometimes ascribed to nineteenth-century Balkan or gypsy musicians coming to perform in the Ottoman center. In the first decade of the twentieth century, we find renowned Aleppan musicians of the Sawwa family arriving in Egypt to play the violin.[85]

The second point concerns the gramophone (the successor of the cylinder phonograph), which can certainly be considered a mass-produced technological device – a medium, as well as a musical instrument.[86] It made its appearance in the Middle East only in the first decade of the twentieth century, but very soon, recordings of local music gained great popularity and were made in large numbers.[87] The new device was, at first, played in coffeehouses and in public at large, and owner-operators also went around with it, but it was not long before it reached the private homes of whoever could afford one. It gave rise to an active indigenous recording industry and frequent advertising in the press, which also attests to its diffusion.[88] It has been said that the gramophone changed Arab music, and possibly Turkish and Persian music as well, shortening the duration of songs, and also elevating the *zajal* poetic genre from its lower class status.[89] Above all, it created a new market and a wider celebrity status for local singers

---

[85] Lagrange, 'Musiciens', pp. 73, 85, 92, 124, 129, 133, 149.

[86] For an image of musical instruments sold, see Kupferschmidt, *Orosdi-Back Saga*, p. 65. Thibouville, the largest violin factory in Mirecourt, France, began marketing gramophones as well; see their catalogues on the web.

[87] Edison's phonograph (1877) and more so Berliner's gramophone (1887) gained wide popularity in the USA, later in Europe, especially due to the transition from cylinders to disks. In 1880 *al-Muqtataf* was still skeptical, writing that it first had to see and hear the new invention before believing it. We do not know what Western music was initially played, but original American records were advertised and probably were soon superseded by local taste.

[88] Ali Jihad Racy, 'The record industry and Egyptian traditional music, 1904–1932', *Ethnomusicology* 20:1 (1976), pp. 23–48, and other articles by the same author; Pekka Gronow, 'The record industry comes to the East', *Ethnomusicology* 25:2 (1981), pp. 251–284.

[89] Ziad Fahmy, 'Media-Capitalism: Colloquial mass culture and nationalism in Egypt, 1908–18', *IJMES* 42:1 (2010), pp. 83–103; cf. idem, *Ordinary Egyptians Creating the Modern Nation through Popular Culture* (Stanford, 2010).

such as Yusuf al-Minyalāwī, and somewhat later, Umm Kulthum and other famous artists.[90]

As in America and in Europe, gramophones were considered domestic musical instruments. This may also be concluded from a few divorce cases in Beirut, in 1910, in which the female partner received the gramophone, it being identified as an instrument of pleasure, for music had no specified value, while the male would take the household utensils.[91]

Though the gramophone continued to play its own role, it lost ground to the radio, which we must leave to a discussion of the post-World War I wave of globalization.

2.8 Incandescent light bulbs and electric household implements

Marc Bloch once chided historians for not taking up electricity.[92] Our research agenda on "small" technologies ought therefore to comprise electrical implements as well. Electricity is generally thought of in terms of power grids, and the books which tell us exactly how many plants were erected, when and where, or how many kilowatts were produced, are legion.[93] But we are interested in additional aspects. Street lighting enabled nightlife, and electric power made night shifts in factories possible. The illumination of mosques for festivals and public celebrations in general (often 'from above' or by interested commercial parties) was intended to impress.[94]

---

90 Orosdi-Back also jumped at the market and recorded al-Minyalāwī; see Kupferschmidt, *Orosdi-Back Saga*, p.40.
91 Toufoul Abou-Hodeib, 'Taste and Class in late Ottoman Beirut', *IJMES* 43:3 (2011), pp. 485–486. The qadi even declared that the gramophone and the records deserved to be broken. This gender point needs further research, e.g. against the background of Victoria de Grazia and Ella Furlough (eds), *The Sex of Things* (Berkeley, 1996).
92 Marc Bloch, *The Historian's Craft* (New York, 1953), p. 66.
93 Istanbul was the first large city to get a power station, the well-known Salihtarağa station near Eyüp coming into operation in 1910 (later than most cities in western Europe); it closed in 1983 and has recently been converted into a museum. Davis Trietsch wrote in 1910 that power plants had 'till recently' been prohibited, but now Damascus, Beirut, Medina (the Holy Mosque), Constantinople, Salonika, Smyrna and Bursa have them. Idem, *Palästina-Handbuch* (Berlin, 1910), p. 217, cf. idem, *Levante-Handbuch* (Berlin 1909), p. 197, and idem, *Levante-Handbuch* (Berlin, 1914), pp. 407–408, on the expansion of electrical grids, also in Egypt.
94 The big mosque in Mashhad was illuminated in 1900 by a generator imported from Russia. Mecca began using illumination around 1926.

However, there is hardly any profound study on the application of electricity in ordinary Middle Eastern homes.[95] The use of electricity in the personal environment, including the gradual transition from oil and gas lighting, has, thus far, remained a stepchild in our social history. *Al-Muqtataf* published several articles on the usefulness of *kahrabā* (the neologism for electricity, possibly introduced by al-Tahtawi) from the 1880s onwards (e.g., on lighting in 1881, cooking in 1906, the household in 1915).[96]

There is a story which comes to mind about Sultan Abdülhamid's alleged opposition to electricity, said to have developed out of his paranoid fear of attempts at his life, confusing "dynamite" and "dynamo".[97] The importing of all electrical apparatus, including electric bells, was repeatedly prohibited.[98] Whatever the origins of the story, or indeed the myth, it did not last. The French journalist Paul Fesch writes that the turnaround came in 1906, and indeed, the remnants of a small power plant can still be seen at Yildiz Palace today.[99]

We have to distinguish between fully-fledged power plants and batteries or generators, which reflect the narrative "from below". By importing private generators, one could get around whatever prohibition there was: A British trader in Istanbul acquired an installation at the cost of £5.000, and it was said that the German consul could do "something" for him on the condition that the equipment was bought from the German Electric Corporation.[100] In Palestine, the Rothschild-owned Rishon le-Zion winery had already installed lighting in 1890, fed by an array of batteries. The Pera Palace Hotel in Istanbul too imported "batteries de cuisine" clandestinely, around 1906.

Some of the Sultan's higher servants had already brought electricity into their own homes. One of them was Uryanzade Cemil Molla, a former governor of Anatolia, who defiantly lighted his own mansion – seizing the occasion of the Sultan's anniversary of accession to the throne – by means of a secretly installed generator.

---

95 Duygu Ulas Aysal is presently working on a PhD dissertation at Bilkent University regarding government-consumer relations after the introduction of electricity in Istanbul.
96 Obviously, electricity was part of other technologies, e.g., the telegraph or telephone or somewhat later the cinema and, more conspicuously, of tramways (and their concessions to foreign companies).
97 Sir W. M. Ramsay, *The Revolution in Constantinople and Turkey: A Diary* (London, 1909), p. 241.
98 The weekly *Electrical Engineer* 25 (1900), p. 645. Exceptions were apparently made for medical and mining equipment.
99 One contemporary source explicitly writes 'He is not wary of innovations such as electricity or the telephone'. See Jacob M. Landau, 'An insider's view of Istanbul: Ibrahim al-Muwaylihi's *Ma Hunalika*, WI 27 (1987), p. 80 (quoting pp. 234ff.); Paul Fesch, *Constantinople aux derniers jours d'Abdul-Hamid* (Paris, 1907), p. 146.
100 Ramsay, *Revolution*, p. 242.

Reportedly, he was summoned to the Palace but convinced the authorities of its usefulness.[101] Around 1906 we hear of a concession given to Zaki Pasha, a former artillery officer, as well as to Abdülrahman, a former Grand Vizier.[102] Said Bey, the public servant in Istanbul made famous by Paul Dumont and François Georgeon due to his notebooks, was apparently also among the first in the city to enjoy electricity from 1908, even before the Silahtarağa plant had been opened.[103]

In a few places we have some data on the rise in numbers of "subscribers" (*abonnés*), albeit mainly for the interwar period, but our questions here are of an anthropological nature rather than inspired by business history.[104] Domestic use of electric light would change sleeping hours and daily schedules, as well as reading habits where the new lamps gave more comfortable light than candles and gas.[105] Light bulbs (greatly improved by Edison in 1879) were, indeed, the first home application.[106] But it would help to know more about uses other than lighting, namely consumer devices. What application came second after the light bulb: electric irons, as was often the case in Europe or the USA, or, electrical fans or stoves? (Fig. 14)[107] In this respect, one should not forget the mediating, but often ignored, invention of plugs and sockets (patented by Hubbel in 1904).[108] In short, there is still a considerable research agenda before us.[109]

---

101 *New Istanbul Times*, 20 February 2009, see www.newistanbultimes.com/mansions-18h.htm (accessed 3 July 2013).
102 Fesch, *Constantinople*, p. 146.
103 Paul Dumont, 'Said Bey – The everyday life of an Istanbul townsman at the beginning of the twentieth century', in Albert Hourani et al. (eds), *The Modern Middle East* (London, 1993), pp. 271, 284. The Silahtarağa plant was built by the Hungarian firm Ganz and opened in 1914.
104 No doubt, consumer bills, etc., can teach us a lot. It may also be useful to ask, for the Middle East, the questions raised e.g. by Dominique Desjeux et al., *Anthropologie de l'électricité. Les objects électriques dans la vie quotidienne en France* (Paris, 1996).
105 Beth Baron, *The Women's Awakening in Egypt* (New Haven, 1997), p. 89. Some conservative 'ulamā' (even Sha'rāwī in the 1970s) saw lighting the night as an unwarranted interference in the Order of Creation, disturbing sleep and upsetting the morning *adhān* (call to prayer).
106 In the USA by the turn of the century, 18 million light bulbs had already been sold. Advertisers in the Egyptian press (probably also the Turkish press) were Osram, Philips-Argenta, Tungsram etc. We do not know what quantity has been marketed in the region. Ultimately, Turkey started producing light bulbs itself.
107 Davis Trietsch, *Levante-Handbuch* (Berlin 1914), p. 384, gives an indication of the increasing imports of Beirut: cables (0.25–4 mm.), lightbulbs (of 16–200 'candles'), switches, transformers, and, owing to the high summer temperatures, ventilators of different sizes and types. The refrigerator came later and washing machines even longer after (needing electricity, water, and sewerage).
108 This is important for hooking up electrical devices other than for lighting. Also, sockets could be used to tap electricity from lamp fittings,

Fig. 14: Advertisement for an electric fan. "Don't wait and say '*bukra fi'l-mishmish*' (literally: "tomorrow when the apricots bloom", the equivalent of "when pigs fly") but buy the best electric fan in the world now, before the onset of the hot season, from the Thomson Houston Company at al-Madabigh Street, today Sharif Street, in Cairo (*al-Lata'if al-Musawwara*, 26 June 1922).

2.9 The automobile

With our focus on accelerated movement of technologies and commodities in the first era of globalization, the diffusion history of motor vehicles in the region calls for more research.[110] We ought to be aware that the world automotive industry from the beginning of the twentieth century was dominated by the United States, and less so by Western Europe, which is also reflected in our secondary sources.[111] Around 1922, when the USA probably already had over ten million officially registered motor vehicles, a mere 4.000 may be estimated to have been registered in Turkey proper, some 2.800 in Egypt, 2.100 in Syria, 800 in Iraq, 700 in Palestine, and far fewer in other countries of the region.

The automobile made its first sporadic appearance in the urban centers of the Middle East in the first decade of the twentieth century, only a couple of years after the manufacturing countries. Not a few local chroniclers marked its first appearance in their cities. The military powers of World War I brought more motor vehicles to the region, and some were even left behind after the war. There is nothing exceptional in the fact that motorcars soon played to local curiosity and imagina-

---

[109] This includes the emergence of specialized stores and electricians. See Rizkallah Hilan, *Culture et devéloppement en Syrie et dans les pays retardés* (Paris, 1969), p. 155.
[110] Literature on America is wide and diverse, complemented by studies on Western Europe. Lewis H. Siegelbaum (ed.), *Cars for Comrades: Automobility in the Eastern Bloc* (Ithaca, 2011), for instance, shows the need and potential for research on other societies.
[111] The USA in 1922 had 84 per cent of the world's motor cars, a rate that declined to 68 per cent in 1939.

tion (caricatures, literature, movies), but also aroused criticism (accidents, speed) (Fig. 15 and 16).

Figs. 15 and 16: The scare of hooting cars endangering pedestrians: caricatures from a critical article entitled "Azrael [the Angel of Death] in the Automobile" (*al-Lata'if al Musawwara*, 10 Jul. 1922).

However, diffusion in the region at large and especially beyond the cities was slow, not only because of the relatively high price to be paid for a motor car but owing to the lack of suitable infrastructures: a lack of paved roads, of a higher quality than desert "pistes", and usable the year round. Even in cities and towns, with their narrow streets and cobblestones, improvements were necessary, as they had been when horse-drawn carriages became more numerous some half-century before.

Because of the lack of suitable roads, even where railways and coastal shipping operated, cargoes had to be transported from the station on camelback to its destination. Most indicative is an anecdote related by Amin al-Rihani to the effect that a new Ford car, ordered by King Ibn Saud, allegedly had to be towed by camels all the way from an Eastern harbor to the Palace in Riyad.[112] Of course, highways and gas stations were then still absent in later oil-rich Saudi Arabia. Yet camel transport, although in decline all over the region, held out till mid-century.

Evidently, the spread of motor vehicles suffered a delay but could not be stopped. This is also clear from legislation and taxation being brought into line with the new means of transport.[113] There is still much to be mapped out in order to understand the process. Roads are one factor, but capital is another. The persistent advertisements of major car manufacturers in the press, however, seem to prove that there was to whom to sell.[114] We know little about the preferences of those who bought automobiles. That there emerged a critical mass of owners from the 1920s onwards may be proven by secondary commercial publicity for tires and lubricants, occasional handbooks and magazines and exhibitions; it would also not take too long before second-hand cars were offered.[115] Secondly, we note the differential rate of diffusion of private cars, trucks, buses (to which motorcycles and tractors may be added) in the various countries of the Middle East, which is another relevant but lacking aspect.

Rarely do our secondary sources go beyond the statistical evidence. In a laconic but inspiring paragraph, a Syrian economist named Rizkallah Hilan has aptly summed up the economic and social effects of the new automotive "industry" which comprised garages, gas stations, commercial and maintenance networks and road construction. It generated a new professional category of mechanics (at the time mostly for motor vehicles), but, in his view, also enabled faster exchanges of goods and ideas, agricultural development, Bedouin settlement – in

---

112 Amīn al-Rihānī, *Tārīkh Najd al-Hadīth* (Beirut, 1928).
113 For Iran, see Willem Floor, 'Les premières règles de police urbaine à Téhéran', in Chahryar Adle and Bernard Hourcade (eds), *Téhéran: Capitale Bicentenaire* (Paris and Teheran, 1992), pp. 173–198. This is also true for the Ottoman Empire/Turkey and other states. The aspect of legislation on new technologies in this first era of globalization is, of course, not limited to automobiles, and needs systematic consideration. This applies also to international conventions to which the Ottoman Empire, including Egypt and Iran became parties.
114 In 2012, the Ford cars importer in Israel re-issued a trade catalogue from the 1930s, which was at the time translated from an Arabic original into Hebrew (the regional dealer worked from Alexandria). It left the pertinent publicity images of the Egyptian countryside and the pyramids intact.
115 Some advertisements also specified prices, which was usually not the case for other commodities. An American trade report asserted that closed sedans and limousines sold better than open cars, as they could withdraw women from sight. Some European producers (Citroën, Fiat, Austin etc.) may have done better business owing to price and credit.

short, regional and national integration. Learning to drive and making car trips added social prestige.[116]

We may consider the automobile as a rather unique case, as it is almost the only technological device, discussed here, which ultimately led to a local industry in Turkey, Egypt and Iran.[117]

2.10 Some conclusions

With reference to those "small technologies" here discussed, we have attempted to show that not all were imported wholesale or blindly imitated from the West: While some were speedily adopted and diffused in one or more Middle Eastern societies, others were delayed, or, in one exceptional case, even rejected. [118]

It might be useful to think of a schematic model of the kind which has been used to explain the diffusion of communication technologies, namely in parameters of "four Cs": Culture, Competence/Capacity, Control and Capital.[119] Thus, the temporary prohibition on the import of typewriters would come under Control, and the slow acquisition of automobiles under Capital and Competence, the need for reading glasses under Competence, but Culture would seem to us to be a predominant factor at play for most.[120]

---

116 Hilan, *Culture*, pp. 154–5.
117 Ford started assembling cars in Istanbul in 1928 with the support of the Turkish regime but gave up in the 1940s. Other manufacturers came much later. Of the consumer goods we discussed, sewing machines "Cleopatra" were for some time manufactured in Egypt and light bulbs in Turkey.
118 Many 'small' technologies still have to be mapped out. A case, also for an earlier period, is the acquisition of rifles as personal arms, not primarily for defense or hunting, but as a status symbol (e.g., beduins). Much attention has been paid to large weaponry for armies including local production. It is equally surprising that the so-called Primus stove, now a generic term but a technological implement developed and marketed by the Swedish firm Hjort & Co. has not received the attention it deserves, as it probably changed home cooking (where kerosene became available). This lacuna also applies to the transition of combustion engines to diesel. We know little about the purchase of different categories of steam engines and generators by private persons, but the number of diesel engines (from combustion engines, the next stage) in the countryside rose sharply from the 1920s.
119 There are several such Four Cs models, we slightly adapt here 'Assessing the impact and technology drivers in region' (which discusses present day NIC aspects): www.au.af.mil/au/awc/awcgate/cia/nic2020/nov6_panel4.pdf (accessed 22 February 2013).
120 While not adopting a rigid deterministic pattern (but working closer to a 'soft' determination model), we remember Robert Heilbroner, who answered the question 'do machines make history?' with the qualifying statement that 'certain cultural aspects' ought to be taken into consideration.

3 The social history of the sewing machine in the Middle East

Can we write the *social* history of an object? It surely has been done, since we have studies dealing with such diverse "things" as certain items of clothing or furniture, the Persian rug, musical instruments, the machine gun, or even the potato. While focusing our investigation on a new mechanical or technological device or method, we must not, of course, neglect the human inventors, distributors and consumers. Even the introduction of small objects or inventions can sometimes have far-reaching consequences and they therefore qualify for a place in social history in the sense of Charles Tilly's "How people lived the big changes".[1]

"Do machines make history?" remains a crucial question. Although we would like to put man (and woman) in the center, we tend to accept the premise that machines (as well) do, with the proviso that "certain 'cultural' aspects" must be taken into consideration.[2] This, indeed, is the point which needs elucidation: labor, gender, and social conditions differ from country to country, from era to era, or from one social class to another. We also have to bear in mind, that there is no point in repeating what is obvious or banal with regard to the introduction of new European or American inventions into the Middle East (which anyway is complex as a region). We have to sort out those aspects which are different and worth giving thought to.

Most historical accounts tend to deal with big technologies (steam power, electricity, aeronautics or computers). However, small machines have no less an impact. With some exaggeration it might be argued that "[n]ext to the plough, the sewing machine is humanity's most blessed instrument"; and "with the exception of the clock, [it] was the first piece of mechanism to be introduced into the home."[3] We would therefore suggest a distinction between "big" technologies" which arrived from Western-Europe or North-America as turnkey projects, often to the benefit of foreign or local elites, and "small" technologies – sometimes what we would call today "consumer durables". Though the latter were perhaps not immediately diffused to all social strata, they became gradually available to and affordable

---

[1] "Retrieving European Lives" in O. Zunz (ed.), *Reliving the Past, the Worlds of Social History* (Chapel Hill 1985), p. 15.
[2] R.L. Heilbroner, "Do Machines Make History?" and "Technological Determinism Revisited" in M.R. Smith and L. Marx (eds.), *Does Technology Drive History? The Dilemma of Technological Determinism* (Cambridge, Mass. 1994), pp. 53–78.
[3] F.P. Godfrey, *An International History of the Sewing Machine*, (London 1982), p. 21. The author reminds us that Gandhi exempted sewing machines from his boycott of western goods.

by large populations and proved their benefit to them. Although Marx understood the revolutionary potential of the recently invented sewing machine and particularly its negative effects on labor relations, he failed to see its more beneficial use in a home environment.[4] One is also reminded of a statement by Mahatma Gandhi: "Every machine that helps every individual has a place, but there should be no place for machines that concentrate power in a few hands and turn the masses into mere machine minders, if indeed they do not make them unemployed."[5]

The importance of the invention, or rather the development of the sewing machine for European industrial and technological development, e.g. the clothing industry, shoe-making, glove-making, saddlery and book-making has been highlighted by David Landes, as well as by Eric Hobsbawm[6] The social consequences of this invention maybe even further stretched: "it made ordinary women seamstresses and seamstresses tailors, and so doing hastened the transformation of what had once been the task of every woman into a professional activity."[7] Undoubtedly, the development of a large scale garment industry in the United States cannot be envisaged without the sewing machine and the availability of cheap female labor, often immigrant women, many of them working in sweatshops or from their homes.[8] Clothes became more refined, and as early as the 1860s-1870s, one discovers a mutual connection between mass-production and enhanced fashion consciousness, which crossed class lines. This break-through differs remarkably from developments in the Middle East.

Of all producers of sewing machines the success of the Singer company stands out. Its efficiency in mass production was unequalled, particularly as it was matched by a network of dealers, showrooms, depots, repair workshops and after-sales services, and sustained by advertising campaigns. Singer itself also provided instruction in the homes of new customers and executed repairs on the spot

---

[4] Karl Marx, *Das Kapital* (Berlin 1951), pp. 494–505.
[5] As quoted by E.F. Schumacher, *Small is Beautiful, Economics as if People Mattered"* (New York 1973), pp. 34–35.
[6] David Landes, *The Unbound Prometheus* (Cambridge 1969), p. 294; Eric J. Hobsbawm, *Industry and Empire* (London 1969), p. 175. The development of the sewing machine, as so many technological innovations, must be seen as a series of inventions and patents, e.g., the needle with a point and eye on one side (around 1755), an abortive breakthrough by Barthélemy Thimonnier (1841), the perfection of the shuttle principle by Elias Howe, and Isaac Meritt Singer (b. 1811) who turned it all into an engineering as well as a commercial success. Cf. on Germany: Karin Hausen, "Technical Progress and Women's Labour in the Nineteenth Century, The Social History of the Sewing Machine" in Georg Iggers (ed.), *The Social History of Progress* (Leamington Spa 1985).
[7] Landes., p. 119.
[8] A.H. Zophy and F.M. Kavenick (eds.), *Handbook of American Women's History* (New York 1990), p. 260.

when needed. This applied to the United States itself and soon also to its agencies abroad.

But Singer's outstanding success was probably due in the first place to its innovative marketing strategy, providing easy long-term credit for the purchase of an implement which was very much in demand, but beyond the immediate cash reach of a family. Machines were sold directly to the customer by the company's own salaried agents. After making the last of an agreed number of monthly installments, the purchaser would receive a final bill of sale. It was Edward Clark, Singer's partner, who seems to have developed the easy payment scheme, which ultimately became successful in the Middle East as well. Initial fears entertained by men that such a payment would make women more independent were apparently soon overcome.[9]

From 833 machines manufactured in until 1855, annual production in the USA rose to 2.564 in 1855 and even more steeply to 13.000 in 1860 and 464.254 in 1870. Indeed, graphic representations show a very steep, almost vertical rise in sales between approximately 1867 and 1872, which was almost certainly due to the famous "new family" model for domestic use which was introduced in 1865.[10] This machine was particularly easy to master, easy to power, and easy to transport.

In 1867 an overseas factory was set up in Clydebank near Glasgow, a location chosen for its iron and cotton industries, as well as for its shipping connections. Somewhat later another plant was built in Kilbowie nearby. By 1913, Singer in Glasgow produced no less than 1.301.851 machines a year, thereby becoming the largest sewing machine factory in the world, in fact in history. This 'overseas' plant – in itself another business novelty –, exported to various parts of the world, including the Middle East, and, for a long time, represented a third of Singer's production capacity.[11] More factories followed in Podolsk, Wittenberg, Monza, Bonnierès, Blankenloch and Campinas.[12] Thus, Singer became an early multi-national. "Sing-

---

[9] R. Brandon, *Singer and the Sewing Machine, a Capitalist Romance* (London 1977, p. 121; A.D. Chandler, *The Visible Hand, The Managerial Revolution in American Business* (Cambridge, Mass. 1995), pp. 302–306 and 402–405. For its mixed blessings in France, where department stores had pioneered new credit systems, see Judith G. Coffin, "Credit, Consumption and Images in Women's Desires *French Historical Studies* 18/3 (1994), pp. 749–783. For Britain see A. Godley, "Homeworking and the Sewing Machine in the British Clothing Industry 1850–1905" in B. Burman (ed.), *The Culture of Sewing* (Oxford 1999), pp. 255–268.
[10] See graph in A. Baron and S.E. Klepp, "'If I Didn't Have My Sewing Machine ...': Women and Sewing Technology" in J.M. Jensen and S. Davidson (eds.), *A Needle, a Bobbin, a Strike, Women Needle Workers in America* (Philadelphia 1984), p. 33.
[11] Godfrey, p. 157.
[12] Godfrey, p. 256.

ers, it may be said, are the Fords of the sewing machine world but much earlier."[13] The Singer company, characteristically, also built one of the first skyscrapers in New York as its headquarters (1908).

The Singer emblem of later years – a capital S embracing the world – was not chosen haphazardly. Indeed, it was the first American industry producing a mass product to seek a global market.[14]

From 1853 onwards the company participated with a spacious showroom of its own in International Exhibitions (e.g. London 1853, Paris 1855, Vienna 1973, Chicago 1893), a fact which was always proudly advertised. In 1893 it distributed to the public colored plates representing Serbs, Hungarians, Swedes, Bosnians, Japanese and Zulus all called "intelligent" and using Singer machines to make their traditional clothing. "The Singer Manufacturing Co., with its factories and offices reaching out and covering every quarter of the globe, is better able than any other company in the world to understand just what is required in a sewing machine for family work ..."[15]

An American – possibly writing on behalf of the Singer company – imagined a universal sisterhood (though also one with ethnic stereotypes), boasting as early as 1880:

> On every sea are floating the Singer machines; along every road pressed by the foot of civilized man, this tireless ally of the world's great sisterhood is going upon its errand of helpfulness. Its cheering tune is understood no less by the sturdy German matron than by the slender Japanese maiden; it sings as intelligibly to the flaxen-haired Russian peasant girl as to the dark-eyed Mexican señorita, it needs no interpreter, whether it sings amidst the snows of Canada or upon the pampas of Paraguay; the Hindoo mother and the Chicago maiden are to-night making the same stitch; the untiring feet of Ireland's fair-skinned Nora are driving the same treadle with the tiny understandings of China's tawny daughter; and thus American machines, American brains, and American money are bringing the women of the whole world into one universal kinship and sisterhood.[16]

Up from an estimated 6 million machines sold in 1880, sales in 1904, indeed, were already estimated at 20 million. Singer claimed to have three quarters of the world market.

---

13 Brian Jewell, *Antique Sewing Machines* (Timbridge Wells, Kent 1985), p. 9–10.
14 E.S. Rosenberg, *Spreading the American Dream* (New York 1982), p. 20. An earlier trademark represented a spool, two needles and a thread modeled in S-shape.
15 *The Singer Manufacturing Company, Sewing Machines*, The Columbian Exposition, ed. May 1893, p. 4.
16 J. Scott, *Genius Rewarded, the Story of the Sewing Machine* (New York 1880), p. 34. Note the absence of the Middle East from this description.

In Russia, for instance, Singer sales started in 1866, to sour to 110.000 machines sold by 1900, and 700.000 by 1914.[17] Also in Japan, Singer machines were introduced in the 1860s, and sales seemed to be successful for several decades, till that country started its own mass production and then restricted imports in 1937.[18] On the other hand, Singer operations in China were not particularly successful, at least not in the 1880s (when intensive efforts were made to penetrate this market), owing to Chinese reservations about the rigid stitching of seams.[19]

3.1 Marketing and consumption in the Middle East

Though the Middle East had not only a long-standing reputation in textile production but also a tradition of its own in needlework, the above quoted Singer company's publicity material of the late 19th century hardly mentions the Near or Middle East. While the European market had become totally covered by Singer and its rivals, the Dutch correspondent of a trade journal, in 1884, saw Turkey still as an interesting new option.[20] Indeed, it would seem that the Ottoman-Turkish market had first been approached in 1881 by Georg Neidlinger, the Singer company's agent in Hamburg.[21] As a German immigrant to the USA he had worked his way up "from carrying water for the man and doing odd jobs", to becoming the key marketing figure for most of continental Europe and the Middle East till his retirement in 1902.[22]

Though the new appliance was advertised in Beirut as early as 1860, it took apparently some time to enter public consciousness: the influential journal *al-Muq-*

---

17 Frederick Vernon Carstensen, *American Multinational Corporations in Imperial Russia. Chapters in Foreign Enterprise and Russian Economic Development*, PhD Thesis, Yale University, 1976, p. 87 et passim. The elevated Singer globe on its headquarters on Nevski Prospect in St-Peterburg (1904) even survived the Soviet Revolution, and is presently (2002) being restored.
18 Susumi Hondai, "Organizational Innovation and the Development of the Sewing-Machine Industry" in R. Minami et al. (eds.), *Acquiring, Adapting and Developing Technologies, Lessons from the Japanese Experience* (London 1995), pp. 193 ff. In Japan the Yasin Co. (since 1934 Brother) produced sewing machines from 1906.
19 R.B. Davies, "'Peacefully Working to Conquer the World:' the Singer Manufacturing Company in Foreign Markets, 1854–1889", *Business History Review*, vol. 43/3 (1969), p. 324.
20 Davies, p. 320.
21 According to *L'Indicateur Ottoman* of 1881 (Istanbul 1881), Neidlinger had established himself as Singer Agent at the Beltazzi Khan, rue Voyvode 69, but E. Eldem's *Bankalar Caddesi* (Istanbul 2000) does not mention him in his survey of this famous street.
22 Scott, p. 32 and p. 38. Davies, pp. 306–307, and pp. 312–313. See regular advertisements (in French !) by the Istanbul Singer office in *The Levant and Eastern Express* for e.g. 1900 and 1901. Also: Nur Akın, 19. *Yüzyılın İkinci Yarısında Galata ve Pera* (Istanbul 1998), pp. 220 and p. 222.

tataf, for instance, on the verge of moving from Beirut to Cairo in the 1880s, published two articles on the new invention.[23]

Singer embarked on a long running advertising campaign in the local press addressed both at the workshop owner and the private consumer.[24] Advertising in women's journals, too, probably contributed much to the Singer sewing machine's diffusion.[25] Typical Singer advertisements in *al-Muqtataf* in 1902 still disclosed an upper-class bias, using Neidlinger's appointment as purveyor to the courts of the Sultan, the Khedive and the Prince of Bulgaria. However, instruction and repairs were to be done at the purchaser's domicile, and special mention was made of the Singer's payment scheme and its moderate prices.[26]

Singer, indeed, seems to have been the first to introduce its innovative payment scheme to the Middle East as well.[27] Of course, here, too, given the low economic potential in the region, the purchase of a sewing machine meant a relatively large expense, and the sort of hire-purchase arrangement as proposed by the Singer agents, definitely advanced sales.[28] With some envy, Ernest Weakly, wrote in his extensive report on Syria to the British Board of Trade in 1911:

---

23 *Ḥadīqat al-Akhbār*, 29 March 1860, where an advertisement with illustrations refer the reader to the editorial office for the name of the sole agent in town of the new American implement (with thanks to my collegue Dr. Fruma Zachs for this reference); *Al-Muqtataf* 1883, vol. 8, p. 634, and more extensively with illustrations 1884, vol. 9, pp. 93–96.

24 E.B. Frierson, "Cheap and Easy: The Creation of Consumer Culture in Late Ottoman Society", in D. Quataert (ed.), *Consumption Studies and the History of the Ottoman Empire, 1550–1922, an Introduction* (Albany, SUNY 2000), pp. 244–245 (note illustration) and p. 250.

25 B. Baron, *The Women's Awakening in Egypt, Culture, Society and the Press* (New Haven 1994), p. 94.

26 E.g. *al-Muqtataf*, vol. 26 (1902) after p. 576. A more systematic analysis of Singer advertisements in the Middle East (esp. their class aspects) could yield interesting conclusions, possibly different from Diane A. Douglas, "The Machine in the Parlor: A Dialectical Analysis of the Sewing Machine", *Journal of American Culture* 5/1 (1982), pp. 20–29. Singer also printed its own instruction manuals in Ottoman Turkish (Fig. 17). I wish to thank Professor Ekmeleddin İhsanoğlu for obtaining a copy for me.

27 This point is worth pursuing, as it may apply also to other USA-made mass products, e.g. Ford motor cars, McCormick reapers, Remington typewriters etc. On buying machines on credit see also Frierson, p. 247+n7, and Quataert, *Manufacturing*, pp. 56–57. See further E. Frangakis-Syrett, "American Trading Practices in Izmir in the Late Nineteenth and Early Twentieth Centuries" in D. Panzac (ed.), *Histoire Economique et Sociale de l'Empire Ottoman et de la Turquie (1326–1960), Collection Turcica*, vol. VIII (Paris 1995), pp. 183; Paul Fesch, *Constantinople aux Derniers Jours d'Abdul-Hamid* (Paris 1907), p. 570; Arthur Ruppin, *Syrien als Wirtschaftsgebiet* (Berlin 1920), p. 214.

28 A poor widow of Zaḥla even stole money to buy a sewing machine to provide for her income, E. Thompson, *Engendering the Nation, Statebuilding, Imperialism and Women in Syria and Lebanon, 1920–1945* (Unpublished PhD Thesis, Columbia Univ. 1995, p. 73; omitted in her *Colonial Citizens, Republican Rights, Paternal Privilege, and Gender in French Syria and Lebanon* (New York 2000).

... although a machine can be obtained for 8s. less if bought for cash, comparatively few can afford or are willing to make cash purchases when payments are allowed to be spread out over a series of weeks or months ... The organisation has been admirably thought out, and, being so thoroughly suited to the conditions of the country, it has very materially helped the wide sale which the American machine has in the country. Local dealers who import German and British machines are also obliged to sell largely on credit, ... and their terms in many cases have to be more liberal in order to obtain some small share of the trade, which would otherwise be monopolized by the American agencies. There does not seem much to choose, as regards price, between the American machine and other makes prices varying between £T. 3 and £T. 3.75 for hand machines.[29]

Also Donald Quataert sees a break-through in Singer's marketing success around 1900 and equally emphasizes in this connection its monthly payment system.[30] On the whole, this marketing policy proved highly successful and is often highlighted by our sources; in the first decade of the 20th century it allegedly pertained to 80% of all sales of sewing machines, overtaking competing English, French and German brands (Fig. 17). However, it also could have its draw-backs. At least, in the early 1930s in Syria, hundreds of machines were re-possessed by the Singer company for defaults of payment.[31]

Since the establishment of a Singer depot in Ma'mūret-ül-'Azīz (Elazığ) in 1900, commercial activities branched out to Diyarbakir, Mosul, Baghdad and Basra. By 1903, 1.230 machines had been sold in what was called Asia Minor.[32] Around 1900, 400 machines were said to have been sold in the Kharput region, up to Diyarbakir, Mosul, Baghdad and Basra, and some 375 in Bursa, 250 in Konya. In Izmir 16.000 were sold between 1903 and 1905 alone. Sivas accounted for 2.000 machines.[33]

In the first decade of the twentieth century sewing machines entered many Ottoman households. In the satirical press criticism could be found against women working hard on the machine, as well as against the machine invasion

---

[29] *Report upon the Conditions and Prospects of British Trade in Syria by Mr. Ernest Weakly* (London 1911), p. 191 [hereafter: Weakly]. Also earlier British trade reports lamented British conservatism with regard to affording advance credit, as the Italians, Germans, Austrians and Belgians did. I would like to thank my colleague Gad Gilbar for lending me a copy of the above source and other trade reports, and for discussing with me some of the issues raised in this study.
[30] Donald Quataert, *Manufacturing and Technology Transfer in the Ottoman Empire, 1800–1914* (Istanbul, Strasbourg 1992), p. 15.
[31] Thompson, *Engendering*, p. 73.
[32] Fesch, p. 570. On Baghdad in 1907, see Ch. Issawi (ed.), *The Economic History of the Middle East, 1800–1914* (Chicago and London 1966), p. 183.
[33] Quoted from Quataert, *Manufacturing*, pp. 23–24. Cf. Chambre de Commerce Français, Istanbul, *La Revue Commerciale du Levant* 1904.

Fig. 17: Frontispiece of the instruction manual for owners of the Singer model 28K with its "vibrating shuttle" (app. 1908), a term here phonetically transcribed in Ottoman Turkish script. On the cover, the New York based Singer company prided itself to have its headquarters for the Ottoman lands, Egypt, Greece, Bulgaria at the Dūzoğlu Han in u Beyoğlu (Rue de Pera), a building today owned by Koç University (courtesy: Ekmelledin İhsanoğlu).

in general.[34] Paul Fesch estimated in 1907 that some ten thousand American and German-made sewing machines were owned in Istanbul, and twelve thousand in Damascus. In the Ottoman capital, the well-known French department store

---

[34] P. Dumont, "Said Bey – The Everyday Life of an Istanbul Townsman at the Beginning of the Twentieth Century" in A. Hourani et al. (eds.), *The Modern Middle East* (London 1993), p. 284; Palmira Brummett, *Image & Imperialism in the Ottoman Revolutionary Press 1908–1911.* (Binghamton 2000), pp. 201–202.

of Orosdi-Back acted as local depot for Singer.[35] This firm also acted as Singer agent in Beirut.[36]

Arthur Ruppin, moreover, reported in 1920 that Singer had been very successful in keeping competitors on the sideline. American sewing machines were alleged to have 60% of the Syrian market; in Damascus alone, by 1912, the annual turn-over was Fr. 130.000, and in Tripoli (Lebanon) even Fr. 150.000. Considering that a machine cost Fr. 80–120, it meant that in each of these places over a thousand pieces were sold every year.[37]

In the region of Palestine, the local agent was Yitshaq Hayutman who was to become one the founders of Tel Aviv in 1909. Upon his arrival in Jaffa in 1905, Yeshayahu Lewin, a friend from Metulla, arranged the Singer job for him: "You can be successful, putting its business in order, adding clerks, salesmen, collectors and technicians from among our brothers, in order to distribute this product in towns and settlements. The main director in Beirut is also a Jew ...". Hayutman's store at Bustrus Street became much more than an outlet for sewing machines, – a focus of information and activity on the Zionist enterprise, and indeed a first point of absorption for arriving immigrants.[38] In Palestine, but probably elsewhere as well, at first, sewing machines were sold to professional tailors, then to housewives. All were still pedal-driven, as electricity would not be readily available for a few decades to come.

Hayutman was a dynamic person, who sold his merchandise far beyond the boundaries of his urban environment. He even ventured into the desert to demonstrate and sell sewing machines to nomads, as a highly illuminating anecdote told by the zoologist Aharoni attests. During the "Italian war in Tripoli" – probably in 1911 – Aharoni, on one of his exploratory excursions, had been arrested and jailed in Karak, apparently after arousing the suspicion of the Ottoman authorities because of his "Italian-sounding" name and lacking a proper travel permit for the Hijaz. It was Hayutman's accidental arrival on a sales mission in Karak, and his intervention with the local governor (and his telegrams to influential Zionist notables in Jerusalem) which ensured his speedy release.[39]

---

35 Fesch, p. 517, and p. 604.
36 *Bulletin Consulaire, Turquie d'Asie: Situation de la Region de Caïffa et de Saint-Jean d'Acre*, no. 1058, p. 82.
37 Ruppin, p. 214, and p. 310.
38 Zekharya Hayut (Hayutman), *'Im Yitshaq Hayutman, Meyyased Metulla we-Tel-Aviv* (Haifa 5728 [1967–1968], p. 143, pp. 157–8, pp. 178–9.
39 Y. Aharoni, *Zikhronot Zo'olog 'Ivri* (Tel Aviv 1946/7), vol. 2, p. 51. I wish to thank Prof. B. Hayutman in Jerusalem for this reference.

Singer gradually established a large network of agents throughout the Middle East. In Syria (in the 1930s) it had 16 offices and 168 employees.[40] In Egypt (1951), its headquarters were located at Sharīfayn Street in the modern business district of Cairo with branches in Alexandria and some thirty branches in provincial towns all-over the country, from Damiette, Rosette and Port Said in the north, to Luxor and Aswan in the south.[41]

Unfortunately, we have no systematic data on the regional imports of sewing machines, or even a total; where sewing machines are specifically mentioned among other hardware, the volume of trade is sometimes given in money values, and sometimes in weight or crates.[42] However, it is clear that rather large quantities – also other brands than Singer from the United Kingdom, Germany and Italy – continued to be imported over the next decades.[43]

3.2 Use of the sewing machine in industry

Jamāl al-Dīn al-Qāsimī, a Damascene *'ālim* who has left us an interesting lexicon of professions in his city at the end of the 19th century, notices – not without misgivings, it seems – the spread of sewing machines, "one of the works of the Franks which dazzle the mind". While stating that this implement, much quicker than manual work, is widely used by Christian tailors, especially those making uniforms for the army or the civil administration, he adds that the use of the sewing machine has spread to some Muslim women as well.[44]

---

40 Thompson, *Engendering*, p. 73, and her *Colonial Citizens*, pp. 34–35.
41 *The Egyptian Directory* (Cairo 1951), p. 610. Cf. *idem* for 1931, p. 993 (21 branches). The Cairo manager for many years was Ezra Rodrigue, G. Krämer, *Juden, Minderheit, Millet, Nation, Die Juden in Ägypten* (Wiesbaden 1982), p. 192.
42 E. g. *Bulletin Consulaire*, p. 82, e. g. in 1909: 9.950 kilograms, 1910: 11.200 kilograms etc. Great Britain, *Report for the Year 1898 on the Trade of the Vilayet of Aleppo*, p. 11; idem for 1902 on Damascus, p. 12; idem for 1903 on Damascus, p. 13; idem for 1905 on Aleppo, p. 10; idem for 1906 on Damascus, p. 10; idem for 1907 on Beirut, p. 5, – all in Pounds Sterling. Cf. *HaHerut*, 12 Sivan 5669 [1909]: 100 sewing machines at value of 6.000 Franks, as quoted by Shelomo Sheva and Dan Ben-Amotz, *Eretz Tzion Yerushalayim* (Jerusalem 1973) [in Hebrew], p. 188. Or 224 crates to Baghdad in 1907, Issawi, p. 183.
43 Cf. Great Britain, Department of Overseas Trade, *Report on the Economic and Financial Situation of Egypt* (June 1925), pp. 32–33; idem for June 1939, p. 83; *Report on Economic and Social Conditions in Turkey*, April 1939, p. 73; idem for September 1947, p. 196; idem for April 1950, p. 157.
44 Jamāl al-Dīn al-Qāsimī, *Qāmūs al-Ṣināʿāt* (Paris, The Hague 1960), vol. I, p. 131. The production of uniforms, not only for the army and the civil service, but also for high school students, and later for para-military youth movements, calls for more research. Many such uniforms were manufactured at home.

Much academic credit for drawing our attention to the significance of the sewing machine in the Ottoman industrial context goes to Donald Quataert. He has shown its importance for the production not only of ready-made clothing and shoes, but also of umbrellas.[45] Even more important is the link which he establishes between the massive diffusion of sewing machines and the revival of Ottoman manufacturing which took place between 1870 and 1900. Thanks to cheap labor, a small ready-made garment industry in Istanbul began to emerge, and the shoe-making branch, threatened by Western imports and tastes, went through a revival, at least for some decades. Quataert has ascribed a new impetus given to the faltering Syrian textile industry to sewing machines in the 1890s; women, for instance in Damascus, working on knitting machines at home, started to produce hosiery.[46] Part of this went on in small sweat-shops, often run by one family, part of it was the work of women as a cottage industry.[47] In Salonika, too, an advanced center of the Ottoman textile industry, some hundreds of women were employed in the female ready-to-wear industry.[48]

3.3 Home industry

Machines, indeed, were purchased by tailors and cobblers, as well as by non-professional women. The sewing machine may have worked as a catalyst to the rise in the formal away-from-home employment of women in the Ottoman Empire, at least where it existed, and on a small scale. Indeed, traditionally, women who contributed to the Middle Eastern textile industries, e.g. spinning, weaving or embroidering, did so usually from their homes rather than from bazaar workshops.[49]

Indeed, one gets the impression that the availability of machines led to the development of a new home industry, e.g. of socks and stockings in Damascus.[50] In Istanbul shoe-making workshops, women would complete the work done on ma-

---

45 Quataert, *Manufacturing*, pp. 22–25. Orosdi-Back in Istanbul operated a large workshop for assembling umbrellas from imported parts.
46 J.A. Reilly, *Women, Property and Production*, MESA Conference Nov. 1991, quoted by Thompson, *Engendering*, p. 50; or his "from Workshops to Sweatshops, Damascus Textiles and the World Economy in the last Ottoman century", *Review* 16 (1993), p. 210. This point needs further research as sewing and knitting machines may have been confused.
47 Quataert, *Manufacturing*, p. 24.
48 D. Quataert in *Salonique 1859–1918*, *Autrement* (Paris 1992), p. 181.
49 M. Shatzmiller, *Labour in the Medieval Islamic World* (Leiden 1994), pp. 358–359, emphasizing the contrast with European practices.
50 Reilly, "From Workshops to Sweathops", p. 210.

chines by men.[51] This "invisible economy" has continued ever since, enabling both urban and rural women to gain some additional family income from the privacy of their homes.[52] A form of pre-mass production, one could say, which exceeded the consumptive needs of the family.[53] Sewing machines also became a much desired element in the dowry of women getting married.

While seamstresses can be found in both urban and or rural settings, possibly in slightly variegating production roles, sewing machines belong also to the regular outfit of nomads.[54]

3.4 Gender aspects

Sewing machines are operated by both men and women, but the exact gender division of labor may differ.[55] Gender roles in factories are often determined by the size, weight, strength and speed of the production tools.[56] Looking at pictures from

---

51 D. Quataert, "Ottoman Women, Households, and Textile Manufacturing, 1800–1914" in his *Workers, Peasants and Economic Change in the Ottoman Empire 1730–1914* (Istanbul 1993), p. 90.
52 A.M. Jennings, "Nubian Women and the Shadow Economy", in R.A. Lobban (ed.), *Middle Eastern Women and the Invisible Economy* (Gainesville 1998), p. 49; B.J. Michael, "Baggara Women as Market Strategists", *ibid.*, p. 67; D.M. Walters, "Invisible Survivors, Women and the Diversity in the Transitional Economy of Yemen", *ibid.*, p. 74 and p. 88 (note that sewing machines are made in Germany and Japan); N. Mustafa M. Ali, "The Invisible Economy, Survival, and Empowerment, Five Cases from Abbara, Sudan", *ibid.*, p. 100 and p. 106. Also Z. Kamalkhani, *Women's Islam, Religious Practice Among Women in Today's Iran* (London, New York 1998), pp. 158–159, and E. Mine Cinar, "Unskilled Urban Migrant Women and Disguised Employment: Home-working Women in Istanbul, Turkey", *World Development*, vol 22 (1994), pp. 369–390.
53 We know near to nothing about the exact distribution of sewing machines between industrial workshops and factories and private homes; even anthropological studies which pay attention to TV-sets, laundry machines or electric fans, fail to mention whether the household owns a sewing machine. One exception, a study on Istanbul (1991), states that 60.6% of the households own one, quoted in M. Sönmez, *Statistical Guide to Istanbul in the 1990s* (Istanbul, n.d.), p. 30.
54 D. Ammon, *Crafts of Egypt* (Cairo, AUC 1991), p. 7. On the expanding household belongings in beduin tents see Shirley Kay, "Social Change in Saudi Arabia" in T. Niblock (ed.), *State, Society and Economy in Saudi Arabia* (London 1982), pp. 173–174. Nomads today may store their sewing machines in a truck, see D. Chatty, *Mobile Pastoralists* (New York 1996), p. 133, p. 157, p. 175. Cf. the anecdote on Hayutman above.
55 S. Rowbotham "Feminist Approaches to Technology, Women's Values or a Gender Lens?" in S. Mitler and S. Rowbotham (eds.), *Women Encounter Technology, Changing Patterns of Employment in the Third World* (London and New York 1995), pp. 44–67. In the New England shoe industry women initially resisted training for machines, but later replaced male workers.
56 A very perceptive overview is C. Overholt et al. (eds.), *Gender Roles in Development Projects, a Case Book* (West Hartford 1985), see for instance p. 72.

the Middle East one often gets the impression that sewing machines have become a feminized implement; men operating such low-tech machinery, on the other hand, may be less willing to be photographed.

Middle Eastern textile and garment manufacturing branches, in their production as well as their marketing aspects, have always been strongly segregated. Indeed, a conspicuous gender aspect (not only) in the Middle East is that tailoring is mostly by gender: women's clothing is fitted by women, and men's clothing by men.[57]

Like elsewhere in the Third World, in Middle Eastern countries with a sizeable industrial labor force it is mainly the textile branch (not only the clothing industry) in which women are overwhelmingly represented. Though "needlework" and embroidery – as in many countries – were a compulsory part of the curriculum of girls in elementary schools, this was not enough, however, to vouch for their capability to be smoothly absorbed in an industrial plant.[58]

For lack of data, one has to be careful with assumptions of large-scale feminization. During World War II, a Syrian newspaper was quoted as portraying women to have given up their sewing machines and typewriters for heavier jobs, held in normal times by males.[59] There could be much European-centered bias in this quotation. We found only one documented case – in Aden (1977) – in which the status of tailoring was allegedly deprecated because of a take-over of tasks from men by women, working for smaller wages.[60]

Though homeworking in the textile and garment sectors is still very popular in Egypt, and probably also in Turkey and other countries of the Middle East, it seems that economic liberalization can reverse the process. With the *Infitāḥ* in Egypt, and with new private investments in the garment industry, women, previously in home industries, were seen moving into industrial jobs.[61]

---

[57] B.K. Larson, "Women, Work, and the Informal Economy in Rural Egypt", in Lobban, p. 152. Cf. the male village tailor setting up business with a secondhand Singer in the 1930s, in Hani Fakhouri, *Kafr el-Elow, an Egyptian Village in Transition* (New York 1972), p. 49.
[58] Randa George Siniora, *Palestinian Labor in a Dependent Economy: Women Workers in the West Bank Clothing Industry*, Cairo Papers in Social Studies 12/3 (1989), p. 59.
[59] *Le Jour*, 22 May 1943, quoted by Thompson, *Engendering*, p. 194.
[60] M. Molyneux, *State Policies and the Position of Women Workers in the People's Democratic Republic of Yemen, 1967–1977* (Geneva: ILO 1984), p. 46.
[61] V.M. Moghadam, "Manufacturing and Women in the Middle East and North Africa: a Case of the Textiles and Garments Industry", *CMEIS Occasional Paper no. 49 (1995)*, University of Durham, p. 33 and p. 36. See also A.M. Abdel-Latif, "The Non-price Determinants of Export Success or Failure: The Egyptian Ready-Made Garment Industry, 1975–1989, *World Development*, vol. 21 (1993), pp. 1677–1684.

Fig. 18: This photo from Kuwayt, 1955, by Jean-Philippe Charbonnier, may be open to divergent interpretations but it drew my attention to women moving about with a sewing machine to earn money (Getty images).

The personal ownership of a sewing machine, however, is eminently important where it gave some measure of initiative, and even economic power, to women. There are ample examples in the literature.[62] Single women sometimes operated an atelier with two or three seamstresses where richer women had their dresses made.[63] Not only could a village seamstress support herself by making clothing for individual customers, but she could in some cases even enter into a business relationship with an urban shop-owner.[64] Enterprising women could become itinerant seamstresses, selling their services from home to home. (Fig. 18)[65] Also Leila Ahmed tells of a spinster, a relative who supported herself by making clothing for entire families in their homes, bringing her own Singer machine along.[66] Another case would be that of refugee women, e.g. Armenian women in Aleppo, who found a new existence by means of their sewing machine.[67]

Women's journals not only carried advertisements for sewing machines but initiated from an early stage onwards columns or pages on sewing.[68] Fashion in general became more important; European and American styles seen at the movies were also imitated for homemade dresses.

3.5 The sewing machine as a tool of development

Already in the 1930s, bourgeois women's volunteering organizations in Syria started providing sewing machines and work to their poorer sisters.[69] Revenues from tablecloths embroidered by village women were channeled into the purchase of sewing machines.[70]

---

[62] However, the interesting recent collection of essays by Barbara Burman (ed.) *The Culture of Sewing*, unfortunately ignores the Middle East.
[63] As captured in the memories of Elias Petropoulos, "Ah Allegra" in *Salonique 1850–1948, Autrement* (Paris 1992).
[64] S. Schaefer Davis, "Working Women in a Moroccan Village", in L. Beck and N. Keddie (eds.), *Women in the Muslim World* (Cambridge, Mass. 1978), p. 420.
[65] Baron, *Women's Awakening*, p. 158, quoting E. Cooper, *Women of Egypt*, p. 121 and p. 123.
[66] L. Ahmed, *A Border Passage* (New York 1999), pp. 163–164.
[67] Ch. Rouyer, *Femmes d'Outremer* (Constantine, 2nd ed., 1941), pp. 13–17 quoted by Thompson, *Engendering*, p. 28.
[68] Cf later *Les Echos de Damas [?]*, 14 March 1934 as quoted by E. Thompson, *Colonial Citizens*, p. 217. Also, the use of paper patterns in the Middle East deserves more investigation.
[69] *Washington NA [?]*, as quoted by Thompson, *Engendering*, p. 73.
[70] Thompson's interview with Nadida Cheikh al-'Ard of Yaqzat al-Mar'a al-Sha'miyya, *Engendering*, p. 380.

Because of the positive advantages of the sewing machines in terms of cost-benefit analysis, the provision of sewing machines and training courses has become a common phenomenon not only in the Middle East – in community centers, village associations, and charitable organizations.[71] Sewing clubs have become a classical type of development project in many Third World countries, providing poor women with the possibility to manufacture family clothing themselves, and possibly even generating some additional income from their homes.

In Yemen, in the 1970s, for instance, the Women's Union set up a sewing cooperative and provided training for the women working there.[72] Similarly, a monograph on an Upper Egyptian village in the 1970s, shows that this aspect is by no means exhausted yet. In Hanya village, a government appointed sewing teacher arrives once a week to train local women in making clothing for their children, as well as for sale at the village bazaar.[73]

The Singer company itself opened in 1982 in the Baqʻa refugee camp in Jordan, a workshop for the manufacturing of ready-made clothing, mainly school uniforms, thereby providing jobs for women within the confines of the camp. In the larger Wahdat camp sewing courses were provided.[74]

3.6 Ready-made clothing

An interesting related issue is the fact that, in spite of the introduction of sewing machines, and in spite of existing potential, no large-scale confection industry for local consumption in the Middle East has emerged till rather recently.[75] Weakly, for example, mentioned in 1911 that a local ready-made clothing industry had sprung up in Beirut, giving employment to some 300 to 400 persons, and ascribed this to

---

[71] E.g., a spontaneous settlement area of Cairo, B. Tekçe L. Oldham and F. Shorter, *A Place to Live, Families and Child Care in a Cairo Neighborhood* (Cairo 1994), pp. 50–51 and 59.
[72] Molyneux, p. 20.
[73] Marileen van der Most van Spijk, *Eager to Learn* (Leiden 1982), p. 26, and p. 45 and p. 77. This is undoubtedly a charitable project, as this source adds that a local social worker also provides cheap cloth.
[74] Rima Yusuf Salah, *The Changing Roles of Palestinian Women in Refugee Camps in Jordan* (SUNY, Unpublished PhD Thesis 1986), p. 286. For the opening of a new commercial training center in Sinak, Baghdad, see large advertisements in *Iraq Times*, e.g. 23 January 1953 ("Make yourself a dress as you learn home dressmaking").
[75] On the limited garment industry (as opposed to textile industry) see E. Longuenesse, "L'Industrialisation et sa Signification Sociale" in A. Raymond (ed.) *La Syrie d'Aujourd'hui* (Paris, 1980), p. 341 and 345. We ignore here E. Ashtor's reference to ready-made clothing in 10th century Iraq, *Social and Economic History of the Near East in the Middle Ages* (London 1976), p. 151.

the proximity of Egypt which had "given people a taste for a certain smartness in dress and appearance."[76]

Imports, however, remained dominant for a long time. With the impact of the penetration of Western manufactures and fashions into the Middle East, trade reports for the end of the 19th and the beginning of the 20th centuries also make specific mention of rising imports of ready-made clothing, predominantly from Austria-Hungary (or rather Bohemia) and Germany. Undoubtedly, these imports were destined only for the fashion-conscious elites. Ready-made clothing was often sold at the luxurious department stores, such as Orosdi-Back and others.[77]

But as a rule, even one generation back, and even in relatively westernized cities, such as Port Said, women went to the dressmaker for lack of a choice of ready-made clothes in stores; cloth (and patterns) were often selected under the impact of American movies they had watched.[78] All women of the then Saudi Finance Minister's household, to cite one more example, were said to be dressed in the same material purchased from the *sūq*.[79] In San'a', Serjeant observed in the early 1970s a growing preference for ready-made clothing over tailored clothing, though women also went on sewing themselves.[80] The Egyptian economist Galal Amin who analyzed social mobility over three generations noted in 1995 that "the figure of the seamstress coming and going" had recently become superseded as even wedding dresses were now ready-bought.[81]

Large popular classes, both urban and rural, still prefer hand-made clothing, either produced at home or in small sweatshops, or in their neighborhood by a familiar tailor or seamstress.[82] Middle Eastern clothing styles (e.g. *galabiyyas*) are relatively simple, without too much variation or whims of fashion, and are therefore easily manufactured at home.[83] Men, on the other hand are more likely to go

---

[76] Weakly, p. 153.
[77] J. Berque, *Imperialism and Revolution* (London, [transl.] 1972), p. 332. At some department stores also local made-to-measure clothing could be ordered. Ready-made clothing was, and often still is more expensive than home-made clothing.
[78] Sylvia Modelsky, *Port Said Revisited* (Washington 2000), p. 127.
[79] Kay, p. 179, quoting from Marianne Ali Reza, *At the Drop of a Veil* (Houghton 1971).
[80] R.B. Serjeant and R. Lewdock (eds.), *Sana'ā', an Arabian Islamic City* (Cambridge 1983), p. 259, cf. pp. 538–539.
[81] Galal Amin, *Whatever Happened to the Egyptians?* (Cairo 2001), p. 110; Cf. E.A. Early, *Baladi Women of Cairo* (Cairo 1993), p. 168, a story on the *baladi* bride's sisters who sewed a dress even cheaper than the seamstress could do.
[82] D. Singerman, *Avenues of Participation, Family, Politics, and Networks in Urban Quarters of Cairo* (Princeton 1995), p. 24, this in spite of the availability of ready-made Islamic dress or even imported jeans and shirts. Cf. Early, p. 79.
[83] This is also true for school uniforms.

to a professional tailor especially where their (western-style) garments require better skills.[84]

Data are lacking on which segments and/or percentages of the diverse populations have their clothing made at home or in the neighborhood, and which purchase ready-made confection. Personal observation at the relatively spacious cloth sections at *sūq*s in many a Middle Eastern country, would suggest that the former populations are still very large. However, Middle Eastern societies are marked by complex stratification, and superficial impressions may be misleading. It all depends on one's vantage point.[85] Unmistakably, however, the purchase of ready-made clothing has been on the rise, a process undoubtedly differing according to country, region, population, gender, class etc.

There are several indications for this, such as the following observation:

> The people who wear ready made are mostly the middle classes including the newly arrived lower middle classes (women) who care about locally defined fashion whether it be a Western style or the latest Islamic dress. Certain mosques sell at subsidized rates or give out the Islamic dress in a fairly conventional form and this creates an industry of sorts for seamstresses.[86]

Trade liberalization and *Infitāḥ* tendencies have probably contributed to this trend.[87] In Istanbul, Tekbir, a small workshop which started with only two sewing machines, has in recent years become a large company with branches in Turkey and Saudi-Arabia, allegedly producing no less than 10,500 garments a week for the "Islamic chic market."[88]

3.7 Conclusions

In societies, or sectors of societies, which entertain strong reservations against women working or trading outside the home, the sewing machine comes as a perfect solution. In the Middle East, the sewing machine remains a relatively affordable "small" technology, one which can be beneficially put to use within the confines of a private home, which can increase family income or savings, and which has the capability of empowering women. While Singer had to close its factories in

---

[84] Men, however, often do the fancy stitch-work. Comments kindly supplied to this author (5 Sept. 2001) by Andrea B. Rugh, author of *Reveal and Conceal, Dress in Contemporary Egypt* (Cairo 1986).
[85] Ready-made clothing was even sold in the village by itinerant peddlers, cf. Fakhouri, p. 54.
[86] Andrea Rugh, see footnote 314.
[87] Singerman, p. 24.
[88] "Fashion Creeps into the Islamic World", http://www.worldroom.com/pages/wo...int_version.

Clydebank in 1969 owing to declining demand by women in America and Western-Europe, the private market for sewing machines in the Middle East, for the time being, does not seem to have been exhausted yet.[89]

---

[89] The Singer company, once a jewel in the American industrial crown, was taken over by new entrepreneurs in Hong-Kong, henceforth called Semi-Tech. Its archives, however, were transferred to the State Historical Society of Wisconsin in Madison.

4 Leapfrogging from typewriter to personal computer in the Middle East

4.1 Introduction

Typewriters, once symbols of technological progress, modernity, bureaucratic efficiency, business profitability ("time is money"), and journalistic speed have today, by and large, become obsolete. They have appeared in novels, movies, and memoirs, and are today fully documented, not to say eulogized, on the internet.[1] Typewriters have become museum objects and collectors' items everywhere, from dozens of cities in North America and Europe to less expected locations in the Middle East such as Istanbul, Eskişehir and Tehran.[2] In this chapter we propose to follow the typewriter's diffusion, particularly in Ottoman and Republican Turkey and in Egypt.[3]

Typewriters were first adopted by commercial corporations and government agencies but after a while also by individual persons.[4] As this process could significantly differ from country to country, we will look into the factors at work in the Middle East.[5] I am aware that compared to Europe (and China earlier), printing

---

[1] We will not go into the technical evolution of typewriters with its thousands of patents, dozens of brands, and hundreds of models, but will try to focus on "the" typewriter's social and cultural impact.

[2] Even the last major factory in India has since closed. The Koç Museum collections in Istanbul and Ankara, the more specialized Tayfun Taylipoğlu Museum in Odunpazarı (Eskişehir), and the Safir Office Machine Museum in Tehran (thanks to Willem Floor for drawing my attention to it) are cases in point. As in a similar museum in Shanghai, it would appear that most items are in Roman script.

[3] This chapter had a long gestation period. A primordial version was presented as a paper to the Eighth Mediterranean Social and Political Research Meeting in Montecatini in 2007. Additional source material was collected during my term as Affiliate-in-Research at the Harvard Center for Middle Eastern Studies in 2013. In particular I wish to thank the library staff of the Harvard Business School there.

[4] Printing presses may be an ambiguous case: Nile Green, "Journeymen, Middlemen, Travel, Transculture, and Technology: The origins of Muslim Printing", *International Journal of Middle Eastern Studies*, vol. 41/2 (2009), p. 212 described the purchase by Levantine individuals of second-hand Stanhope hand presses at the beginning of the 19th century when steam-powered presses came into use in Great Britain. Whether the latter may have signified a transition from "small" to "big" in our terms deserves further research

[5] Also Jared Diamond in his *Guns, Germs, and Steel* (NY: Norton 1997), pp. 247–249, has argued that not all societies are equally receptive to all innovations, ascribing this to relative economic advantages, social value and prestige, compatibility with vested interests (where he discusses the Qwerty

had also been held up for a variety of reasons, mainly cultural it seems, at least for its Muslim populations until the 18th century, but the printing press then became accepted rather quickly in the 19th century.[6]

4.2 The typewriter's genealogy

Even on the American market the first mass-produced commercial typewriter of 1874, as developed by Christopher Latham Sholes and Carlos Glidden, and manufactured by the Remington gunsmiths was not an immediate success. It was awkward to use, and at $125 apiece expensive, and therefore not immediately "persuasive" in Everett Rogers' terms.[7] It was felt that copyists and printers could still do the work more efficiently and cheaper. Initially, it was even an insult not to write by hand. Typewritten letters were considered as a breach of good manners; for some time, it remained an offense to address somebody personally with a 'machine-made' letter.[8] But after a hesitant start, and technical improvements, the 1880s began to show a steep upsurge. In the USA, Remington was followed by other manufacturers such as Underwood, Hammond, Smith Corona, and Royal, each with their own innovations and improvements. [9] Within two decades – with a contemporary expansion of commercial business –, a revolution in office management evolved in Government agencies as well as corporate offices. By 1896, one researcher estimated that 450.000 typewriters had been produced of which 150.000 were in use in the United States.[10] Remington, at the time the largest producer, took pride in that between 1893 and 1899, it had doubled its output to 800

---

keyboard), or ease of observation. Michael Adas, *Machines the Measure of Men* (Cornell UP 1989) while dwelling on the divergent impact of big Western Imperialist technologies, saw such consumer products (including typewriters) as enhancing awareness of ongoing technological transformations, p. 142.

6 Ami Ayalon, *The Arabic Print Revolution, Cultural Production and Mass Readership* (Cambridge: Cambridge University Press 2016), esp. pp. 1–17. This applies also to the Ottoman lands.

7 See Introduction, pp. 28–29.

8 Cynthia Monaco, 'The Difficult Birth of the Typewriter", *Invention & Technology Magazine*, vol. 4/1, Spring/Summer 1988.

9 Remington, initially an arms manufacturer, after the Civil War at first shifted to sewing machines, and then bought the typewriter patent from Sholes and Giddens. Typewriters became more or less standardized only in 1910. Meanwhile European competitors had emerged in Britain, Germany, and Italy. For an economic history see George N. Engers, *The Typewriter Industry: The Impact of a Significant Technological Innovation* (PhD Thesis, Univ. of California, Los Angeles 1969).

10 Edward W. Byrn, *The Progress of Invention in the Nineteenth Century* (NY: Munn and Co. 1900), p. 182. The same source concluded that 75.000–10.0000 were then made annually.

a week, and that 203.000 of their machines were already in use in what they called "the Universe". For several decades the United States remained the predominant manufacturer, but with competing brands elsewhere, its global market share would gradually decline.

Meanwhile, the growth of the corporative sector and public services in the USA led to new clerical openings for women, a process with which the typewriter, promoted as enhancing efficiency and saving time, became associated. Already in 1881, as the common narrative runs, the Young Women's Christian Association acquired its first six typewriters to train women and, within five years, 60.000 women entered this new field of employment, both in businesses and factories, and from the 1890s onwards, also government service. By the turn of the century, 200.000 young women had been trained as typists and stenographers. [11] According to Margarie Davies, female stenographers and typists in the US by 1890 had already become a majority (63.6%, and by 1930 they would be 95.4%).[12] The new female labor force may have enhanced the sale of typewriters, even though the emancipatory effect of the typewriter can be disputed: However respectable such employment appeared for young American ladies, it was rather poorly paid and pushed men out. Western Europe followed suit, though at a slower pace. In Germany, new, faster methods of handwriting such as the Sütterlin script of 1911 were still being introduced in the twentieth century.[13]

We propose a double comparison. The typewriter story in the Middle East contrasts with the successful diffusion in the Middle East of its "older sister", the sewing machine (Singer and its competitors). In any event, their narratives were quite different. Neither had initially been destined for private use but the multi-tasking

---

[11] Remington and other companies, too, conducted training courses for typists ("Remington is "an educator as well as a labor saver" (*Remington Notes*, vol. 2 no. 5 (1912), p. 8, and stenographers in particular were in demand. By 1930 this number had grown to two million. See also Daniel J. Boorstin, *The Americans, the Democratic Experience* (NY 1974). p. 126, p. 55, and esp. pp. 398–400. On the slow development of a standardized keyboard, see Delphine Gardey, "The Standardization of a Technical Practice: Typing (1883–1930)", *Reseaux, The French Journal of Communication*, vol. 6/2 (1998), pp. 255–281.

[12] Margerie M. Davies, *Women's Place is at the Typewriter, Office Work and Office Workers 1870–1930* (Philadelphia 1982), table 1 and 4 et passim; Sharon Hartman Strom, *Beyond the Typewriter, Gender, Class, and the Origins of Modern American Office Work, 1900–1930* (Urbana 1992); Donald Hoke, 'The Woman and the Typewriter: A Case Study in Technological Innovation and Social Change', *Business and Economic History* 8 (1979): 76–88; Francisca de Haan, *Gender and the Politics of Office Work, the Netherlands 1860–1940* (Amsterdam 1998).

[13] According to the German media theorist, Friedrich Kittler, who wrote perceptively on the history of the typewriter, Germany entered the field late. The Ministry of Commerce and Trade there endorsed the typewriter only in 1897, *Gramophone, Film, Typewriter* (Stanford: Stanford UP 1999), p. 218.

sewing machine celebrated its most conspicuous triumphs in households, while the typewriter remained mainly a specific office technology, and entered homes (and private offices) much later and in far smaller numbers.

Secondly, in comparison to the USA or Europe, the diffusion of the typewriter was much slower in the Ottoman and Persian lands. Was it primarily a lack of usefulness owing to lower literacy which had a negative impact on potential consumers and manufacturers alike? Or was the new device then still understood as another "Frankish" invention, aimed at breaking traditional calligraphic or clerical conventions, in yet sharper words, a colonialist plot against local habits by economically interested western manufacturers?[14] Or, alternatively, was the diffusion of the typewriter held up by technological incompatibilities in language and script, which had to be solved before it could make headway?

It is not certain to what extent American typewriter manufacturers from the beginning expected large sales beyond the sphere of western languages. Rather it appears from their publicity campaigns that they were out to prove technological supremacy, and the universal benefit of their innovations. In the Ottoman Empire, Remington, for instance, advertised in local foreign-language papers, that "all" [sic] governments, railway and navigation companies, telephone companies, banks, armies, insurance companies and, more specifically, the Emperor of Russia, the Queen of England, the Queen of Spain, as well as the Khedive of Egypt possessed one.[15] Another jubilant claim (in 1900) needs similar reflection: "All over the world it [the typewriter] has already traveled – from the counting house to the merchant to the Imperial Courts of Europe, from the home of the new woman in the Western hemisphere to the Harem of the East [sic] – everywhere its familiar click is to be heard, faithfully translating thought into all languages, and for all places." One may note here the mention of royals, and of women too, but whether this supposed ownership implied regular use needs further verification.[16] Yet, the Remington

---

14 Gandhi had reluctantly endorsed sewing machines in India, and though having used a typewriter in South Africa, remained ambivalent on its implementation, see Vikram Doctor, "History of Typewriters in India: MK Gandhi's Love-hate Relationship with the Machine", *The Economic Times*, 9 March 2012 (quoting a 2012 lecture by David Arnold in Mumbai). In his book, *Everyday Technology, Machines, and the Making of India's Modernity* (Chicago: Chicago University Press 2013). Arnold has extensively discussed the impact of the "immigrant typewriter" for India.
15 It should be understood that Roman-script keyboards were meant. See *Levant Herald*, 26 Feb. 1899. Also, Abrevaya Stein, *Making Jews Modern, the Yiddish and Ladino press in the Russian and Ottoman Empires* (Bloomington, Ind. 2004), p. 197, notes that Remington after 1902 was a regular advertiser in the Ladino newspaper *El Tiempo* (Istanbul), even though Ladino itself was written in Hebrew so-called Rashi script.
16 If the Khedive indeed owned one (in Roman script), this was before the "invention" of the Arabic typewriter.

Company itself was aware of its over-statement, writing in 1908: "There is one institution of the West which the East has not yet duplicated and that is the typewriter girl. However, even her advent may be expected in good time".[17]

4.3 Techno-linguistic adaptation

Unlike the sewing machine, here was an invention which was not immediately ready for use in the region because it was a priori incompatible with the indigenous languages and scripts. Thus, the typewriter's stage of "Persuasion" (in Rogers' terms), – if we adopt a determinist approach assuming that sooner or later it would enter most national and linguistic regions in the world –, went through a phase of awareness in the local press. The invention reached local consciousness by means of mere descriptions as a remarkable new item, not initially by commercial advertisements. As was the case with other technological innovations, readers still had to become familiar with its very existence before they could see it in reality and be persuaded of its practical uses.[18]

The prominent journal *al-Muqtataf* (founded in 1876) published its first descriptions of a typewriter from 1884 onwards.[19] Taking this self-styled "scientific and industrial" magazine as a criterion for the "Knowledge" stage of the new invention, and then still reaching only a segment of the literate population, it would still have to cross an additional "chasm" of two decades before reaching the stage of "early adoption" locally. This cannot be solely ascribed to geographical distance or slow communication as several other technological innovations of the

---

17 *Remington Notes*, vol. 1. No. 7, p. 4.
18 By 1909 Remington claimed to supply typing machines in 84 languages, including Armenian and Hindi, Greek and Russian (Cyrillic) script typewriters. Hebrew and (slightly different) Yiddish machines entered the market soon after. A Thai typewriter was developed by a British dentist Macfarlane (1913) and manufactured by Smith Corona. Thomas Mullaney has admirably unraveled the vicissitudes between the early Chinese and Japanese (and Korean) typewriters, a narrative which takes into account the use of different characters (Kanji, Katakana etc.), see his "Controlling the Kanjisphere: the Rise of the Sino-Japanese Typewriter and the birth of CJK", *Journal of Asian Studies* vol. 75/3 (2016), pp. 725–753, and his book *The Chinese Typewriter, a History* (Cambridge, Mass.: MIT Press 2017). However, Japanese and Chinese typewriters remained cumbersome till the 1980s when the personal computer offered more sophisticated solutions. Mullaney's work indicates, in our view, how much detail is still missing from the typewriter's Middle Eastern social history.
19 Consider that the first mass-produced typewriters reached the American market in 1874, ten years earlier. Initially it was called *alat al-kitaba* or *al-ala al-katiba*, though in later years it was still referred to as *taybraytir* as well. Japanese also retained the word *taipuraita*. Neither Turkish (*yaz makinesi*), nor Arabic had the ambiguity of English word typewriter, which could be a person (a typist) as well.

west were meanwhile making headway.[20] A competing magazine, *al-Hilal*, which began to appear in Cairo in 1892, made a highly premature mention of a Chinese typewriter in 1897 but hinted that an Arabic machine was, in fact, in the making.[21] Both magazines, and – we assume without a systematic search – various parallel Arab and Ottoman journals, continued to report from time to time on the development of typewriters, albeit not with great impatience for a vernacular adaptation.[22] Neither was the supposed "relative advantage" of speed typing, implying the typically American maxim, "time is money", brought forward as a compelling incentive.

Yet during those decades, while typewriters had become a success in America, and more slowly in Europe too, one assumes that experimenting with Arabic-script keyboards must have taken place, or as Titus Nemeth in his impressive book on Arabic typography writes "... the inventive and entrepreneurial spirit of the late nineteenth century – in particular mechanization and the nascent consumer market – probably gave rise to parallel developments of such a machine by numerous people and companies, without knowledge of each other's work".[23] One assumption is that the later "inventors" Wakid and Haddad (see below) had been inspired by a typewriter which was exhibited at the Chicago World's Columbian Fair in 1893, but that no one could as yet develop an Arabic model.[24] Moreover, we suppose that strongly progressing Ottoman Turkish, Arabic and Farsi printed newspapers and books were making readers acquainted with the typographic look of their languages.

The forces converging toward the development of an Arabic-script typewriter are difficult to unravel. The designing of Arabic types and keyboards for typing machines and printing presses and the development of a workable typewriter are sometimes confused. On the one hand, there developed commercial-industrial incentives, supported by the technical know-how and successes of the big manufac-

---

20 Compare the photographic camera and the gramophone (phonograph). Periodicals such as *al-Muqtataf* played a crucial role in making foreign technologies known to local readerships.
21 *Al-Hilal*, vol. 6 (November 1897), with further articles on typewriters in the following years.
22 However, see *al-Muqtataf*, vol. 13 (1888), p. 710 where the question is raised as to where one could purchase one is still referred to New York. For earlier entries, see *al-Muqtataf*, vol.9 (1884), p. 79; vol. 13 (1888), p. 67. *Al-Hilal*, vol. 8 (15 Sept. 1900) claimed that Bishara eff. Shihada in the US and Selim Haddad were near solving all problems.
23 Titus Nemeth, *Arabic Type-Making in the Machine Age, The Influence of Technology on the Form of Arabic Type, 1908–1993* (Leiden: Brill 2017), p. 75. Also H.S. Deuchar in *IJMES* 51/1 (2023).
24 Zeina Dowidar and Ahmed Ellaithy, "The Invention of the Arabic Typewriter" who argue that e.g. they were too young; the machine did not work well; and nobody at the factory who made it knew enough Arabic, https://medium.com/@kerningcultures/the-invention-of-the-arabic-typewriter-a6d26e0554a (accessed 8 Aug. 2022).

turers and, on the other, there were individual local inventors who felt challenged by the new technology.

New models of typewriters were being constantly developed and marketed. Soon they would consist of hundreds of parts, but technological intricacy was not the main difficulty in adapting them to the non-Roman script.[25] The reversing of the typewriter's carriage from left to right for Arabic (Ottoman, Farsi) script was arguably the least engineering hurdle. Anyway, this had already been overcome by Hebrew script models.[26] It was the Arabic script, in particular the character keys and the keyboard itself, which needed "domestication". In practice this meant that the number of characters as used in typographic printing had to be decreased. From twenty-three Roman characters on American keyboards, at least twenty-eight basic characters for Arabic had to be added (and several more for Ottoman script). A major new challenge lay in the fact that until then no cursive script had been adapted to any typewriter. Most complicated was the design of the necessary first, middle, and end characters, some with kerns (the part of a metal type projecting beyond its body or shank), and of the ligatures (*kashida*) with differential spaces.[27]

The original QWERTY keyboard had been re-calculated on the basis of the frequency of the single Roman characters in English so as to prevent the annoying jamming of the arms which then had to be separated before typing further.[28] It would take time until practicable local language keyboards appeared but, in the

---

[25] In 1918 Royal had 388 parts, Underwood had 400, Remington had 969 etc., *A Salesman's Manual of the Royal typewriter* (NY 1918).

[26] In fact, Hebrew-script keyboards were at the time mostly used by Yiddish writers and later underwent some changes to facilitate easier typing in Hebrew.

[27] Arabic has 17 basic characters; the others are variations with diacritical points. We are leaving aside the many technical improvements of the Arabic keyboard, in order to focus on the typewriter as such. But for some of the technical elucidations, I wish to thank Amnon Acho, my own former "typewriter man", who had learned the now extinct craft in Cairo (interview 3 Sept. 2001).

[28] Qwerty, patented by Sholes and Glidden in 1873 and sold to Remington, was endorsed by an international typewriter congress in Toronto, 1888, see Kittler, p. 229. The French use a re-calculated Azerty keyboard, the Germans Qwertz, the Italians Qzerty etc. However, the Qwerty set-up has given rise to major controversies, most of which revert to the economy of speed typing. Qwerty has been called "one of the biggest confidence tricks of all times", or even a curse. Research proved that the so-called Dvořák arrangement of 1932, equally scientifically designed, is more efficient. On the ongoing debate see Everett Rogers, pp. 8–11, and many others, e.g. Paul David, 'Understanding the Economics of QWERTY: The Necessity of History' in W.N. Parker (ed.), *Economic History and the Modern Economist* (NY 1986); S.J. Leibowitz and Stephen E. Margolis, 'The Fable of the Keys', *Journal of Law and Economics*, vol. 33 (1990; and Michael Shermer, 'Exorcising Laplace's Demon, – Chaos, and Antichaos, History and Metahistory', *History and Theory* 34 (1995), pp. 74–75.

end, a calculation of each of the ensuing keyboards for Arabic, Ottoman-Turkish or Farsi proved not to be an unsurpassable obstacle.

The first practical break-through in developing a workable keyboard is ascribed to two Lebanese "inventors", Philippe Wakid and Salim Shibli Haddad who by 1889 had overcome the main technical difficulties, assembling a typewriter with 75 keys (later increased to 83) called "al-Hilal", meaning the crescent moon. [29] However, it seems not to have gone immediately into production. Neither did the Arabic keyboard designed in 1899 by Ernest Tatham Richmond, a British-educated architect who later shifted to a prominent military and administrative career in Egypt and Palestine. [30] In the end, Salim Shibli Haddad, a painter of Lebanese descent who lived in Cairo, was most successful.[31] He had been working on a keyboard from at least 1898. In 1899 he obtained in the USA two subsequent patents, one for the notation itself and one for the slightly differing Arabic, Turkish, and Persian types. In a detailed eight-page explanation with illustrations, he claimed to have reduced the necessary 638 characters to 58 distinguishable types, including the indispensable *ta marbuta*, and the *madda*, and *hamza* signs. A connection bar enabled the typist to make alignments.

The *Muqtataf* journal, which had meanwhile relocated from Beirut to Cairo, came in 1902 with a detailed story on Haddad:

> We announce to the Sons of the Arabic-speaking community [*abna al-'Arabiyya*] and their educated [members] that an Arabic typewriter has been produced which writes connected Arabic characters ....The inventor is *khawaga* [the Christian gentleman] Salim Haddad ... He has registered his invention in his name, and has proceeded to America, [... where] he has entered into an agreement with a factory to produce machines on demand. After much effort and continuous expenses, Mr. Sirri Idris Bey Raghib has returned with such a machine. Yesterday he invited a group of Arab journalists and notables of the capital to his home. There he showed them the machine and demonstrated its use and even invited them to use it. We also wrote on it immediately. We had never written on a machine before,

---

29 Mohammed Sadid, *L'Alphabet Arabe et la Technologie* (Rabat, ISESCO 1993).
30 Richmond was the son of a well-known British painter, who from 1900 first served in the Antiquities Department in Egypt, and thereafter in various capacities in Palestine. In Jerusalem, he wrote an important book on the Dome of the Rock, and facilitated its restoration, in close cooperation with the Grand Mufti Hajj Amin al-Husayni, see Uri M. Kupferschmidt, *The Supreme Muslim Council* (Leiden: Brill 1987), pp. 20, 27, 38–39, 129–130. For Richmond's patent see United States Patent Office, no. US 637109 A: "Types for Type-Writers or Printing-Presses", 14 November 1899 (See Google Patents: on internet). However, it expired in 1916.
31 According to *al-Hilal*, vol. 12 (15 May 1904), see Fig. 10, which carried a detailed description and an illustration of the machine, he was funded by Idris B. Raghib. Robert Messenger discovered a painting from 1897 by Slatin Pasha, the famous Austrian explorer and later administrator of the Sudan, https://oztypewriter.blogspot.co.il/2014/10/the-arabic-typewriter (accessed 11 May 2018).

either in Arabic or in Frankish [European language] .... Inevitably, this machine will arouse some controversy in the beginning, as was the case with [other] Frankish machines. We heard that European diplomatic offices used to reject any document written by means of a typewriter. However, nowadays, the situation is reversed as the rule has become that anything submitted to them has to be written by typewriter if it is not to be disregarded.[32]

Around the same time, there was another surprising attempt geared to the Ottoman side. Theodor Herzl, the founder of the Zionist movement, may be ascribed a role in the development of a Remington typewriter with Arabic-script (though it is not clear whether in this case it was specifically designed for Ottoman Turkish). Being a journalist by profession, who regularly used a typewriter, he conceived the idea in the autumn of 1901 that presenting Sultan Abdülhamid on his upcoming visit to Istanbul with a token of the latest technology, namely a typewriter, would expedite the obtaining of a Zionist charter for Palestine. He accordingly approached Richard Gottheil, the President of the Federation of American Zionists, who was a professor of Semitic languages at Columbia University. Knowing that Hammond was working on an Arabic typewriter (and possibly had filed an application in 1901), Gottheil contacted Hammond as well as Remington. The latter firm must have known that one or more patents were under consideration but agreed to make one – by hand – within a relatively short time. Though the costs of such a typewriter would amount to $300, Gottheil proudly reported that he had succeeded in bringing down the price to $150, with the understanding that it would become a marketing success for the Remington works.[33] In his enthusiasm he even suggested to Herzl that "...we shall probably get a rebate on all that we have contracted to pay. I have a clear understanding on that subject with the man in charge in the New York Office." One of Gottheil's former students was sent to the plant to assist in its production: "Though I can write the Arabic characters quite fluently, I cannot write them well enough to use my writing as a copy for the engravers. Of course, I might have called in some Arab or Turk here, but he would never have kept the secret." [sic][34] Soon, Gottheil wrote to Herzl that the prototype model worked beau-

---

32 *Al-Muqtataf*, vol. 26 (1902), p. 575.
33 A similar one had also been made by hand for the new Queen Wilhelmina of the Netherlands.
34 Raphael Patai (ed.), *The Complete Diaries of Theodor Herzl* (New York: Herzl Press 1960), p. 1224. Also Central Zionist Archives (CZA), H1/1282, letter Gottheil to Herzl, 7 Jan. 1902. The file also contains the other letters by Gottheil on the Remington machine. Herzl's letters are found in his *Briefe, Ende August 1900-Ende Dezember 1902* (Frankfurt 1993), vol. VI, pp. 324–325, 337, 348, 352, 387, 418, 427, 446, 460–461, 466, 467, 480, and 488. However, Gottheil urged Herzl to have utmost care taken of the typewriter and even to send a student along to instruct the Sultan in its proper use. See also Mordechai Naor, *Ha-Rishonim: Sippurim min ha-'Aliya ha-Rishona le Eretz-Yisrael* [The Forerunners] (Tel Aviv: Am Oved 1983), pp. 102–108. There may still be more details in the Remington re-

tifully and even added "I want to call your attention to the fact that I have heard of one or two machines being in use in Egypt. As far as I know, these are most unsuccessful, and they write the Arabic and Turkish characters in a manner which makes them practically unrecognizable by Orientalists [sic]. All the machines that have been attempted heretofore fail to give the proper ligatures to the letters."[35] Remington would later speak of "a veritable triumph of mechanical skill".[36]

Herzl, however, had not reckoned with Sultan Abdülhamid's ban on importing typewriters. It is not entirely clear when this order had gone into force, but it certainly was effective from 1901 and it had generated pressure from manufacturers in the US and Britain, as well as protests by the ambassadors of their countries in Istanbul. Maybe, as a journalist, Herzl could have known better, because the quibbling of the ever-suspicious Ottoman censors and customs authorities had become famous in Europe.[37] Ottoman authorities were said to fear that "in the event of seditious writing", they would not be able to trace the "operator of the machine". The embargo appears to have been caused by a misunderstanding that a typewriter was a sort of printing press.[38] It might be that the ban was not strictly enforced on Roman-script typewriters. Regular advertisements for these had already begun to appear in the local foreign-language press.[39] In 1895, the *Journal de la Chambre*

---

cords (boxes 3 and 5) kept at the Hagley Museum and Library; information by the archivist Marjorie G. McNinch.
35 CZA, H1/1282, letter Gottheil to Herzl, 7 Feb. 1902.
36 *Remington Notes*, vol. 1 no. 1 (1907), p. 2. Issue no. 3 has a more elaborate description of the Arabic machine, pp. 8–10.
37 The motive was apprehension of insubordination. A similar ban applied to firearms, ammunition, hand printing presses, and telephones. G. des Godins de Soushesmes, *Au Pays des Osmanlis* (Paris 1894), p. 260, relates that electrical instruments for physics education were barred too (for some time). Only a few later cases of banning typewriters in the Middle East are known to us, such as happened in some communist countries. It would seem that the Imam of Yemen, at least around 1939, had banned typewriters which had been ordered from India for local merchants, see Yosef Tobi, *Yehudi beSherut ha-Imam, Ish ha-'Asaqim ve-Soher ha-Nesheq Yisrael Tzubayri* [A Jew in the Service of the Imam, the Businessman and Arms Dealer Yisrael Tzubayri] (Tel Aviv: Afiqim 2002), pp. 141–142. See also below on Iraq, Najm al-Wali, p. 138.
38 *Scientific American*, July 1901, quoted by Phil Wolf, 'Ban Typewriters'. The article also mentions that a consignment of 200 typewriters at the customs would have to be returned.to the sender.
39 *Levant Herald and Eastern Express*, 26 February 1899 (Remington); ibid.,12 March 1899 (Hammond); ibid., 21 Oct. 1901 (Remington), all in French. This is only a random sample; advertising might have started earlier. We deduce this also from the *Revue Commerciale du Levant*, organ of the French Chamber Commerce, reported by 1904 – with some jealousy, it seems – that the local market was dominated by Remington, which was said to be no match for the French machines, Dec. 1904, no. 213. The journal also investigated the marketing of Johannet typewriter paper, p. 794.

de Commerce de Constantinople, after recommending the new invention in the USA, and explaining how it had led to the shortening of letter-writing time (using American slogans such as "struggle for life" and "time is money") concluded: "La machine à écrire et ses adeptes les dactylographes sont donc des inventions utiles et se répandront assurement de plus en plus dans tous les pays civilizés" [sic].[40]

The first-ever Remington with an Arabic (or Ottoman) keyboard arrived in Istanbul after some logistical delays. But in the end Herzl did not meet the Sultan; thus, his strategic schemes came to nothing. Neither Herzl's local representative, Soma Wällisch, who was Chief Sanitary Inspector of the Ministry of Public Education, nor other high-placed connections, succeeded in releasing it from customs. Herzl himself, in a letter to the Ottoman diplomatic representative in Vienna, admitted that the plan he had in mind did not materialize.

The ban was revoked following the Young Turk Revolution in September 1908.[41] The machine itself, however, had apparently, after all, reached the Palace, as becomes clear from *The Remington Notes* of 1909: "After the dethronement of the former Turkish Sultan…a thorough search was made in his old palace… and among his personal belongings … one of the most interesting was a Remington typewriter."[42]: If it still survives, it deserves to be a museum piece.

We do not know whether this particular prototype was ever patented, but before long, Remington began manufacturing Arabic-script machines.[43] The first issue of *Remington Notes* proudly mentioned that the Shah of Persia had been presented with an Arabic-script (!) Remington by his Minister in Washington, and that even the Mikado of Japan had also been given one in *katakana* script by the company's local representative. The magazine adds a questionable anecdote relating

---

40 27 April 1895, p. 196.
41 *Papers Relating to the Foreign Relations of the United States*, 1902 (Washington 1903), pp. 1026 – 1041 and p. 1045; idem for 1905 (Washington 1906), pp. 883 – 885; idem for 1910 (Washington 1910), pp. 1073 – 1074; idem for 1908 (Washington 1912), p. 756. The ban also affected Remington typewriters in Farsi which had arrived at the customs in Trabzon for transit (1907). See further L.J. Gordon, *American Relations with Turkey, 1830 – 1930* (Philadelphia 1932), pp. 174 – 175.
42 Vol. 2, no. 5 (1909), p. 14. See my "On the Diffusion of 'Small Western Technologies'", first published in 2015, and reprinted here as Chapter 2, which served Robert Messenger's blog very well: https://oztypewriter.blogspot.com/2015/05/ataturk-sultan-his-harem-remington-7.html (first accessed 1 June.2018). The blogger has yet to ascribe it to me as the author. See further his "The Arabic Typewriter Keyboard and the Syrian Artist".
43 Competing pioneering manufacturers of Arabic-script typewriters, as listed by Messenger, were Smith (1904), Underwood (1914), and Müller in Dresden (1925).

that the Shah's predecessor had already owned one, but that his first act upon receiving it had been to give it a bath in boiling water to cleanse it from evil spirits.[44]

No less a scholar and prominent intellectual than Edward Said claims that his father, Wadi', had designed the Arabic keyboard for the Royal brand, albeit without mentioning a year.[45] His father was the proprietor of the Standard Stationary Company in Cairo as well as in Jerusalem. Said graciously gives credit to both his father and his mother: "With my mother's help, he developed – 'invented' would not be wrong – the Arabic typewriter with Royal, whose aristocratic owners, the John Barry Ryans, he came to know quite well". From leading museum collections, it would seem that around that time there existed at least four commercial versions of Arabic keyboards and probably several more. [46]

We also have to consider here the technological interaction between Arabic Linotype typesetting machines, with their (more or less) conventional keyboards, and the typewriters themselves. One outstanding name here must be that of the Lebanese-Maronite Mokarzel brothers in New York, who founded in 1898 the *al-Hoda* newspaper, and in 1910 adapted the linotype keyboard to Arabic script.[47]

It is difficult to pinpoint the year in which non-Roman typewriters made their commercial entry into the Ottoman or the early Turkish market, and which were the American or European brands which did so. European manufacturers such as Olivetti in Ivrea, Seidel & Naumann in Dresden, and Hermes in Yverdon, may have had a marketing advantage (with credit installments, as pioneered by the Singer

---

44 Vol 1, No. 1 (1907), p. 2. This means that the first model must have reached Mozaffar al-Din Shah in 1906 as he was succeeded by Mohammad 'Ali Shah on 3 January 1907. The Japanese katakana script is syllabary; and the machine was beautifully decorated with mother-of-pearl.

45 Edward Said, *Out of Place* (New York 1999) p. 94. His father owned these well-known office supply stores in Jerusalem and Cairo. The Said brothers had stationary stores in Jaffa and Alexandria as well; and in Jerusalem and sold Arabic, English and Hebrew typewriters, see Ami Ayalon, *Reading Palestine, Printing and Literacy 1900–1948* (Austin 2004), p. 9 and pp. 83–84.

46 Wilfred A. Beeching (the director of the British Typewriter Museum), *The Century of the Typewriter* (Bournemouth, new ed. 1990), pp. 49–50. One specific Arabic keyboard was used in Morocco, there were also two keyboards for Turkish, two for Persian, two for Hebrew and different French and English ones for Egypt, pp. 53–54, p. 58, p. 63, p. 70. In 1979, a Palestinian researcher, Qustandi Shomali, published a proposal for yet a new keyboard, *al-Bahith* (special issue 1980) (my thanks to the author).

47 On *Al-Hoda* see Wikipedia. The Mergenthaler Company in New York began in 1898 to market the revolutionary Linotype machine (later also Monotype, Intertype) which replaced typesetting by hand, see Lyn Gorman and David McLean, *Media and Society in the Twentieth Century, a Historical Introduction* (Oxford 2003), p. 7, and Nemeth, *passim*. We disregard here teleprinter keyboards which came earlier but rather remained "big technologies" and in Roman script.

sewing machine company) over Remington and other American manufacturers.[48] It is impossible to make a cost-benefit calculation, but we assume that most local byers acquired the machines for practical purposes, not as status symbols.

The Koç Museum for Technology in Ankara owns a Seidel & Naumann typewriter (model Erika of around 1911, we think) with a rare Ottoman script keyboard including the characters which are not in the Arabic alphabet (Figs. 19 and 20). It was previously owned by Fuat Bayramoğlu, a well-known diplomat, poet, and researcher; this case proves that the new device was somehow on its way to become a genuine "small" technology.[49] Looking at the said Ottoman typewriter, one observes an ingenious arrangement in three, not four, rows of black upper and red lower-case letters, with red and green shift keys, but a lack of reading marks. In fact, traditionally, Arabic manuscripts used a different marking system to that end.

4.4 Typography and calligraphy

Supposedly, this pace of adoption was slow because early Arabic or Ottoman typewriters were still difficult to handle, and because the typescripts which were produced were not always legible.[50] But similar reservations had also been voiced with American typewriters in the beginning.

This brings us to a major cultural or psychological impediment to their diffusion, namely the relation to Arabic script and the sacred character of the language of the Qur'an: the reason why Islamic calligraphy remained a beloved art form.[51]

Moreover, traditional handwriting was not necessarily strictly horizontal as westerners are used to, but sometimes slightly vertically tilted; in principle, type-

---

[48] Not only a matter of geographical proximity, it would seem, but also owing to the American dislike of allowing credit terms. Seidel und Naumann in Dresden manufactured bicycles from 1872 and also made Singer sewing machines under license but became well-known for their Erika typewriters.

[49] Seen in the Koç Museum in Istanbul, now in the Çengelhan branch of the museum in Ankara. Apparently, it was donated by the owner who was befriended by the Koç family. However, it is unlikely that its date of production was 1900 as a sign states. The said typewriter carries the nameplate of Carl E. Halbarth (an agent in Berlin). An illustrated catalogue of items from Abdülhamid's era, *Sanayi Devrimi Yıllarında, Osmanlı Saraylarında Sanayi ve Teknoloji Araçlari* (Istanbul: Yapı ve Kredi, Yayıncılık, n.d). has a short section on typewriters (pp. 62–64) but no specific data on its use at the Court.

[50] See Gottheil's remark on the art of handwriting above.

[51] Even until quite recently, linotype-set newspapers would still proudly maintain their hand-styled headlines (sometimes even signed by the calligraphic artist).

4.4 Typography and calligraphy — **129**

Figs. 19 and 20: A rare Seidel & Naumann typewriter, model Erika (app. 1911) with an Ottoman-Turkish keyboard, from the Rahmi M. Koç Museum for technology at Çengelhan, Ankara (formerly exhibited in the Istanbul branch), sold by Halbarth in Berlin, and allegedly owned by Fuat Bayramoğlu, a well-known diplomat and poet, who was befriended by the Koç family. The keyboard has ingeneous lower red and upper black keys for the various Ottoman characters and their begin, middle, and end positions in words (permission: Rahmi Koç Museum, Ankara).

setting could never match this model. In fact, in the realm of printing, lithography (developed by Alois Senefelder from 1796) could preserve some of this calligraphic ideal, and for quite some time competed with typography. On the other hand, the latter part of the nineteenth century, with more and more type-set books and periodicals in the different Arabic, Ottoman, and Turkish styles, had made readers familiar with modern typography, which in turn must have facilitated the "Confirmation" phase of the typewriter's diffusion.[52]

4.5 From foreign use to local business

Though it has been suggested that, ultimately, the "sublime quality" of typescript made it the standard for state bureaucracies and commercial business, this would still take time in the Middle East.[53] For several decades, typewriters remained an alien technology, used solely for foreign-language correspondence. This was true even for certain government offices. Consider the following anecdote on the Customs office in Alexandria (somewhat before 1909). Following a leak to the local press of an imminent change in tariffs, an official investigation discovered a typewritten document on official stationery. Since each of the few machines in use there had its specific peculiarities, the machine on which it was written could easily be detected. However, the civil servant regularly working on it had been in Europe at the time of this premature publication, and the culprit was never caught.[54]

Our assumption is that before World War I, in government offices, typewriters using the local language must still have remained a rarity. It is a moot question when the Ottoman government acquired its first typewriters. At the Sublime Porte there was possibly only one in use for Roman script, and a few more were used in other official agencies around the Empire. To what extent this

---

[52] Raja Adal has emphasized the appearance of aesthetics of printed and typewritten texts, see "The Print Sublime", a lecture delivered in Toronto, 26 September 1919. Nemeth's book shows how decisive the aesthetics of typography (including the design of new types) were and still are today.

[53] This applied also to full stops, commas, question and exclamation marks, brackets etc., see Dana Awad, "The Evolution of Arabic Writing Due to European Influence. The Case of Punctuation", *Journal of Arabic and Islamic studies* 15 (2015), pp. 117–136. Arabic script keyboards initially had Arabic numerals, but even before Microsoft keyboards they had already given way to Indian numerals.

[54] Arnold Wright and H. A. Cartwright, *Twentieth Century Impressions of Egypt* (London 1909), p 213. The story reminds us of Sherlock Holmes, who was the first literary hero to solve a mystery by identifying an impostor's typewriter: Conan Doyle's *A Case of Identity* (1882).

slow implementation was caused by reluctance in the conservative bureaucratic *kalemiye* environment calls for further research.[55] One may even hypothesize that the printing of circulars was more common than typewriting.

Ernest Giraud, the President of the French Chamber of Commerce in Istanbul was well-aware that France had "fallen asleep", and was strongly lagging behind sales of Remington and other leading American brands, as well as some British and German competitors. In a lengthy report in 1910, he not only summed up his own conservative reservations regarding the new device, including doubts about its speed, but also the typographical shortcomings of Remington's Arabic or Ottoman typewriter (" not a great success", he says). It was his view that it would be impossible to properly link the characters. Haughtily, in a twist of his country's *mission civilisatrice*, he stated: "C'est, du reste inutile: Grâce à l'essor donné à l'instruction publique, dans un certain temps, tous les habitants de l'Empire sauront le Français". But there were objective delaying factors too, such as the high price, transport costs, and the lack of demonstrators.[56] The lack of skilled repairmen, which he mentioned, was an additional factor, not only in Istanbul.[57]

Turning the pages in Edhem Eldem's lavishly illustrated history of the Ottoman Bank Archives, which bank allegedly was one of the first to introduce typewriters, one gets some idea of the transition from handwritten to typewritten business: the earliest local typewritten documents reproduced in that volume date from the 1890s.[58] Without exception, all of these are in French or in English, the primary working languages of the bank.

---

55 *Kalemiye* from *kalem*, pen, means the scribal division of the administration. Consulting Carter V. Findley on this question, he kindly answered (18 July 2000) from his recollection that he had seen one beautifully typed government contract in Ottoman, probably dating from WWI, but added that it might have been written by the other party. Typewritten documents appeared to be already more common in the Cumhuriyet Arşivi in Ankara. A student of ours recently asked the personnel of the Başbakanlık archives the same question, and was told that the Porte, even after WWI, possessed only one typewriter in Roman characters. Suraiya Faruqhi, in an interview with the journal *Tarih*, issue 1 (2010), related that though carbon copies from the 1930s in *riq'a* script (!) of the Maliyeden Müdevver revenue register existed, a more legible typescript became available only in the 1970s.
56 E.g., "Machines à Écrire", *Revue Commerciale du Levant* (1910), pp. 159–176. It was not only his preference for stylish handwriting but also his aversion to blue-and-red ribbons.
57 Douglas Sladen, *Queer Things About Egypt* (London: Hurst & Blacket 1910), pp. 124–125, where the author relates that he finally found an Armenian who "understood" how to repair his Williams (an American brand) typewriter.
58 Edhem Eldem, *A 135-Year-Old Treasure, Glimpses from the Past in the Ottoman Bank Archives* (Istanbul 1998), e.g. p.230 (Banque Abraham 1891); p. 240 (Pera Palace 1994); p. 231 (Insurance Co. 1896); p. 176 (Selim Pasha 1898). No typewriters are seen in the many pictures in the volume.

Therefore, as to business use, typewriters remained mainly confined to western oriented enterprises such as big European banking and foreign trading companies. The tilt toward foreign interests and minorities can equally be deduced from commercial almanacs. It is indicative that in Alexandria in 1915, for instance, half of the typewriter agents were Jewish.[59] For the sake of comparison, in Palestine, somewhat later, from the 1920s onward, typewriters were used in all banks and government departments, but this did not refer to Arabic (or Hebrew) machines.[60] On the other hand, Bayan Nuwayhad al-Hout remembers that in 1948, upon their flight from Jerusalem, her father, a prominent Palestinian who had served in a series of public offices, took his two personal typewriters, one English, the other Arabic, along with him.[61]

4.6 Typing (dactylography) and shorthand (stenography) as ancillary skills

Given the use of typewriters for foreign business, it is not surprising that the training of personnel also remained in the orbit of foreign interests and foreign residential communities. At the time, it was commonly thought that the time-saving skill of touch-typing required systematic training, which in those days should include its "twin art" of shorthand as well.[62] Closely following examples abroad, local schools of commerce, in many cities pioneered by the French, gave predominance to the French language (conforming to the foreign residential and minority

---

For both handwritten and typewritten envelopes and postcards around WWI, see Yavuz Çorapçıoğlu (ed.), *International Ottoman Postal History Exhibition* (Ankara 1999), p. 39 and p. 69.

59 Thomas Gerholm, 'Economic Activities of Alexandrian Jews on the Eve of World War I, According to *Le Mercure Egyptien* in Shimon Shamir, (ed.), *The Jews of Egypt, a Mediterranean Society in Modern Times* (Boulder 1987), p. 101.

60 *al-Nafa'is al-'Asriyya*, August 1922 (and also *Filastin*, 4 January 1929) as quoted by Ayalon, p. 9. We have seen that at least some Shari'a courts in Mandatory Palestine began writing their *sijills* (records) by typewriter from the 1920s.

61 "Evenings in Upper Baq'a: Remembering Ajaj Nuwayhed and Home", *Jerusalem Quarterly*, no. 46 (2011) p. 19.

62 Sir Isaac Pitman's shorthand was first publicized in England in 1837 and subsequently introduced in the US in 1852 (before the typewriter), but it was adapted to other languages much later. In 1905–6, the Toynbee Hall in London, well-known for its free reading room, organized men's evening classes in Pittman's shorthand (for a fee!), see Asa Briggs and Anne MacCartney, *Toynbee Hall, the First Hundred Years* (London 1984). In the US, touch typing had been developed from 1888 but became common in business schools around 1910. Remington and other manufacturers supplied their own training classes, as Singer had done for sewing machines.

elites).[63] The French Chamber of Commerce in Istanbul propagated typing skills and shorthand as follows: 'Ce nouveaux progrès rend de grands services aux gens pressés, c'est-à-dire à toute la catégorie des gens d'affaires, commerçants et industriels, qui ont chaque jour à répondre au courrier qui leur arrive de tous côtes...'.[64] The École Commerciale Française in Kadiköy prided itself on very good exam results.[65] Other Commercial Lycées followed, all apparently including typing in their curricula.[66] In Alexandria, in 1901, a stenography course was offered by the Italian-sponsored Popular University.[67] In Beirut, it was the four-year curriculum of the Syrian Protestant College which included the English, French and Turkish languages, as well as typewriting and stenography.[68] In Palestine, these skills were taught at the Anglican Girls Schools and several others.[69] These are just a few examples.[70]

---

[63] In 1850, Gervais Pigier had founded a first École Pratique de Commerce et de Comptabilité, and his son, Emile, added courses for girls in 'dactylographie.' They were followed by Frères (Jesuit) schools, which expanded into the Middle East. Elisabeth Longuenesse, 'Système Éducatif et Modèle Professionel: le Mandat Français en Perspective, l'Exemple des Comptables en Liban' in N. Meouchy and P. Sluglett (eds.), *The British and French Mandates in Comparative Perspectives* (Leiden: Brill, 2004), pp. 544–545: She mentions commercial schools for boys in Jerusalem 1878, Tripoli 1886, Beirut 1890, and Alexandria 1911.
[64] 27 April 1895, p. 196.
[65] 30 June 1908.
[66] *Istanbul Ansiklopedisi*, "Ticaret Liseleri', vol. 7, pp. 268–269. See also Y. Surmen et al., 'Higher Education Institutions and the Accounting Education in the Second Half of the XIXth Century in the Ottoman Empire', https://www.researchgate.net/publication/24113130_ (last accessed 6 March 2023), p. 6 who state that the "Ticaret Mektebi" was based on a French curriculum for which it was initially difficult to find lecturers.
[67] Anthony Gorman, 'Anarchists in Education: the Free Popular University in Egypt (1901)', *MES* 41 (2005), p. 314.
[68] *Remington Notes*, vol. 1, no. 10 (1908 [?]), pp. 6–7 (see cover page), see also Arthur Ruppin, *Syrien als Wirtschaftsgebiet* (Berlin, Vienna, 2nd. ed. 1920), pp. 372–374 (data for 1912–3). Ruppin recommended similar schools for Aleppo, Haifa, and Jaffa.
[69] Note the training of women. Inger Marie Okkenhaug, "She Loves Books & Ideas & Strides Along in Low Shoes Like an Englishwoman: British models and graduates from the Anglican girls' secondary schools in Palestine, 1918–48", *Islam and Christian-Muslim Relations*, vol. 13(4): 2002, p. 467
[70] In Turkey, American modernization programs imparted typewriting skills on conscripts, however without the expected civic benefit, as they often returned to their villages without typewriters, Andrew James Birtle, *US Army Counterinsurgency and Contingency Operations Doctrine 1944–1946* (Center of Military History US Army 2007), p. 328. In Egypt too, some learnt typing during military service, see "Typewriters Still in Use in the Public Sector", *al-Masry al-Youm*, 24 February 2015.

It is not clear when Arabic and Turkish typing became part of regular commercial curricula.[71] A teacher at the Alliance Israélite Universelle schools, Avram Benaroya, is said to have developed a system for Turkish stenography, which was adopted by the government in 1929, the year after the alphabet reform (discussed below).[72] After World War II, the number of (non-missionary) commercial schools significantly increased – from Cairo and Alexandria, Beirut, and Tripoli, to Baghdad and Aleppo –, and all of these taught typewriting in both Arabic and English (or French).[73]

4.7 From early adopters to an early majority: another time chasm

Before going back to our main question on the relatively slow spread of typewriters in the Middle East, we will attempt to examine the evidence on their marketing. As is the case with so many "small technologies" in the region, one has to piece together the information from a variety of sources, as distinct from the quality of the statistical data which served researchers of the American or European markets.

If we are right in assuming that a typewriter cost $ 125 at the time, an alleged $1.902.153 of exports in 1898 might point to 15.000 machines, but only a very small number can have gone to the Middle East.[74] Unlike North America and Western Europe, where national, corporate, and even household statistics or, at least, estimates exist, trade reports for our region are confusing: Brands, sorts, quantities, prices, and weight of exported typewriters were confused, and there is no specification of script and category. Slightly less problematic is the lack of a breakdown of office versus the portable models which were probably acquired by private persons.

Trade statistics on typewriter sales give us only a very limited insight into differences between countries, in particular, but we would like to know more about those who acquired them and for what purpose. According to Robert Messenger's

---

[71] Higher School of Commerce (Ticaret Mekteb-i Alesi) was founded in 1882, and a first Ticaret Lisesi had been founded in Istanbul in 1894.
[72] Rifat N. Bali, *Avram Benaroya: Un Journaliste Juif Oublié Suivi de ses Mémoires* (Istanbul 2004).
[73] R.D. Matthews and M.Akrawi, *Education in the Arab Countries of the Near East* (Washington 1949), p. 67, pp. 178–179, p. 377, p. 463 and p. 480. In 1964, the well-known Egyptian journalist, 'Ali Amin, proposed a Higher Institute for Secretarial Studies, see Arthur Goldschmidt, *Biographical Dictionary of Modern Egypt* (Boulder 2000), p. 21.
[74] Cf. Byrn, p. 182.

blog, the United States exported in October 1921 – one randomly chosen month, but toward Christmas, as he says – typewriters to the total value of $ 956.240 of which $14.807 to Egypt and $ 2.832 to "Turkey in Europe". To this amount, smaller exports from the United Kingdom and probably Germany must be added.[75] These figures remain elusive as we do not know who acquired them and for what purpose. Low overall rates of acquisition must be offset by urbanization, composition of the populations, and foreign involvement. Another computation – of mediocre reliability – in 1930–1931 pertains to Mandatory Palestine with its mixed Arab-Jewish population, which makes a difference. It shows that 6.760 typewriters were imported, as against 4.330 in Syria and Lebanon, 859 in Egypt, 168 in Iraq and 14 in Iran.[76] In 1935 and 1936, Egypt was stated to have imported respectively 2.095 and 2.144 typewriters, overwhelmingly of American manufacture.

Indeed, a British trade report ascribed the United Kingdom's poor performance to three factors: Most Egyptian typists were used to American machines, British typewriters were more expensive, and the latter could not compete with certain "well-known American makes" which are "fitted with Arabic script".[77] Nevertheless, it seems that typewriters in the Arabic language did steadily advance. Around 1950, according to a knowledgeable person in the trade in Cairo, most typewriters he handled were in Arabic rather than in Roman script.[78]

Surprisingly, in Ibn Sa'ud's palace in Riyadh in 1922, the Lebanese-American journalist and writer Amin Rihani relates, there had already been an Arabic typewriter. The king's secretary had not only asked him to instruct him in its proper use, but he had an additional request. Among other acts of harsh maltreatment, such as users who had pounded the keys with force, Rihani discovered that its ribbon had once been broken, and had then been stitched together in such a way that it could not move. The anecdote does not relate whether the typewriter was thereafter properly used.[79]

As to outlying Yemen in 1934, we can learn a little from the delivery by a Jewish agent in Tel Aviv, Israel Subeiri, in Tel Aviv, of one single typewriter to a local

---

[75] "Overseas Exports for US Typewriters: Christmas 1921", OzTypewriter.blogspot.com > 2012/12. Even rough population estimates, supposedly 13 m. for Egypt and 14 m. for Turkey, or the low literacy rates of that time, are not helpful.
[76] D. Gurevich, *Foreign Trade of the Middle East* 1930–1931 (Jerusalem: Jewish Agency 1933), p. 215. No script is indicated.
[77] Department of Overseas Trade, *Report on the Economic and Financial Situation of Egypt* (London, May 1937), pp. 86–87.
[78] Amnon Acho, see n. 346. See further *The Arab Directory 1957–1958* (Beirut 1958 [?]), which lists 17 typewriter dealers in Cairo and 6 in Alexandria.
[79] Amin Rihani, *Ibn Sa'oud of Arabia, His People and his Land* (London 1928), pp. 148–149.

merchant, Sulayman Bukhari. In 1939, however, it would seem that the Imam refused, for some reason, a batch of typewriters which had been sent via Aden.[80] From Brinkley Messick's book on the *Calligraphic State* – a study focusing on writing documents by hand –, one discovers that typewriters began to be used regularly after Yemen had become a Republic (1962). From then onwards, government offices, as a rule, would type their documents.[81]

Advertising in newspapers helps us reconstructing the process of (at least) the potential diffusion of many small technologies, and in this case of typewriters. Remington and some other manufacturers locally advertised their Roman-type typewriters long before they came up with models in the local languages. One supposes that consecutive advertising indicates the existence of a potential market consisting of at least government agencies and large corporations, and not so much private consumers as one might think. In fact, various new brands made their entry in the following decades.[82] German, French and Italian manufacturers tried to capture a share of the market.[83]

In Egypt, as early as 1914, "the only Arabic typewriter" was advertised under the brand name of Monarch "as being used in the Khedivial Palace, in all Egyptian government offices, in some banks, businesses and stores, as well as in Ottoman, Moroccan and foreign lands." The agent was none other than Salim Haddad, the inventor mentioned earlier, whose American Warehouse now had stores in Cairo and Alexandria.[84] Later, Smith Premier regularly propagated an Arabic machine for "Egyptians aspiring to progress and advancement", called "Misr" (Egypt) adding that it was used in government services (Fig. 21).

---

80 Josef Tobi, *Yehudi beSherut haImam*, (Tel Aviv: Afiqim 2001–2002), p. 60, and pp. 141–142.

81 Brinkley Messick, *The Calligraphic State, Textual Domination and History in a Muslim Society* (Berkeley 1993), pp. 244–245. Kindly answering a question of mine, the author wrote (10 Jan. 2007) that at the time of his research in Ibb (1974–1976) he met there a former minister who owned an Arabic typewriter.

82 The Press Museum in Istanbul exhibits old Torpedo and Siemag specimens.

83 By 1913, Hammond, Underwood, Adler, Lambert were represented in Istanbul, followed by Royal, and Olivetti, AEG. Later, Hermes from Switzerland, Erika and Continental from Germany and Halda from Sweden joined the ranks. For Egypt see various brands in *Egyptian Directory* (Cairo 1951).

84 *Al-Hilal*, vol 22 (Nov. 1914) (with thanks to Fruma Zachs). Monarch was part of the Typewriter trust in which Remington was also a partner. The said advertisement has a drawing of the "Visible" model which dated back to 1907. It is noteworthy that office machines were advertised in the general press.

4.7 From early adopters to an early majority: another time chasm — 137

Fig. 21: Advertisement for the "Misr" (Egypt) Arabic typewriter, manufactured by the Smith Corona Company, and here said to be the choice of the Egyptian government (*al-Lata'if al-Musawwara*, 24 Jan. 1927).

Unfortunately, my investigation shows how little we know about the modernization and growth of smaller firms.[85]

The growing frequency and size of advertisements for typewriters in the Arabic, as well as in the Turkish press, does show that the market, or at least the potential market, gained momentum in the 1950s.[86] An additional interesting angle using advertising as an indication, might be the screening of small advertisements for personnel, but this would be a project in itself. The impression is that in Egypt

---

[85] One "Misr" model (1922) forms part of the British Museum collection on modern Egypt. See also Yunan Labib Rizq, 'Enter Illustrated Adverts', *al-Ahram Weekly*, 29 March-4 April 2001. Also, *The Egyptian Directory* (Cairo: Max Fischer 1931), p. 994 listing Remington sales agents in Cairo, Alexandria, Sudan (*sic, no place*), Damascus and Aleppo. The brands Royal, Olivetti, Smith, Monarch and Adler were represented, too.

[86] See Olivetti in *al-Ahram*, e.g., 29 March 1956.

and Turkey, from the early 1940s onwards, demand for skilled typists and stenographers was on the rise.[87]

4.8 No local typewriter celebrities

However, we did not discover advertisements with well-known users (except royals) to convince buyers and increase sales. In American and European literature, anecdotes on famous authors and journalists and their typewriters are frequent. Famously, Mark Twain used a Remington model II writing *Life on the Mississippi* in 1882 (not *The Life of Tom Sawyer* as Remington advertised). Friedrich Nietzsche had adopted a typewriter to overcome his eye problems, or others such as Lev Tolstoy and Sigmund Freud both had daughters who typed their manuscripts.[88] Arabic writers and intellectuals appear to have been reluctant to use one, be it out of respect for a resilient manuscript culture or even a calligraphic preference.[89] An Egyptian journalist, Mursi Saad El-Din, wrote as late as 2003 (we realize that this is not exclusively typical for the Middle East): "Typewriters and computers are cold metal structures that have none of the warmth a pen or pencil can give. Come to think of it, the greatest classical literature – fiction, poetry, drama – was produced long before the typewriter or the computer was known."[90]

This is not to say that relevant references are totally lacking in memoirs. The renowned linguist, Pierre Cachia, is one of the few Middle Eastern scholars to mention the topic at all, remembering that he learned typing from an Italian teacher (provided that he would write exclusively in Italian on his Olivetti) and that, as a student at the American University of Cairo during WWII, he had to exchange his German brand for an American one for lack of spare parts.[91]

We do not know which prominent Arab writers possessed typewriters at all. Najib Mahfuz, stabbed by an Islamist radical in 1994, was limited in the use of his

---

[87] *Cumhuriyet*, 22 March 1940 (for a lady typist [*bayan daktilo*] in a bank; idem, May 1940 (for an auditor's office); *Son Posta* 19 February 1943 (a POB in Ankara). It is interesting to compare this to *l'Égypte Nouvelle* (in French!), e.g. of 17 Oct. 1925, where two typists (one male, one female) look for employment, while one (female) is required.

[88] Sigmund Freud bought a typewriter for his secretary in 1913; Franz Kafka wrote his love letters on a typewriter, but not a single manuscript, see Kittler, p. 215 and p. 223.

[89] Titus Nemeth drew my attention to the need to distinguish between a manuscript culture and a calligraphic culture.

[90] *Al-Ahram Weekly*, 30 Oct. – 5 Nov. 2003.

[91] Anna and Pierre Cachia, *Landlocked Islands, Two Alien Lives in Egypt* (Cairo 1999), pp. 127–128. We add that his Russian-born mother, Anna, worked as a linotype typist at the *Egyptian Gazette*, pp. 74–75. At the time they were still rare in the region.

right arm for quite some time, and did not write anything, which hints at his writing in longhand. Yusuf Idris's writing habits are described as using a pen and a large sheet of brown paper. [92] Apparently, for some writers, this began to change only in the 1960s, but it could also be for worse. Under Saddam Husayn, in Iraq, all typewriters had to be registered with the authorities as a form of control and censorship. The Iraqi writer, Najm (Najem) al-Wali, upon being honored by the Frankfurt Book Fair in 1984, even prided himself on never having possessed one before going into exile.[93] In Turkey, more writers and journalists may have preceded their Arab colleagues, but systematic information is difficult to come by. For instance, Oktay Akbal, a journalist and writer (1923–2015) was known to have used a typewriter, and novelist Selim İleri (b. 1949) who already fully belonged to the post-World War II generation also did. But this was not a publicity aspect as elsewhere.[94]

4.9 Transitions: QWERTY and Turkey, and de-colonization in Egypt

The diffusion of typewriters in the languages of the Middle East was not only held up by technical hurdles, but no less by the prevalence of foreign economic routines and interests. This situation began to change when a "Control" factor from above began to work in the form of new legal and bureaucratic measures, increasing the sale and use of typewriters in local languages: Under the dynamics of the nationalist evolution in the region, the predominant use of French or English in banks and other commercial establishments would give way to Turkish and Arabic. In Turkey, the former anomaly was formally terminated by the Law on the Compulsory Use of Turkish in Economic Enterprises of 6 April 1926.[95] The said law predated the script reform of 1928, which became an even more determining landmark for the use of Turkish typewriters. In the following year, the government ordered

---

92 P. M. Kurpershoek, *The Short Stories of Yusuf Idris, a Modern Egyptian Author* (Leiden 1981), p. 49 and p. 47. It is unclear why it was brown paper.
93 Eamonn Fitzgerald's Rainy Day blog, "The Crime? Possession of a Typewriter", http://www.eamonn.com/2004/10/the_crime_possessio (site inactive 2023).
94 Information on Turkey from our friend, Prof. Filiz Özbaş but there must be more information. Neither Egypt nor Turkey, it seems, manufactured typewriters locally. This contrasts with India with its larger literate population, and English as one of two official languages, see Arnold, pp. 112–114.
95 İktisadi Müesselerlerde Mecburi Türkçe Kullanması Hakkında Kanun, with thanks to Rifat Bali for this information.

6.000 Roman-script typewriters from the USA and from Germany. Between 1928 and 1930, to give one example of the impact, branches of the ruling party made the transition from handwritten to typewritten reports.[96] Now, more private firms as well began using typewriters.[97] In the period of transition, one could still see advertisements in Ottoman-Arabic script, propagating Roman-script typewriters.[98] It could be hypothesized that it ultimately facilitated a faster acceptance of personal computers than in the Arab world, an aspect to which we will come later. Atatürk's reform not only introduced Roman characters, but from the very beginning also foresaw a standard model for a "national" Turkish typewriter keyboard, scientifically calculated according to the frequency of Turkish characters as had been the Qwerty design.[99]

Still, there were several new hurdles to be cleared.[100] After trials and errors, an advanced Standard Turkish Keyboard was officially adopted on 20 October 1955 and duly patented. A major contribution was made by Ihsan Sitki Yener, a business administration expert with a higher American academic degree, who taught at various commercial schools and experimented with new keyboards in order to enhance the speed of typing in Turkish. He cooperated with the Turkish Language Institution and the Ministry of Education.[101] Yener himself won several prizes in highly popular speed typing championships in Turkey and abroad. It was said that the Turkish keyboard enabled even faster typing than by Qwerty. Yet, under the impact of Microsoft, Turkey reverted to Qwerty in the 1980s which did not mean the end to debates regarding an ideal keyboard.[102]

---

96 James D. Ryan, *The Republic of Others. Opponents of Kemalism in Turkey's Single Party Era, 1919–1950* (PhD Thesis, University of Pennsylvania 2017), p. 88.
97 *Levant Trade Review*, vol.17/1 (Jan. 1929). See also Fig. 22. Also K. Krüger, *Kemalist Turkey and the Middle East* (London: Allen & Unwin 1932), p.78. Literacy in Turkey in 1935 was 29% for males, and 10% females, in Egypt in 1937 23.4% for males, and 6.1% for females. The process of alphabetization was slower than what was hoped but it had its impact also on (type-)writing.
98 Orhan Koloğlu, *Reklamcılığımızın İlkyüzyılı 1840–1940* (Istanbul 1999), p. 270.
99 Recommendation by the *Dil Encümeni* as quoted from Ulkutasir, by Vrolijk, *Een Turks Alfabet* (Leiden 1998), p. 33.
100 Modern Turkish orthography has an ı without a dot, as well as a so-called soft ğ, in addition to the characters ç, ş, ö and ü which exist also in either French or German. Some of these are lost again in contemporary word processing.
101 The Turkish keyboard runs FGĞIOPR. The Koç Museum has one older Remington typewriter with an AIERTY keyboard. On Yener see http://www.sampiyon-kurslari.com.tr/kurucumuz_eng.htm.
102 Asım Egemen Yılmaz and Emrah Ciçek, "Optimized Rearrangement of Turkish Q and F Keyboards by Means of Language-Statistics and Simple Heuristics", *Hittite Journal of Science and Engineering*, vol. 3/1 (2016), pp. 23–28; Tuncay Kayaoğlu, "Keyboard Warriors: Battle for Turkish Typists Heats Up" (3 March 2015) http://aa.com.tr/en/turkey/keyboard-warriors-battle-for-turkis (accessed 14 Dec. 2017).

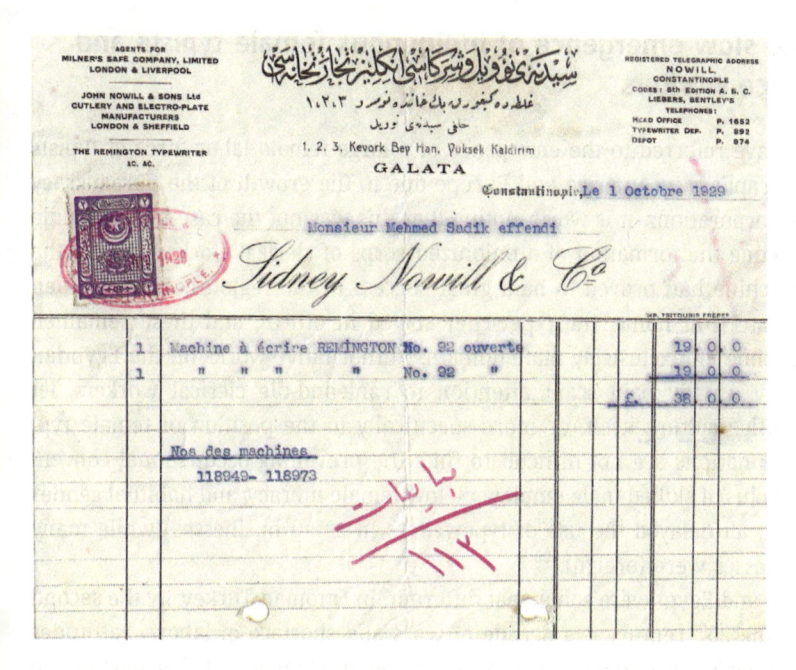

Fig. 22: A typed invoice of the Sidney Nowill firm in Istanbul for the delivery of two Remington machines dated 1929, a year after the transition to Roman script. The stationary is in English and Ottoman, the specification is in French, and a handwritten notation is still in Ottoman script (author's collection).

In Egypt as well, in a different political setting, notably following demands by students of the Commercial Faculty in Cairo, and probably effendi-led movements such as the Muslim Brothers and Misr al-Fatat, the Government promulgated Law no. 62 of August 1942 which made the use of Arabic compulsory in commercial dealings.[103] Further Egyptianizing steps came in the late 1940s and in the 1950s, followed by wholesale nationalizations of former foreign companies a decade later.[104] One assumes that these must have had an impact on the typewriter market.

---

[103] Banque Misr had from its founding in 1920 adopted Arabic, but we do not know how much business was handled by typewriters.
[104] It is noteworthy that India went through a similar nationalist process as in the Middle East. Gandhi urged the replacement of English as the business language by Hindu or Hindustani (being interested himself in typewriters and in stenography). Although typewriters with a locally pioneered Hindi keyboard were being manufactured in India, they did not make any headway till 1948. In that year of independence, a government committee was set up which submitted its recommendations in 1951 for a standard keyboard, as well as a related stenographic method.

4.10 The slow emergence of indigenous female typists and office workers

Since we have referred to the emergence of a large female labor force of typists and stenographers in America and Europe due to the growth of the bureaucracy and large corporations, it is worth noting that this was not the case in the Middle East. Let alone the formation of a unionized corps of clerical workers. While the sewing machine had proved to be a great success in the region because women could operate it at home, the typewriter stayed in offices, and these remained for a long time an exclusively male domain. Authoritative works on the Egyptian bureaucracy usually paid scant attention to rank-and-file clerical workers, let alone to office culture itself, or more specifically to the position of female typists.[105] Explanations are not difficult to find: the prevailing occupational conventions, a surplus of skilled male employees, low female literacy, and habitual gender segregation, all delayed the use of typewriters in small businesses (while many large businesses were foreign).[106]

Turkey and Egypt were somewhat different. In Ottoman Turkey, by the second decade of the 20th century – a decade of wars and shortage of labor –, attitudes toward women's work appeared to engender some change, – a development to which women's magazines – it seems – also contributed their share.[107] But it was far from a lasting momentum with regard to clerical occupations.

Let us review a few specific examples. In Istanbul, soon after World War I, five department stores (a commercial branch identified with European commodities and habits) were said to employ some female typists, almost certainly for corre-

---

[105] Berger, *Bureaucracy and Society in Modern Egypt, a Study of the Higher Civil Service* (NY 1957); Frederick Harbison and Ibrahim Abdelkader Ibrahim, *Human Resources for Egyptian Enterprise* (New York: McGraw Hill 1958); Nazih N.M. Ayubi, *Bureaucracy and Politics in Modern Egypt* (London: Ithaca 1980); Monte Palmer et al., *The Egyptian Bureaucracy* (Syracuse: Syracuse UP 1988); or even Laura Bier, *Revolutionary Womanhood, Feminisms, Modernity, and the State in Nasser's Egypt* (Stanford, Stanford UP 2011), esp. pp. 90–100. See also Arlene Elowe Macleod, "Transforming Women's Identity, the Intersection of Household and Workplace in Cairo" in Diane Singerman and Homa Hoodfar (eds.), *Development, Change, and Gender in Cairo, A View from the Household* (Bloomington: Indiana UP 1969), pp. 27–50, on under-employed typists in the bureaucratic system.
[106] Piano playing and typewriting were often compared as a typically female keyboard skill; see Kittler, pp. 194–195. However, in the Middle East, pianos came late and then for the select elite only.
[107] Nicole van Os, "'Müstehlik Değil Müstahsil", Not Consumers but Producers; Ottoman Muslim Women and Milli İktisat', in *The Great Ottoman-Turkish Civilisation* (Ankara 2000), vol. II, pp. 269–275.

spondence in European languages.[108] A few indigenous Muslim women worked in the Ottoman telegraph and telephone offices, or as secretaries in banks, insurance companies, hotels etc.[109] But in 1923 not a single Muslim female stenographer could be found in Turkey, wrote Ruth Woodsmall, a Christian feminist, all were Greek, Armenian, or Jewish. Demand for female stenographers, rather than mere typists, now exceeded supply, a fact which Woodsmall in her optimism about women's emancipation, ascribed to the abandoning of the veil, and the adoption of the Latin alphabet which supposedly gave a push to business.[110] However, in Palestine, in 1936, the rule was that the few women in clerical positions in government service, upon getting married, could not continue to be employed. But there, in the Jewish environment it had become possible to obtain competent shorthand typists without difficulty.[111]

In foreign companies in Egypt, as described by Flora Karanasou for the year 1947, just before the first stringent steps of nationalization, all the typists, secretaries, clerks and telephone operators still were foreigners or minorities, mainly Copts.[112] Widening the scope to the early Nasirist 1950s, the American sociologist Morroe Berger wrote on the civil service: 'The first thing he [the Western visitor] notices is the presence of men, rather than women, in the reception rooms... The outer office presided over by a female secretary or two, so familiar in the West, is never seen in Egypt. Nor can one usually see the large office with scores of desks at which typists, clerks, and machine operators are working.'[113] The Suez Canal Company employed no women at all until 1941, and even after rules were changed

---

[108] Lawrence S. Moore, 'Some Phases of Istanbul Life' in Clarence Richard Johnson (ed.), *Constantinople To-Day or the Pathfinder Survey of Constantinople, A Study in Oriental Social Life* (NY 1922), p. 187. Though this source does not mention male typists, it indicates that wages were lower for women than for men, a phenomenon known also in the West.
[109] Serpil Çakır, *Osmanli Kadın Hareketi* (Istanbul 1993), pp. 261–300.
[110] Ruth Francis Woodsmall, *Moslem Women Enter a New World* (New York: Round Table Press 1936), p. 258. Note that her statement refers to 1923, and possibly things had already changed when she published her book.
[111] Ilana Feldman, *Governing Gaza, Bureaucracy, Authority, and the Work of Rule, 1917–1967* (Durham NC: Duke University Press 2008), p. 117.
[112] F. Karanasou, *Egyptianisation, the 1947 Company Law* (unpublished PhD Thesis 1992), p. 335. Cf. Angelos Dalachanis, *The Greek Exodus from Egypt, 1937–1962* (New York: Berghahn 2017), p.103 on "Egyptiot" (meaning local Greek) "touch typists" hired by the British after the war to try and prevent repatriation.
[113] Berger, p. 12. In Riyad, Saudi Arabia (1996), when the Institute of Public Administration had one opening for a secretarial job, it reportedly received 500 applications, Eleanor Abdella Doumato, "Women and Work in Saudi Arabia: How Flexible Are Islamic Regimes?', *Middle East Journal* 53/4 (1999), p. 570.

(1954) no more than more than half of the office workers could be women, all of them hired only after taking exams in typing and stenography.[114]

Aspirations to advance clerical professionalization for women (apart from training courses mentioned above) may have been more evident than reality warranted. For the time being, these came to expression in specialized periodicals. One short-lived magazine called *Sténo-Dactylo* seems to have appeared in Cairo as early as 1912, and another, *l'Orient Sténographique*, in 1926, both in French, which is not surprising.[115] So far, I have not been able to trace issues. Anyway, the readership could not have been large. But a few Arabic textbooks to instruct aspiring typewriting secretaries were published.[116] One similar handbook appeared in Lebanon; it was clearly translated from a European example: Though originally meant for women typists, the writer inadvertently here and there addressed the user in the male Arabic form (Fig. 23).[117]

In Istanbul, a new periodical *Sekreter Daktilograf* was published from 1956 to 1960 (Fig. 24). It claimed to be "a professional magazine to enhance success in the fields of dactylography, stenography, communication, book-keeping, fiscal legislation, foreign language and general secretarial occupations".[118] The *Daktilograf*'s editor was Ihsan Yener, whom we have already met as the developer of the Turkish keyboard. He tried to propagate his own innovations, devoting much space to the Turkish participants, and indeed champions at international speed typing contests (female and male). The journal, though not particularly captivating, contains some other information on the training of typists, on the secretarial occupation in general, as well as photographs and advertisements. We do not know the reason for its disappearance, but a successor journal called *Sekreter Dergisi* (1992) – defined as a 'monthly specialized women's magazine' – became defunct too, which makes us think that the market was not yet ripe for such publications.[119]

---

114 Barbara Curli, "*Dames Employées* at the Suez Canal Company: The 'Egyptianization' of Female Office Workers, 1941–56", *International Journal of Middle East Studies* 46 (2014), pp. 553–576.
115 Jean-Jacques Luthi, *Lire le Presse d'Expression Française en Égypte, 1798–2008* (Paris, l'Harmattan 2009), p. 305.
116 *Die Presse in Aegypten, Internationale Press-Ausstellung* [sic], (Cologne 1928), p. 43. The publisher was a certain Edouard Gargour.
117 *al-Uslub lil-Sahih fi al-Darb 'ala al-Ala al-Katiba al-'Arabiyya, 'Ilmi-'Amali bi-Tariqat al-Lams* (Beirut: al-Khayat 1966).
118 Cf. in the USA *The Typewriter Operator* (from 1887) and similar magazines.
119 Asli Davaz-Mardin, *Hanimlar Alemi'nden Roza'ya, Kadın Sureli Yayinlari Bibliografyasi; 1928–1996* (Istanbul 1998), contains basic data on the first journal, pp. 120–121, but not on the second, of which we have seen a later issue dated August 2002 (no. 59) which means that it appeared every two months.

Fig. 23: A translated text book " *Al-Uslub al-Sahih fi'l Darb 'ala al-Ala al-Katiba al-'Arabiyya, 'Ilmi-'Amali bi-Tariqat al-Lamas* (The Right Way to Type on an Arabic Typing Machine, Scientific [and] Practical, Based on the Touch Typing Method), published in Beirut in 1966 (Nazarian Library, University of Haifa).

In Turkey but also elsewhere for a long time, one could still see letter writing in the streets as a male occupation, particularly in front of government offices. The erstwhile *katip*s of government offices duplicated themselves in the form of *arzuhalcıs*, filling out forms and writing petitions against a fee for the illiterate public.[120] The skills gap had grown from writing by hand to typewriters acquired by a great many of them.[121]

As to sources which can complete the picture, systematic checking of photographic collections could be helpful to reach more definite conclusions, though

---

120 Public scribes have often been depicted, e.g. Grehan, *Damascus*, p. 183, an etching by William Thomson, 1886.
121 Kemal Karpat, *Studies in Ottoman Social and Political History* (Leiden: Brill 2002), pp. 265–266. By 1957, a developing small Anatolian town also had two "sidewalk typists", Joseph S. Szyliowicz, *Political Change in Rural Turkey, Erdemli* (The Hague, Paris 1966), p. 74.

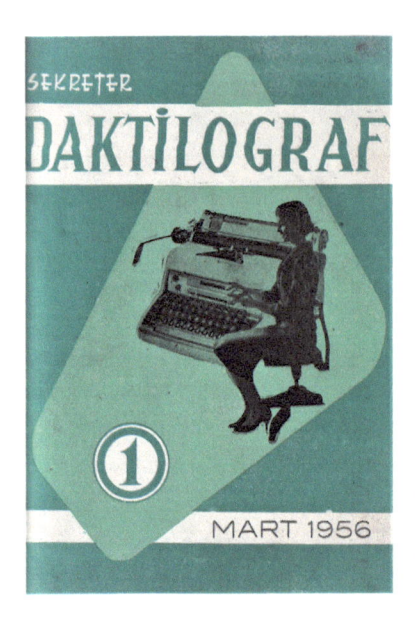

Fig. 24: Cover of the first issue of the monthly *Sekreter Daktilograf* (Typist Secretary) for office workers, a magazine which appeared in Istanbul from 1956 to 1960 (author's collection).

our impression is that photographs of offices only rarely show female clerks.[122] This applies also to movies in which female secretaries play a role; only a few Egyptian instances can be mentioned in this regard (Fig. 25).[123]

In parentheses, it is to be noted that the authoritative *Encyclopedia on Women & Islamic Cultures* (2003), which contains entries on women in a host of occupations, by and large ignored clerical employees (such as secretaries, typists, and computer operators). Within a relatively short time, owing to fast changes, its short section on women and information technologies had become out-of-date for most countries in the region.[124] This, in spite of a proven increase of the share of women in non-agricultural occupations such as "stenographers and typists" especially in non-OECD countries which had begun in the preceding de-

---

[122] Sarah Graham-Brown, *Images of Women* (London 1988), p. 164 has only one such picture of Palestinian refugee women in Kuwait. Orhan Pamuk, *The Innocence of Objects* (New York: Abrams 2012) shows his father's male office staff with two women, pp. 70–71 (and there Daktilo Demir's "famous" typewriter, p. 209).
[123] Among colleagues consulted on this point, Eyal Sagui Bizawe, mentioned the movies *Mar'ati Mudir 'Amm* (1966), *Tharthara Fawq al-Nil* (1971) and *al-'Ar* (1982), and Asaf Dar, *Ana Hurra* (1959).
[124] Suad Joseph (ed.), *Encyclopedia of Women & Islamic Cultures* (Leiden: Brill 2003), vol. 4, pp. 386–389. On this issue it is not significantly updated in the on-line edition.

4.10 The slow emergence of indigenous female typists and office workers — 147

Fig. 25: Street-based public scribes in Istanbul, with typewriters, could still be seen in the late 1970s (photo by author).

cades.[125] In fact, while the literature on women in the Middle East has greatly expanded in recent decades, little has been written on the development of office culture, in particular on women entering secretarial and clerical jobs in co-ed workfloors, and their increasing educational and vocational backgrounds and skills. Also, in the few collections of "self-stories" from the Middle East, relating to a variety of occupations, to the best of my knowledge, none refers to typists, secretaries, or other office personnel.[126]

---

[125] Cf. Richard Anker, *Gender and Jobs, Sex Segregation of Occupations in the World* (Geneva: ILO 1998), esp. pp. 269, 289, and 392–395. At the time, the ILO still considered computer skills under the heading of bookkeepers and cashiers.
[126] E.g., Edmund Burke,III (ed.), *Struggle and Survival in the Modern Middle East* (London 1993); Donna Lee Bowen and Evelyn A. Early (eds.), *Everyday Life in the Muslim Middle East* (Bloomington, 2nd ed. 2002).

4.11 Leapfrogging to the personal computer

In comparison with Northern America and Europe, as well as in comparison with parallel "small technological" inventions and devices, the diffusion of typewriters in the Middle East had lagged behind. Before typewriters could have reached Everett Rogers' stages of an "early" and even "late majority", personal computers – a technological revolution in itself which began around 1981–, was making them obsolete. This happened within a surprisingly short span of time: the diffusion of personal computers went much faster and was more complex than that of typewriters had been. Soon, even the maintenance of old typewriters was becoming a problem too and contributed to their phasing out. Surely, larger percentages of clerical workers (and private persons) are working on personal computers today than have ever worked on their predecessors, the typewriter.[127] In this sense, typewriters were skipped over.

However, this was not the end of the typewriter technology because personal computers themselves signified in a sense a "dynamically continuous innovation".[128] Though often considered as superseding typewriters, it has to be realized that personal computers retained some of the basic features of typewriters, such as the keyboard (or even revived Qwerty in Turkey) and preserved at least part of the terminology of typewriting.[129] An electric typewriter had been patented already by Edison in 1872, but it was too expensive and cumbersome to operate.[130] The advanced IBM Selectric which was launched in 1961 and became a great global success (13 million of its subsequent models had sold by 1986) became a chain in this development. The idea of a changeable typeface, with a "golf ball" shaped head, may be traced back to the Blickensderfer typewriters of the early 1900s, and is today a regular feature of our word processors. This innovative electric typewriter which enabled an expert typist to reach 90 words per minute, also introduced some early features of a computer terminal. But the transition from a global success to widespread use in the Middle East was not self-evident. The Selectric

---

127 In Egypt, however, typewriters were still used occasionally for filling out official forms, see al-Masri al-Yawm, 24 February 2015.
128 Term more or less as used in *Oxford Reference* [on internet] even though new manufacturers took over.
129 Also, some of the word processing terminology clearly continues the typewriter era: cc for carbon copy, etc. Even the @ symbol (fi in Arabic) reverts to the Lambert typewriters (a Frenchman who was the first to manufacture a keyboard of one piece (1902). Word processors, moreover, carry over erstwhile offices management techniques, e.g., files and folders.
130 R.T. Gould, "The Modern Typewriter and its Probable Future Development", *Journal of the Royal Society of Arts*, vol. 76/3940 (1928), p. 728.

was adapted for Arabic, Hebrew, and Farsi around 1977, and became used in the region from the 1980s. Experts were full of praise for the Arabic fonts on the "golf ball".[131] However, considering that the production of this office machine was already discontinued around 1986, its impact in the countries of the Middle East was probably limited.[132]

Personal computers began to make a difference. With regard to the criterion of Competence, new skills had to be mastered (such as software and word-processing) but, on the other hand, methodic typing courses which had once been considered essential *rites-de-passage* to enter secretarial, journalistic, or academic positions were made increasingly redundant by education, as well as by the use of cellphone keypads and the like.[133] The ability to handle a keyboard is nowadays expected from anybody who writes for his or her existence.[134] The popularity of personal computers and the use of the internet went together.

With differences between countries and social sectors in mind (e.g., in literacy, economic development, and globalization), attitudes are, here and there, changing also with regard to Middle Eastern women working outside their homes. More and more women, many conspicuously in Islamic attire, are seen today in offices and other workplaces, operating computer keyboards: In government offices and agencies, commercial establishments, banks, tourist agencies, schools, universities, and the like.

Statistics on the number of personal computers in the Middle East are in certain respects deficient. The research on "connectivity", "internet penetration", and "digital divide" has shifted away from the question of diffusion of ownership of personal computers, in particular, among private individuals. Though, once more, a "small technology" in the Middle East is lagging behind, it seems that great leaps forward are being made.[135] The countries which we discussed are rap-

---

131 Nemeth, pp. 76–77, mentions that its typographical qualities in spite of shortcomings were widely acclaimed.
132 The Selectric was, no doubt, very expensive for the countries of the region. We do not know how many were sold during this short period.
133 Though the *International Herald Tribune* (29 December 2011) remarked on the lack of a researched history of word processing, there are several pertinent studies, e.g., Jeanette Hofmann (1999) and Jacques Derrida (2001), but, indeed, a narrative of word processing in Arabic and Turkish, and an analysis of how it impacted office culture is still lacking.
134 Jon Alterman has mentioned "typing" ability as a factor, counting among the delaying factors: costs (whereby richer oil countries are in an advantageous position), the need for English, technological drawbacks, "The Middle East's Information Revolution", *Current History* 2000, p. 22. Shorthand, too, has disappeared as an occupational specialization.
135 The *Arab Human Development Report* of 2003 had shown a diffusion of 18 computers for each 1.000 persons in the region, with a world average of 78.3; with 1.6% of the population then having

idly catching up but at a different pace.[136] Here also the criterion of affordability (Capital) comes into focus again: Data for 2006 showed that Turkey had a 6.1% household ownership of personal computers, Egypt 2.65%, but the UEA 33.8% (as compared to 78.57% for the USA).

Supposedly, Turkey's higher literacy rate (and its earlier transition to Roman-script typewriters) maintained a lead over Egypt, but the wealth of the UAE, and probably the Gulf states in general, assured a lead over other Arab countries and even over Turkey.[137]

My main argument is that the personal computer has, by and large, in Middle Eastern countries jumped over the typewriter. This phenomenon is an interesting case of "leap-frogging", which had not been recognized as such by Everett Rogers, – we suppose, because he did not deal with non-western countries. There, this seems to happen more frequently. [138] If the full diffusion of a certain "small technology" is delayed for too long, it can be overtaken and eclipsed by a newer one. Similarly, for large populations in the Middle East, in our era of accelerated globalization, the transistor radio has jumped over the home radio receiver, and the cellular telephone over the landline telephone, and, in some places, credit cards over payments by check, to mention but a few technological developments, each of which justifies more thorough research in its national or regional, social, and socio-economic context.

---

access to the internet, the Arab region came almost at the bottom of the list. Three years later, in 2006, internet usage in the Middle East as a region (2.9% of the world population) was estimated at 9.6% as against 15.9% in "the rest of the world". Worldwide Internet Penetration (29 March 2006, source Miniwatts Marketing Group): worldwide penetration estimated at 15.7% the Middle East (no definition) has the second lowest internet "penetration", after Africa, and the second lowest usage (1.8%) after Oceania. However, the usage growth for 2000–2006 was the fastest of the seven regions. While statistics often diverge, another source for 2006 states 22.7% "penetration" in Turkey and 7.0% for Egypt ("usage" being estimated respectively at 64.0% and 15.3%, which indicates access to internet cafes, schools etc.). But here too, even considering a low take-off, the increase is impressive, respectively 193.7% and 1,011.1%, see *Internet World Statistics*

136 Arab Republic of Egypt, Ministry of Communication, *Measuring the Digital Society in Egypt: Internet at a Glance* (2015) p.8: 39.7% of households use the internet, including a ratio of 43% females). According to World Bank data, in 2016, 56% of households in Turkey had a personal computer.

137 Cf. A. Almowanes, "History of Computing in Saudi Arabia: A Cultural Perspective", *International Journal of Social Science and Humanity*, vol.7/7 (2017), pp. 437–441.

138 The phenomenon of leapfrogging in the sense of technological development and consumerism, is used here for skipping over a phase of empirical, linear evolvement.

5 From Ibn Haitham to Abou Naddara and Al-Misri Effendi: eyeglasses in the Middle East

Eyeglasses (or, less formally, glasses, or spectacles, eye-ware, vision aids, whatever we call them) conform to our definition of a "small technology".[1] Compared to other devices discussed in this volume, they are relatively simple artifacts, with few parts: and maybe they are the most widespread mass-produced item here discussed: devices to help the vision for those who want or need them. [2] Eyeglasses have become universal, yet their diffusion differed according to various regional and social factors.[3]

Allegedly, it was in Pisa around 1286 that a craftsman "invented" eyeglasses, not a theoretician of optical science, a monk, or a philosopher. As with other technological innovations in history, there are several hypotheses as to who exactly he was. But his achievement was soon made known by a Dominican preacher in the early years of the fourteenth century.[4] The new device was a breakthrough because it enabled people with presbyopia (defined as the age-related gradual loss

---

[1] The term "eyeglasses" is probably derived from the Italian *occhiale*. In English, *spectacle* (from Latin *spectaculum* is usual, not *speculum* which is a mirror. The German *Brille* comes from *beryl* (a mineral), and the French *lunette* from *lune* (moon or a half-moon aperture). For different names in various languages, see W. Bohne, *Handbook for Opticians* (New Orleans: Griswold 1895), pp. 202–205. In Arabic *naẓẓāra* is the usual word from a verb which means to see, and in Turkish *gözlük* from *göz* for eye. In modern Hebrew initially the word *batei-eynayim* (sort of "eye shelters") was used, but according to the linguist Rubik Rozental, since 1890, at the suggestion of a schoolteacher from Grodno, Russia, *mishqafayyim* (a dual form derived from a verb which means to see) became accepted.

[2] It is not our intention to survey the evolution of types of eyeglasses, of which descriptions are abundant (e.g., *monocles* developed in Germany, *pince-nez* types, as well as *lorgnettes*, also from the 18th century, which mainly served women. The use of each type had its European or American societal context, but this followed for the Middle East as well.

[3] The literature and internet items on the evolution and diffusion of eyeglasses in Europe are ample, with specific material on North America and some on China and Japan, but little has been written on Middle Eastern countries.

[4] An earlier assumption ascribed the invention to Pope John XXI's physician, Petrus Hispanus, in 1276. For the most authoritative research see Edward Rosen, "The Invention of Eyeglasses", *Journal of the History of Medicine and Allied Sciences*, part. I, vol. 11/1 (Jan, 1956), pp. 13–46 and part II, vol. 11/2, pp. 183–213; Vincent Ilardi, *Renaissance Vision From Spectacles to Telescopes* (Philadelphia: American Philosophical Society 2007). pp. 3–6, and Chiara Frugoni, *Le Moyen Âge sur le Bout du Nez: Lunettes, Boutons et Autres Inventions Médiévales* (Paris: Belles Lettres 2011, transl from Italian, 2001), pp.3–33. See also Frank Joseph Goes, *The Eye in History* (New Delhi; Jaype Bros 2013).

of the ability of the eye to focus on an object) to read more easily than with a magnifying glass.[5]

The statement "Spectacles still await their historian" cannot be upheld since research in social, cultural, and medical history, and especially the internet, have increasingly filled lacunae in our knowledge of eyeglasses and their diffusion, at least as far as Europe is concerned.[6] If some dates in the early history of eyeglasses remain speculative, this is challenging rather than disturbing.

The earliest category of people who began using eyeglasses, from around 1300 onward, were, indeed, those suffering from presbyopia. Our contemporaries generally start to realize that between the age of 38 and 43 glasses are needed. Incidentally, we remark that the term reading-glasses is so common that we tend to forget that artisans and craftsmen doing fine work, e. g., tailors, toolmakers, decorators, painters, weavers, as well as women embroidering or in other fine crafts, also require a remedial device. Today, even elderly tea-pickers in India, or other agriculturalists anywhere who do delicate work, benefit by them.[7] A resulting, more or less coerced, division of labor between nearsighted scribes, scholars, judges, goldsmiths, weavers or the like, and long-sighted farmers, livestock holders or military men, must remain an intriguing historical speculation.[8]

People suffering from other visual disabilities, mostly hyperopia (longsightedness), myopia (nearsightedness), or astigmatism (blurred vision), had to wait until the 15th century for corrective eyeglasses with convex lenses. By then, they too, could be helped by advances in the transition from quartz to optical glass, and accumulated optical know-how.

Looking at the world-wide diffusion of glasses in different categories, it was not until after WWII that the 40 % of people who objectively needed them, actually acquired them.[9] The global prevalence of visual impairments is geographically uneven today, and probably has been so for long. It has been estimated that in our days, at least a quarter of all readers regularly need glasses, and a sixth of humankind is myopic and would permanently need them too.[10] Actually, experts think

---

[5] We will not deal here with such "reading stones" (convex lenses mostly made of quartz), or other magnifying devices held over the object.
[6] Quote from David Vincent, *The Rise of Mass Literacy, Reading and Writing in Modern Europe* (Cambridge: Polity 2000), p. 103.
[7] Tea-pickers in Assam, tested, and given glasses in 2017, improved their daily yields by 20 %, *BBC News*, 19 November 2019.
[8] Tomás Maldonado, "Taking Eyeglasses Seriously", *Design Issues*, vol. 17/4 (2001), p. 38.
[9] James Smith Allen, *In the Public Eye: a History of Reading in Modern France, 1800–1940* (Princeton: Legacy Library 1991), p, 166.
[10] Alberto Manguel, *A History of Reading* (New York: Viking 1996), p. 291, quoting Trevor-Roper's *Blunted Sight*.

that myopia, is on the rise in East and South-Asia. They ascribe this to increasing stress – ever more education – in particular to mastering Chinese characters, and to less outdoor activity. On the international scale, these regions therefore score worse, while countries in North America and Europe (with high incomes) do much better. In not a few countries, such as India, El Salvador and Guatemala, non-governmental organizations distribute free eyeglasses or sell inexpensive models.[11]

MENA (Middle East and North Africa) countries seem to be doing better: Myopia (near-sightedness) today is on the average estimated at 15%, which is considerably lower than in East Asia, though higher than in East and South Africa.[12] Still, the percentage of people with an unmet need for presbyopia correction is more than 20% higher than its incidence.[13]

Not much can be deduced from contemporary estimates with regard to the evolving need, over centuries, for eyeglasses in the (broadly differing) countries of the Middle East. Surely, the area is today not at the bottom of the list. Nevertheless, historically speaking, the diffusion of glasses in that broad region has been slow in comparison with Europe, and we propose in this chapter to investigate why.

5.1 Evolution in Europe

Europe is the continent where eyeglasses had been invented, and where diffusion was relatively fast. A coalescence of optical knowledge, experience in polishing appropriate lenses, trial and error on applied materials and designs, and mass-production, enabled people who needed reading and working glasses, to acquire them on their own initiative, and at a relatively cheap price.

---

11 The Scojo Foundation in New York estimated in 2006 that 1.6 billion people in the world need reading glasses, but only 5% have them. See further: Case Western Reserve University, Weatherhead School of Management, 10 February 2006 on the internet. A British charity, Vision Aid Overseas (VAO), recycles glasses for Burkina Faso and other countries in Africa. Yet another charity has distributed free eyeglasses in Northwestern China (2004). An Oxford physics professor, Joshua Silver, has pioneered cheap fluid-filled "adaptive glasses" of which the Centre for Vision in the Developing World has distributed around 100.000 in 20 countries, *CBC Radio*, posted on internet, 25 June 2021.The same source estimates that three billion people in the world still need corrective glasses. In all the above cases there is no specific mention of the Middle East.
12 For statistics, see various reports by the American Academy of Ophthalmology, and the WHO *World Report on Vision* (Geneva 2019).
13 Timothy Fricke et al. "Global Prevalence of Presbyopia and Vision Impairment from Uncorrected Presbyopia", *American Academy of Ophthalmology* (2008), p. 1496.

For a long time, "primordial" rivet glasses with – more or less – colorless lenses, had to be held by hand, or loosely on the nose, while reading or doing handwork. Rivet frames were made of wood, leather, or bones, and later had somewhat awkward cords. It took time before they were replaced by more sophisticated models with fixed temples or arms, to keep them in place on the ears, or metal bridges on the nose, which date from the first half of the 18th century.[14] Since then, basic models have not changed much though materials and fashions have.

Sometimes we forget to focus on such later innovations which signify a more important breakthrough than the initial invention itself, turning a clumsy auxiliary artifact into an easily useable object.[15] Pinpointing the advent of eyeglasses in the year 1300, would therefore disregard the process by which the actual *invention* became diffused as a common *in-use* device. It was a complex interplay of factors of awareness, need, adaptation, marketing, and acquisition. Once, the benefits had been proven ("Persuasion" in Everett Rogers' terms), it was a matter of availability on the open market, albeit with time gaps between the various countries and continents.

For some time, eyeglasses had been literally *manu*factured, made by hand, by crystal makers in Venice, and even more so in Florence, both building up a wide reputation among notables who could order their glasses from these cities. However, with the migration to Western Europe of some Murano glassmakers, the Italian domination of the market was gradually lost. In the 15th century, mass manufacturing, particularly in Frankfurt, Nuremberg, and London, was already overtaking Italy.[16] This development also lowered prices and made cheaper types affordable. Though we have no exact idea of prices and price parities, costs ranged from a reasonably low two to eighteen shillings, while certain pairs with crystal lenses in gold frames amounted to a ducat (82 shillings).[17] As in Italy, the occupation of eyeglass manufacturers in many places was regulated by guilds. Regensburg in Germany had a first specialized guild of *Brillenmacher* in

---

14 Nicolas Veyrat et al., "Social Embodiment of Technical Devices: Eyeglasses over the Centuries and According to Their Uses", *Mind, Culture, and Activity*, vol. 15 (2008), pp. 185–207.
15 The crucial invention of side arms is ascribed to London opticians, Edward Scarlett and /or James Ayscough, respectively in 1730 and 1752.
16 David S. Landes, *The Wealth and Poverty of Nations, Why Some Are So Rich and Some So Poor* (New York: Norton 1998), pp. 46–49. Landes has been criticized for his Europa-centrism, but in our view, it can hardly be disputed that eyeglass remained for three to four hundred years a typical European manufacture. Landes also made a lot out of mechanical clocks as European, but waterwheels, printing presses, and gunpowder were invented in China.
17 This is the estimate of Antique Spectacles, "Eyeglasses Through the Ages", https://www.spectacleoptometry.com/blog/eyeglasses-through-the-ages. Calculations can also be made from Italian account books, see Mozzato below.

1535, followed by other countries. In France (1581) they were called *lunetiers*. England which produced eyeglasses from the 15th century, had a Spectacle Makers Company from 1629. Guilds would be dissolved after the French Revolution, and soon thereafter also disappeared in other West European countries.[18]

In fact, throughout western Europe, the acquisition of eyeglasses for disabilities other than presbyopia now became more widespread. The corrective capability of new convex lenses improved distant sight, be it for hunting or warfare and, above all, enabled some people to wear them permanently.

The quality of glass lens polishing improved over time, and the sixteenth and seventeenth centuries saw significant advances in optical science in London and other centers of Western Europe. Lens grinding for telescopes and microscopes also developed, with great names in the Netherlands: Zacharias Janssen, the brothers Constantijn and Christiaan Huygens, and Antony Van Leeuwenhoek. One may remember that the philosopher, Baruch Spinoza, after being banned in 1656 by his community, chose to make a living from lens grinding (allegedly he also made one pair of glasses for the German philosopher, Leibniz).[19] Germany, in particular, with pioneering companies such as Zeiss, would achieve a lead in quality lens-making by the nineteenth century.

For lack of proper trade statistics, researchers have mapped the process of diffusion, at least as far as Western Europe is concerned, with the help of iconographic evidence. At first, only monks and religious scholars, and some other dignitaries, were portrayed with glasses, the earliest of whom appear to be depicted on a fresco by Tommaso da Modena preserved in Treviso (1352). It represents Cardinal Hugh de St. Cher holding a pair of rivet glasses, which he could not have had at the time because they had not then been available when he passed away in 1263.[20] Such an anachronism was often applied to scenes and portraits of biblical, ecclesiastical, and other historical figures to emphasize learnedness (e.g., Vergilius). In general, the list of frescoes, oil paintings, and miniatures of persons with eyeglasses is noteworthy and varied, and their symbolism could in some

---

[18] Though guilds in Middle Eastern countries have enjoyed much research interest, I have not found the occupation there.
[19] Anita McConnell, *A Survey of the Networks Bringing Knowledge of Optical Glass-Working to the London Trade, 1500–1800* (The Whipple Museum of Science 2016), p. 89 et passim.
[20] This was followed by a fresco in Assisi of a philosopher confronting St. Catharina (from around 1360), and a painting now in Innsbruck dating from around 1370, see Ilardi, p. 18–23, p. 69, and a detailed appendix of 15th-17th century paintings, etchings, and sculptures of bespectacled personalities, many by famous artists such as Van Eyck, Brueghel the Elder, Caravaggio, El Greco and Velasquez, pp. 261–305. Also, Gotthold Prausnitz, *Das Augenglas in Bildern der Kirchlichen Kunst im XV. und XVI. Jahrhundert* (Strassburg: Heitz & Mindel 1915). For miniatures see Frugoni, pp. 20–23.

later etchings and gravures not only suggest eminence and wisdom, but sometimes also an unfavorable characteristic such as avarice, deceit, or evil.[21]

At the same time, a market for spectacles emerged as can be seen from more popular artistic representations. From the 16th century onwards, when more people wanted eyeglasses, out of objective need or to show importance and status, there are multiple etchings, engravings, woodcuts, and even stone, metal, and porcelain sculptures, which enable us to get a glimpse of eyeglass diffusion. As an artistic genre this proves that eyeglasses had by then become everyday objects. This was mainly a trade of itinerant peddlers and small shopkeepers.[22] It is said that the first specialized store for spectacles opened in Strasbourg as early as 1466. Not much is known about the initial diffusion of eyeglasses in North America, but it was allegedly a Mayflower Pilgrim, Peter Brown, who had been the first to introduce them.[23]

The growing diffusion of eyeglasses in Europe has been linked to a specific factor of the same era, namely printing, the invention of which is ascribed to Johannes Gutenberg in Mainz in 1440 (or to his first commercial printing in 1453).[24] The diffusion of Gutenberg's press, and of printed books, accelerated the growth of a public of readers, no longer restricted to privileged people consulting manuscripts. To lower the price of books, printers now began to use smaller types (resulting in fewer pages) which necessitated glasses for older people (or even younger ones with congenital disabilities). One may add that Protestantism, more than Roman Catholicism, elevated one's reading of the Bible to an obligation. Thus, beginning with the Renaissance, reading habits were progressively changing and would in the next centuries also include newspapers in rather small print. The fast spreading of literate people became a critical mass, which in turn led to a so-called Readers' Revolution of the 18th century. However, in contrast to Europe, where glasses gradually became a normative need owing to the spread of books

---

21 Grace A.H. Vlam, "The Calling of Saint Matthew in Sixteenth Century Flemish Painting", *The Art Bulletin*, vol 59/4 (1977), p. 563 and p. 566.
22 Laurence Fontaine, *Histoire du Colportage en Europe, XVe-XIXe Siècle* (Paris: Albin Michel 1993) remarkably has a print of a spectacle peddler on the frontispiece. See also *Antique Spectacles*, curator David A. Fleishman, https://www.collectorsweekly.com/hall-of-fame/view/antiquespectacles-com. (last accessed 6 March 2023).
23 For the sake of comparison, a certain John McAllister opened the first optician's store in Philadelphia in 1799. See also a lecture handout by David A. Goss (2003) on the internet. No American narrative fails to mention Benjamin Franklin as the inventor of bifocal glasses (in the 1780s), and William Beecher as having introduced silver and blue steel frames (from 1826).
24 John Dreyfus, "The Invention of Spectacles and the Advent of Printing", *The Library* 10/2 (1988), pp. 93–106.

and papers, development of printing in the Middle East was much slower and picked up steeply only as late as the 19th century.[25]

5.2 Why did the Middle East not have them before Europe?

These European head starts are all the more intriguing because the Middle East had – in theory – the potential to lead the process of developing eyeglasses. The civilization of the Middle East (as a generalization) could pride itself on early accomplishments in optical sciences, as well as on skilled glassmaking. The Italian-American historian, Vincent Ilardi, who wrote extensively on eyeglasses, asked the same question but tended to ascribe the answer to the technological decline of the Middle Eastern glass industry.[26] There might have been additional, more complex causes.

Several outstanding medieval Muslim scholars worked on the anatomy and physiology of the eye, on the refraction of optical lenses, and on the related working of mirrors. Al-Kindi (d. 873), a multi-talented philosopher from Baghdad, was among the first scholars to write on the working of the eye. Around the same time, an all-round scholar and scientist in Cordoba, called ʿAbbas ibn Firnas (d. 887) was credited with the fabrication from quartz of a sort of colorless magnifying glass which enhanced the capacity of scholars with impaired vision to read manuscripts.[27] His experiments with lenses, if they took place at all, could have led to eyeglasses but probably remained on the theoretical level. In Baghdad, somewhat later, Abu Saʿd al-ʿAlāʾ Ibn Sahl (ca. 940–1000) also excelled in theoretical knowledge on lenses and the refraction of light, as well as on burning-mirrors.

Again somewhat later, it was a great Arab scientist, who moved from Basra to Cairo, Abu ʿAli al-Hasan ibn (al-)Haitham (965–1040), who is said to have conceived – in theory – of eyeglasses (as he did of military burning-mirrors, and a sort of camera obscura).[28] Known in Europe as Alhazen, his original *Kitab al-Manazir* (Book of Optics, 1012) was lost, but a Latin translation survived and became widely known. No matter that he was wrong on several theories, e.g., the physiology of the eye, then still believed to send out beams, an antique concept, instead of

---

[25] The reasons have often been debated, Ami Ayalon, *The Arabic Print Revolution, Cultural Production and Mass Readership* (Cambridge: Cambridge University Press 2020), esp. pp. 1–17
[26] Ilardi, p. 124
[27] He also tried an Icarus-like implement to fly and luckily escaped death.
[28] It has been said that, becoming old, he designed a convex lens to be able to read; this was possibly a magnifying glass rather than reading-spectacles.

receiving them as he himself concluded.[29] The great impact of his theoretical explorations can be shown, in particular, on Roger Bacon (d. 1292), a Franciscan philosopher.[30] Bacon manufactured and recommended magnifying glasses, though not reading spectacles as they would become known in Italy.

Neither did a revival of the study of optics in Persia toward the end of the 13th century make the transition from theoretical (and some experimental) physics to an applied practice of manufacturing eyeglasses as was the case with making mirrors. This has again been explained by a lack of understanding of the exact physiology of the eye, even though it applies equally to Europe, at least before the famous astronomer, Johannes Keppler in 1604.[31]

Traditional Arab ophthalmology, as a branch of medical knowledge focusing on eye diseases, was on a high level as well, and only in the 18th or 19th century had to give way to European science. It appears that it did not identify presbyopia as an eye ailment, and this might possibly be one further explanation why Islamic scholars failed to apply their knowledge towards developing spectacles.

In addition, Islamic glass craftsmanship, in part a Roman and Byzantine heritage, and in part the authentic refining of glass-blowing techniques, notably the use of molten glass and of coloring and enameling, was on a high level. It found expression merely on the artistic level, e.g., in lamps, bowls, and other vessels. At some point, the Venetians brought Islamic master craftsmen to the island of Murano which led to a transfer of knowledge and techniques, but from the fourteenth century the said Islamic art declined. On the other hand, Mamluk and Ottoman traders must have remained aware of Italian glassmaking as they regularly sold essential alkali ashes from Syria to Venice.[32] Most of this supply was used

---

29 Born in Basra, considered to be a genius in mathematics, physics (esp. burning mirrors) and astronomy (beyond the usual *qibla* and Ramadan time fixing), he was invited to Cairo by the eccentric Sultan al-Hakim (d.1021) due to his ideas on how to regulate Nile waters, but feigned madness until after the ruler's death, and then resumed teaching at al-Azhar. He wrote over 50 treatises, see John S. Badeau et al., *Genius of Arab Civilization, Source of Renaissance* (Oxford: Phaedon 1976), pp. 138–139, also pp. 125–126, 132–134 and 151. Though most of it was very theoretical, in practice, he also worked successfully on burning-mirrors.
30 *Encyclopaedia of Islam*, 2nd ed., s.v. Manazir" (by A.I. Sabra). In the 13th century, two other Franciscan theologians, the Polish Vitello and the English John Pecham, also relied on Ibn Haytham's ideas on optics.
31 This is Ilardi's explanation for Europe, pp. 28–30. In contrast to Bacon, Keppler, three hundred years later, did understand how glasses worked.
32 Regarding the end of the 13th century, see Eliyahu Ashtor and Guidobaldo Cevidalli, "Levantine Alkali Ashes and European Industries", *Journal of European Economic History*, vol. 12/3 (1983), pp. 488, 503, 506, 513 and 518.

for artistic vessels and beads, but not, however, as far as we know, for making lenses.

Thus, the aggregate of unique heights of knowledge in optics and ophthalmology, and a well-developed glass industry, did not converge into the invention of reading glasses. And more surprisingly, when Italian and other European eyeglasses became available in the countries of the Middle East, as far as we know, the people there also did not immediately generate an urge to imitate them. Such urge to imitate was the case, for instance, in China and Japan after eyeglasses had become known there after the sixteenth century.[33]

5.3 A small technology known to select elites

The wealth of Arabic poetry sometimes yields surprising hints, such as a few verses by the Sicilian (later Andalusian and Tunisian) poet, 'Abd al-Jabbar ibn al-Hamdīs (1055–1132). The relevant passage on what could be either just a magnifying glass or real eyeglasses had been rediscovered by the Egyptian scholar, Ahmad Taymur, who discussed it in *al-Hilal* in 1919.[34] It reads as follows: "This is a frozen brook held with the hand ….what an excellent aid for the old man whose vision has become weary, and whose age has made the letters small to his eyes, with it he sees the forms of the lines becoming bigger…"[35] The poetic words used were

---

[33] Spectacles had apparently been introduced into China by Christian missionaries in the sixteenth or seventeenth century (though foreign traders might have brought them earlier) but were regularly imported from the 18th century. On copying eyeglasses by local artisans, see Frank Dikötter, "Objects and Agency, Material Culture and Modernity in China" in Karen Harvey (ed.) *History and Material Culture, a Student's Guide to Approaching Alternative Sources* (London: Routledge 2009), p. 164, and on the "spectacle culture" in China, his *Things Modern, Material Culture and Everyday Life in China* (London: Hurst 2007), pp. 215–216 *et passim*. As to Japan, usually more associated with imitations, consider Francisco St. Xavier, a Christian missionary, who might have brought the first spectacles as gifts in 1551 (or 1549), one pair of which still survives in a temple in Kyoto, though possibly earlier eyeglasses might have been introduced by the Chinese in 1530, J. William Rosenthal, *Spectacles and Other Vision Aids, a History and Guide to Collecting* (San Francisco: Norman 1996), pp. 96–97. The Japanese began to manufacture their own lenses and frames in the 17th century and borrowed more western technologies in the 19th century. A first lens-making factory dates from 1875.
[34] "'Uyun min Zujaj am Hiya al-Nazzarat?" (Eyes from glass or are they spectacles?)", *al-Hilal*, 1919, issue 12, pp. 236–239. The periodical referred to it also in a later article on the history of eyeglasses 1923, issue 2, pp. 596–597. See also below.
[35] For an excellent analysis and translation see Farid Benfeghoul, "Through the Lens of Islam: A Note on Arabic Sources on the Use of Rock Crystals and Other Glass as Vision Aids" in Cynthia Hahn and Avinoam Shalem (eds.), *Seeking Transparency, Rock Crystals Across the Medieval Medi-*

hajar for stone, and *jawhar* for jewel cannot be decisive. Both an Iraqi scholar of the 1950s, Mikha'il 'Awwad, and Lutfallah Gari, a Saudi expert on the history of Arab-Islamic technology, deduced that it must have been a pair of spectacles.[36] However, it does not seem likely that Ibn Hamdis's verses hinted at genuine spectacles for the elderly. If it were true, there is no explanation why this invention did not spread over the next two centuries, prior to its claimed appearance in Pisa.[37] On the basis of his meticulous research, Farid Benfeghoul reached in any case the conclusion that it might be the earliest Arabic reference to a rock crystal lens as a reading aid.[38]

In their article (2013), Amir Mazor and Keren Abbou Hershkovits have provided strong evidence that a certain Shams al-Din al-Tibi (1289–1355), a volatile man of many domiciles ranging from Egypt to Syria, and of many occupations, might have been the first Muslim spectacle wearer on record.[39] Tibi is mentioned in a chronicle as using "eyes of glass" (*'uyun al-zujaj*) to see small handwriting (*li-ru'yat al-khatt al-daqiq*).[40] This, indeed, was two centuries later, and is therefore more convincing. Another plausible case which the above researchers found is the mention in a chronicle of the Hanafi qadi who in 1472 asked the Mamluk Sultan Qaytbay for one of the pairs of spectacles of European origin which had come into his possession. It is referred to simply as "eyes" (*'uyun*) to see the writing (*allati yanzuru biha lil-kitaba*).[41] Whoever has tried to read Arabic, Ottoman, or Farsi manuscripts, official, or fiscal documents, and the like, hardly needs to be convinced of the benefit of spectacles.[42]

---

terranean (Berlin: Gebr. Mann Verlag 2020), pp. 237–249, esp. pp. 245–248. The words *jadwal jamid*, or frozen brook, can be taken to mean crystal. See also his "Through the Lens of Islam: the Pre-History of Eye-glasses according to Arabic Sources." In *Zeitschrift für Geschichte der Arabisch-Islamischen Wissenschaften* 23 (Sonderdruck 2022): 259–315. I am very grateful to Benfeghoul for his help and advice.

36 Benfeghoul, p. 247

37 Lutfallah Gari, "The Invention of Spectacles between the East and the West", *Muslim Heritage* (first published 12 November 2008), https://muslimheritage.com/people/authors/9725-2/ (accessed 16 December 2021).

38 Benfeghoul, p. 246.

39 Amir Mazor and Keren Abbou Hershkovits, "Spectacles in the Muslim World: New Evidence from the Mid-Fourteenth Century", *Early Science and Medicine* 18/3 (2013), pp. 291–305.

40 Mazor and Hershkovits, resp. p. 301 and p. 298.

41 Mazor and Hershkovits, pp. 297–298, quoting Nur al-Din 'Ali ibn Dawud al Jawhari al-Sayrafi (d.1495).

42 Derek Hopwood, *Tales of Empire* (London: Tauris 1989), p. 61 quotes a butler of the Khedive, John Young, speaking of a scribe as follows: "Being terribly short-sighted, as befits an oriental scholar, he raises the paper to within exactly one inch of his left eye, holding his head to one

We can only speculate that these individual items had arrived by trade, or as gifts, as was the case with many a precious European object.[43] The significance of their findings is that the lapse in time between Europe and the Middle East is not as big as was sometimes thought. These rare cases are enlightening because they are mentioned by authentic local sources. The suggestion of the two above researchers that quantities of imported lenses at the time may have been framed locally, calls however for more explicit literary or other evidence.[44]

We owe to Lutfallah Gari a few additional, equally noteworthy, references to reading glasses in historiographical and poetic sources. One concerns a vague allusion to familiarity with these glasses (if they were indeed eyeglasses), ascribed to the poet Ahmad ibn al-'Attar al-Masri (15th century) who wrote "After my youth, came my grey hairs and my (old age) / which caused my straightness to become distorted. / It was sufficient that I (once) had a sharp vision / (but) now my eyes have become of glass".[45] Another, more cryptic reference by the historian Shams al-Din al-Sakhawi, pertains to a calligrapher, Sharaf Ibn Amir al-Mardani (d. 1447): "He died at an age that exceeded a hundred years, he had good senses and continued writing even without using a mirror".[46] Did he mean eyeglasses or a mirror? We do not know.

More explicitly, the famous Cairene scholar, Jalal al-Din al-Suyuti, (d.1505) wrote in a poem "My eyes were above my cheek/ and today I put them above my nose".[47] Bernard Lewis quotes the famous Khorasani scholar and poet, Nur al-Din 'Abd al Rahman Jami, supposedly toward his death around 1492, who lamented that his eyes had become deficient "unless, with the aid of Frankish

---

side, and fixing his right eye on the carpet, throwing it as it were out of gear. He then proceeds to move his head forwards along the paper".
43 His full name was Muhammad b. Badi b. Abi Bakr b. 'Uthman b. Badi, Shams al-Din-al Tibi as mentioned in a biographical dictionary of notables from the fourteenth century, a well-known genre, by Khalil al-Safadi (d. 1363). Allegedly al-Tibi was born in Cairo and died in Beirut, having worked as a perfumer, a poet, a teacher, a reciter of *hadiths*, and a silk trader, Mazor pp. 299–302.
44 Mazor and Hershkovits, pp. 303–304. Their suggestion seems to be corroborated by Andrea Mozzato, "Luxus und Tand: Der Internationale Handel mit Rohstoffen, Farben, Brillen, und Luxusgütern im Venedig des 15. Jahrhunderts am Beispiel des Apothekers Agostino Altucci", in Christof Jeggle et al. (eds.) *Luxusgegenstände und Kunstwerke vom Mittelalter bis zur Gegenwart, Produktion, Handel, Formen der Anneignung* (München: UVK Verlag Konstanz 2015), p. 396, referring to a party of 5.000 lenses (*spechietti*) sent to Damascus in 1464.
45 With thanks to Farid Benfeghoul for helping me with the translation and pointing out the Arabic puns.
46 See Gari, above.
47 *Al-Hilal* 1923, issue 2, p. 592 (see note 491)

glasses, the two become four."[48] A study by Seyyed Hadi Tabatabaei, quoting the same poetic lines, makes clear that already in the Aq Quyunlu Turkman period (14th-15th centuries), Venetian ambassadors had made Iranians familiar with eyeglasses.[49] To what extent, in the case of Jami, awareness also meant actual possession, is a matter of speculation.

From the European perspective, an Italian researcher, Andrea Mozzato, has been able to scrutinize the detailed account books of a Venetian pharmacist-turned-trader of the 1460s and 1470s who exported large quantities of eyeglasses of varying qualities, some 90.000 altogether, to Damascus, Tripoli, and Beirut.[50] After all, Italians had also been the inventors of modern bookkeeping. Vincent Ilardi, too, used account books of tradesmen, particularly from Florence, and found impressive sales to their counterparts in Pera (Istanbul), then still a district of mainly foreigners in Istanbul where, it seems, "a great craving" for them existed: 1.100 pairs of spectacles in 1482, 529 pairs in 1484, 2400 pairs in 1540, and unknown numbers in between.[51] Even the friars of the Holy Sepulcher Church in Jerusalem knew about eyeglasses, requesting the visiting Florentine Ambassador Luigi di Angelo della Stufa in 1488 to send them a specified quantity, graded according to their ages.[52] Venetian cargo lists of the last decades of Mamluk rule and trade mention crystal lenses and other artifacts without specifying their purpose.[53]

With the changes in Mediterranean trade, we assume that thousands of pairs of glasses continued to be delivered to the Ottoman, possibly also to the Safawid, and the more distant Moghul Empire.[54] These seem to have been formidable quantities, and there is no reason to believe that the re-oriented European trade with the Ottoman Empire stopped eyeglasses, as these were in demand, and not manu-

---

48 Bernard Lewis, *What Went Wrong? The Clash Between Islam and Modernity in the Middle East* (Oxford: Oxford UP 2002), p. 127.
49 Seyyed Hadi Tabatabaei, "The Introduction of the Telescope into Iran before the Nineteenth Century" in Christopher Neil et al. (eds.) *Scientific Instruments Between East and West* (Leiden: Brill 2020), p, 142, quoting Hossein Mahboubi Ardakani.
50 Mozzato, pp. 377–405. He supplied not only Venetian and Florentine eyeglasses but also cheaper German models, and sometimes bartered with them in the countries of the Levant.
51 Ilardi, pp. 118–123.
52 Ilardi, p. 92 and p. 225. As far as we know, Florence had introduced a scale conforming to five year strengths; a first scientific table (before the usual diopters) is ascribed to the Portuguese Benito Daza de Valdes in 1623.The term, diopter, was introduced by a Frenchman, Ferdinand Monoyer, in 1872.
53 Benjamin Arbel, "The Last Decades of Venice's Trade with the Mamluks – Importations into Egypt and Syria", *Mamluk Studies Review*, vol. 8/2 (2004), p. 56.
54 Cf. Toby E. Huff, *Intellectual Curiosity and the Scientific Revolution* (Cambridge 2011), pp. 120–122, 130, and 206.

factured locally.[55] Existing statistics or customs accounts are not helpful for this sort of too trivial articles.[56] Yet we should always be alert to any incidental mention such as one in a random German source which shows that eyeglasses (and other small European items (*Kleinigkeiten*), were traded as far as Afghanistan at the beginning of the nineteenth century. [57]

If eyeglasses were imported by the thousands, it is enigmatic that so little pictorial, literary, or even archeological evidence remains.[58] While in Europe, iconographic evidence for the diffusion of eyeglasses is quite abundant, it is scant for the countries of the Middle East. Owing to tradition, or even to a religious prohibition, representations of human beings have been much rarer in the Islamic world than in Europe or Norh-America. This certainly applies to the *'ulama*, the intellectual counterparts of the church people often represented in European paintings. But it is not entirely absent. We do have a few rare images such as a portrait by the famous Safawid miniaturist, Mu'in Musawwir, of his master, Ridha 'Abbasi of Isfahan, which shows him at work with a *pince-nez*.[59] We can only hope that more similar images will be found (Fig. 26).

Here we are already in the era of the Ottomans and the Safawids, during which glasses definitely had become spread in large numbers in Europe. From the Venetian chronicler, Marino Sanudo, we know that in 1532, the Ottoman *defterdar*, the chief accountant of the Ottoman court, requested the Venetian diplomat, Pietro Zen, (with the title vice-*bailo*) to procure him a pair of spectacles, and in-

---

55 Paul Masson, *Histoire du Commerce Français dans le Levant au XVIIe Siècle* (Thèse, Faculté des Lettres, Paris 1896), p. 517, mentions 700–800 dozen eyeglasses annually toward 1700.
56 A. Mesud Küçükkalay and Numan Elibol, "Ottoman Imports in the Eighteenth Century: Smyrna (1771–72)", *Middle Eastern Studies*, vol. 42/2 (2006), p. 727 on customs registers which do mention spectacles.
57 A.C. Gaspari et al., *Volkstümliches Handbuch der Neuesten Erdbeschreibung* (Weimar: Hassel 1821), vol II, p. 730. Not long ago, a pair of spectacles in a jewelled frame and with lenses of Columbian emerald of the 17th century, coming from Moghul India, was auctioned by Sotheby's, see: https://www.sothebys.com/en/buy/auction/2021/arts-of-the-islamic-world-india-including-fine-rugs-and-carpets-2/a-pair-of-mughal-spectacles-set-with-emerald. It must have been a piece of jewelry rather than a means to correct vision. Also consider the Sultan of Mysore in southern India, around 1790, though an enemy of the British, who collected eyeglasses, as well as clocks, scientific instruments, and also possessed a printing press.
58 For archeological discoveries of eyeglasses see Ilardi, pp. 306–308, who adds occasional finds in libraries, and relics or votive offerings, pp. 311–316. In some cases, eyeglasses had been forgotten between book pages.
59 See Lutfallah Gari, above. Also, Tabatabaei, pp. 143–144, who claims on the authority of an article by Farzan Haghighi that by 'Abassi's time (1565–1635), glasses were already widely used. However, the drawing, kept at Princeton University, is dated 1673 and might be anachronistic.

Fig. 26: Earliest known Persian Safawid miniature of a painter with eyeglasses (1673): A posthumous portrait by Mu'in Musavvir of his teacher Riza Abbasi (Garret Collection, Princeton University Library, public domain).

deed received them, – a crystal specimen in a silver frame.[60] Two later similar requests are known from 1576 and 1577 which were apparently also honored with pleasure by the Venetians.[61] The usefulness of the device must have been proven;

---

60 Marin Sanudo, *I Darii*, vol. 56, p. 402. Also quoted by Otto Kurz, *European Clocks and Watches in the Near East* (London: Warburg Institute 1975), p. 21.
61 Julian Raby, "The Serenissima and the Sublime Porte: The Art of Diplomacy, 1453–1600" in Stephano Carboni (ed.), *Venice and the Islamic World, 828–1797* (New York: Metropolitan Museum of Art 2007), p. 96. The later requests seem to have been for the Sultan's tutor, and for the Grand Vizier Siyavuş Pasha.

possibly, the *bailos* or other foreigners in the capital were themselves seen in possession of a pair.[62]

Nevertheless, for a long time, eyeglasses in the countries of the Middle East may have remained exclusive valuable items. This is our conclusion from reading glasses mentioned in inheritance inventories of prominent persons. With growing research into Ottoman inheritance registers in recent decades, as well as greater attention to material culture, one discovers that eyeglasses were objects valuable enough to be specifically listed; we learn, for instance, of a high-placed Ottoman scribe, Mustafa Effendi, whose estate (1755) was analyzed by Joel Shinder. The scribe had owned spectacles with cases (as well as clocks, rifles, locks, telescopes, and a multitude of other Western objects).[63] Fatma Müge Göçek, who also investigated the diffusion of western goods among high officials and military men during the 18th century, found mention of several valuable eyeglasses.[64] One later specific case is from Karaferya (today Veria in Greece) where the well-educated Ottoman local commander of the 1870s left two pairs of glasses (as well as other precious items such as field glasses, a mirror, a watch, firearms, and books).[65] The nature of this sort of source is, of course, biased towards the elites. Thus, it is important to keep in mind that eyeglasses in the Ottoman Empire then were still exclusively owned precious objects.[66]

---

62 Huff, pp. 120–122, 130, and 206 (relying on Ilardi).
63 Joel Shinder, 'Mustafa Efendi: Scribe, Gentleman, Pawnbroker", *International Journal of Middle East Studies*, vol. 10 (1979), pp. 416–417, but we have no reason to assume that the glasses might have been bribes or in lieu of payment as the researcher hypothesizes.
64 Fatma Müge Göçek, *Rise of the Bourgeoisie, Demise of Empire: Ottoman Westernization and Social Change* (Oxford: Oxford UP 1996), p. 99 and p. 107. Admiral Gazi Hasan Pasha (d.1789) apparently owned several pairs. Estimated values varied from 960 to 300 aspers (the silver *akçe* coin), the more expensive ones had ostensibly been owned by military men who also had binoculars and telescopes.
65 Meropi Anastassiadou, "Livres et 'Bibliothèques' dans les Inventaires après Decès de Salonique au XIXe Siècle", *REMMM* 87–88 (Sept. 1999), [OCR], n. 24. To the contrary, there is no indication that an 18th century watchmaker or a qadi with many possessions left behind eyeglasses, M. Şükrü Hanioğlu, *A Brief History of the Late Ottoman Empire* (Princeton: Princeton UP 2010), pp. 30–33. For a 17th century Pasha who in his life owned field glasses, see Amanda Phillips, *Everyday Luxuries, Art and Objects in Ottoman Constantinople, 1600–1800* (Berlin: Vetler 201x), pp. 151–152. It probably remains an exception, because not in all detailed studies on estates such as by André Raymond, Colette Establet and Jean-Paul Pascal, Joyce Hedda Matthews, or Pinar Ceylan, have we found eyeglasses.
66 Neither are eyeglasses mentioned by Eminegül Karababa, "Investigating Early Modern Ottoman Consumer Culture in the Light of Bursa Probate Inventories", *Economic History Review*, vol. 65/1 (2012), an article with much attention to the gradual "democratization" of consumer goods, or by James Grehan, *Everyday Life & Consumer Culture in 18th Century Damascus* (Seattle:

In more outlying areas, eyeglasses must have remained rare for a long time. When, in 1827, the French explorer, Victor Fontanier, visited a village near Trabzon, an inhabitant, a local dyer by profession, offensively snatched the glasses of the "infidel", and smeared his face with blue paint.[67] Eyeglasses remained uncommon in more places: In Yemen as late as 1928, children of a northern tribe, upon seeing a visitor wearing them, cried out "He has two jewels in his eyes!"[68]

While archeological evidence on the use of glasses exists in Europe, where urban excavations have more than once discovered antique eyeglasses, this is thus far lacking in the Middle East. In this respect, the Mamluk and Ottoman eras are studied less than those of antiquity in the region and have so far not yielded comparable finds.[69]

5.4 From "knowledge" to "persuasion"

The question is also who in the region, or at least in the big trade centers, among those who actually needed eyeglasses, were aware of their existence and benefit. By the nineteenth century, which is our main focus in the present volume, reading and working glasses did gradually become coveted necessities for whoever – in the bigger urban centers – could get hold of them. But if they knew about their existence – even by hearsay – they still had to afford, to find, and to acquire a pair. We do not know how this worked for "the common man". No shopkeeper or peddler in the Middle East has left us a description of his merchandise, – and visual representations of people wearing eyeglasses came later.

University of Washington Press 2007). The latter found writing implements only in 16% of male estates which is an indication, p. 178n.

[67] Michael E. Meeker, *A Nation of Empire, the Ottoman Legacy of Turkish Modernity* (Berkeley: U. of California Press 2002) p. 70, quoting V. Fontanier, *Voyages en Orient Entrepris par Ordre du Gouvernement Français de l'Année 1821 à l'Année 1829* (Paris, Librairie Universelle 1829) where he discusses the incident as a contrast between the perception of technology and ignorance.

[68] Brinkley Messick, *The Calligraphic State* (Berkeley: U of California Press 1993), p. 296, quoting the traveler, 'Abd al-Wasi al-Wasi'i.

[69] With most attention going to Mamluk and Ottoman buildings and large pottery, smaller artifacts have rarely drawn attention. Bethany J. Walker found two lenses of derelict spectacles which – she thinks – might have been sold to Bedouins in Jordan by Palestinian or Syrian peddlers in the nineteenth century; see her "The Late Ottoman Cemetery in Field L, Tell Hisban" *Bulletin of the American Schools of Oriental Research*, no. 322 (May 2001), pp. 61–62 (with thanks to Tasha Vorderstrasse for the reference, and for answers by other experts, Uzi Baram, Joanita Vroom, Edna Stern, and Yoav Arbel).

With the multiplication of cultural and scientific Turkish and Arabic journals from the second half of the nineteenth century, literate (or otherwise interested) people became exposed to advertising, as well as to scientific or medical descriptions. It is difficult to follow this process. Although magazines such as *al-Muqtataf* and *al-Hilal* occasionally published on eyeglasses, interest in telescopes appears to have been greater.[70]

On the other hand, we exclude religious objections which could have impeded or delayed the adoption of eyeglasses. Or why should they have been raised at all? As far as we are aware, there is no injunction such as *bid'a* (unwanted innovation) against wearing them. However, in modern times, we find *fatwas* discussing whether spectacles inhibit the believer during prayer from touching the ground with the forehead, which is the sort of questioning which offers contemporary evidence of increased use. The permissibility of gold or silver frames, in fact a prohibition, however, is an older Shari'a issue.[71] Around the turn of the century, a British missionary, Napier Malcolm, relates that in Yazd, Iran, till 1895, Zoroastrians were forbidden by the authorities to wear eyeglasses.[72] If this is true, it needs deeper investigation: we assume that the prohibition was not primarily issued against glasses as such, being a western device, but as a discriminatory measure against the said minority.

5.5 Lesser prevalence of visual disabilities?

A further question is whether the relative rareness of eyeglasses in our historical sources of that time might be related to a possibly lower prevalence of visual disabilities in the Middle East than in Europe. If we consider that life expectancy for centuries remained quite low in the Middle East, many people might not have reached the critical age for presbyopia. A lesser need would have affected the degree of diffusion. It is, of course, a tricky argument because physical degeneration

---

[70] See further Ami Ayalon, *Reading Palestine, Printing and Literacy (1900–1948)* (Austin: Univ. of Texas Press 2004), p. 2.
[71] See *Islamweb* on the internet which has various relevant fatwas. The use of gold frames is in principle forbidden for men but allowed for women.
[72] Umbrellas and rings, too, were forbidden for them, Napier Malcolm, *Five Years in a Persian Town* (New York: Murray 1908), pp. 45–50

may then also have set in at an earlier age. On the other hand, constant exposure to sunlight is supposed to cause presbyopia somewhat earlier in life.[73]

Although we would like to think that there is no reason to assume that prevalence of presbyopia or myopia (nearsightedness) in Middle East countries differs significantly from world averages, we know today that age, urbanization, and indoor work do impact the first, and heredity the second disability. Yet, in the past, a few studies reached the conclusion that Turks and Egyptians did have better eyesight than Europeans which might suggest that glasses were, indeed, less needed. We note that such comparative questions already occupied medical specialists in the 19th century, some of whom sought to quantify such physiological conditions – correctly or not. This applied to attempts to measure eyesight in Europe and comparatively in the Middle East, – data from which we might learn something about the need for glasses. A few interesting references:

Lorenz Rigler, a prominent Austrian ophthalmologist who published a survey of the Ottoman Empire in 1852, found a significant differential of myopia between Europe and Turkey. "There are few nearsighted people in the Orient. Wearing spectacles has so far not become fashionable".[74] He ascribed this lower prevalence to a more natural upbringing as well as to the following specific factors: "... the early mental education of children, the freedom which is on the whole allowed to them till the age of fourteen, the rare use of vests and neckties, the favorable location of most living places as far as spaciousness, airing, and the view on the environment is concerned, [and] the rare employment of girls in fine handwork." Another expert, Johann Beger, the German specialist on myopia whom he cites, adds the less frequent smoking of cigars as a factor. "The wearing of glasses has thus far not yet become in fashion, and since the number of myopes is very small, also the number of wearers of glasses is very small. When the accommodation capacity of the eye begins to fail, one observes significantly more frequent presbyopia."[75]

---

73 Manuel N. Miranda, "The Environmental Factor in the Onset of Presbyopia", in Lawrence Stark and Gerard Obrecht (eds.), *Presbyopia* (Haiti: Third International Symposium on Presbyopia 1985), p. 24

74 Translations are mine. Lorenz Rigler, *Die Türkei und seine Bewöhner in ihren naturhistorischen, physiologischen, und pathologischen Verhälltnissen von Standpunkt Constantinopel's* (Vienna 1852), vol. II, p 540, this apparently on the authority of Van Millingen, *Centralblatt für Augenheilkunde,* vol. 7 (1883), p. 125. Rigler arrived in Turkey to reorganize the military hospitals and even operated on Sultan Abdülmecid, see Hirschberg, vol 10, pp. 303–305.

75 Rigler, pp. 540–541. Feza Günergun (whose paper at the CIEPO conference in London was not published) told me that a German physician who examined Ottoman recruits found a low incidence; it may have been Rigler, or Pruner somewhat earlier.

Four decades later, in 1898, Hermann Cohn, an ophthalmologist from Breslau, who was equally interested in vision acuity, especially of German school pupils, was allowed to take measurements in Egypt. Using an E-forked device in the open air as we would use our eye charts of today for literate people, he examined Bedouins, military recruits, and Nubian women, and concluded that, overall, they had better eyesight than the average person in Germany. This could, indeed, point to a lesser need for eyeglasses. Among Cairene high school pupils, however, he found more myopes, a percentage which resembled European schools.[76] It seems that today's statistics, however, do not lead to the conclusion that reading Arabic script or print increases the need for glasses, as compared to Roman script.[77]

5.6 The use and the marketing of eyeglasses in the Middle East

In one particular sense, the Middle East did not differ from Europe in that a certain division emerged between the peddling, or similar free marketing of cheaper eyeglasses, on the one hand, and the procurement of medically prescribed glasses, on the other. Obviously, differences between continents and countries, between urban and rural populations, and between Europe and America, and the Middle East continued to be significant.

From the 19th century onward, the prescription of quality glasses, not just to help presbyopia, but to correct other visual handicaps, began to shift part of the business from the market to specialized opticians' stores and medical clinics.[78] In some European and North American countries, the latter became a matter of regulation or legislation.[79] But not so soon in the Middle East.

With the changes in supply and in consumerism in the course of the nineteenth century, eyeglasses appeared in larger quantities on the market, along

---

[76] Herman Cohn, 'Untersuchungen über die Seheleistungen der Ägypter', *Berliner Klinische Wochenschrift*, 16 May 1898. See also Wolfgang Raff, *Deutsche Augenärzte in Ägypten – von Franz Ignaz Pruner bis Max Meyerhof (1831–1945)*, MD Thesis (Technische Universität München, 1984), pp. 107–112.

[77] For this we need more comparative research findings, e.g., Turkey (which has had Roman script since 1928) or China and Japan.

[78] See the fascinating article by Meltem Kocaman, "Scientific Instrument Retailers in Istanbul in the Nineteenth Century, and Verdoux's Optical Shop", in Neil Brown et al. (eds.) *Scientific Instruments between East and West* (Leiden: Brill 2020), p. 241, quoting Ami Boué, an Austrian geologist-traveler of the first half of the century (in French).

[79] Interestingly, ready-to-wear spectacles are back in supermarkets today.

with many other imported commodities.⁸⁰ Frames and lenses, now increasingly mass-produced in factories, were imported from Europe, and were procured (mainly) by newly established opticians in the big urban centers. Local commercial directories – a medium which catered to the upper classes and other elites, generally in French – began to list opticians and carried their advertisements. Clearly, glasses were an urban commodity, and professional opticians were still rare. Initially, many of these opticians were Armenians, Greeks, Jews, or other minorities and foreigners.⁸¹ At least one of them, the Armenian family firm Foyagian, sold glasses in elegant leather cases with his own imprint.⁸² Yet more systematic research is necessary on the development of this professional branch, in particular on their cooperation with medical practitioners (Fig. 27).⁸³

Trade statistics of the 19th century do not usually specify quantities of eyeglasses (though they probably do include them under a rubric of optical instruments). Nevertheless, any occasional consular report on optical goods can be helpful. Such is the elaborate USA world-wide survey for 1911, albeit with inadequate, and probably biased, coverage.⁸⁴ American eyeglasses – it says – were hardly anywhere known or available, at least not in sufficient quantities to compete with French, German, Austrian, and some British brands. It is interesting to read about the

---

80 Kocaman, p. 244, mentions an import of about a 1000 dozen pairs of eyeglasses every year for the 1870s.
81 These were usually in French and began to appear in a few large cities from the 1860s onwards, mostly annually. For Cairo see, for instance, François-Levernay (ed.), *Guide Annuaire d'Ègypte, Année 1872–1873* (Cairo, Alexandria), p. 227 and unnumbered advertising pages, where we find at least four opticians, and on pp. 329–330, with two *medicins-oculistes*. For Istanbul, for instance, see the *Annuaire Orientale (Ancient Indicateur Oriental)* which appeared from 1881; it lists at least sixteen opticians (on Verdoux see p. 168 n. 535 and below). Equally, in the *Indicateur des Professions Commerciales* for Izmir (1896) we find six opticians.
82 An undated artifact which surfaced in 2011 in a blog by an unknown collector, http://mavi boncuk.blogspot.com/2011/09/p-foyagian-constantiople-galata.htm. We suppose that it was not exclusive.
83 Gudrun Krämer in her PhD Thesis, *Minderheit, Millet, Nation? Die Juden in Ägypten 1914–1952* (Wiesbaden: Harrassowitz 1982), p. 46, cites reports of the Alliance Israélite Universelle (1881) on the Muski quarter in Cairo, listing several Jewish opticians (together with clockmakers). See further Arnold Wright and H.A. Cartwright, *Twentieth Century Impressions of Egypt* (London: Lloyds 1908), p.376, mentioning Davidson & Regenstreif, a firm of opticians from London which opened a branch in the Continental Hotel premises of Cairo in 1903. The particular entry in this handbook emphasizes cooperation with local opticians and hospitals. Prescription or recommendation by "oculists" is also hinted at in the USA Department of Commerce and Labor, Bureau of Manufacturers, A.H. Baldwin, *Optical-goods Trade in Foreign Countries*, (Washington, Government Printing Office 1911), p. 63 (hereafter: *Optical-goods Trade* 1911).
84 *Optical-goods Trade*, pp. 32–80.

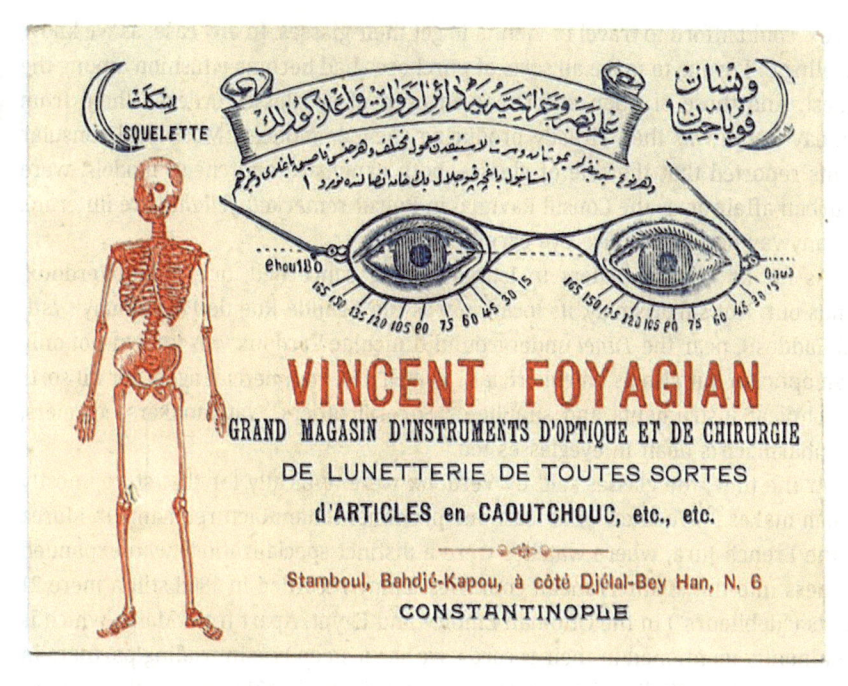

5.6 The use and the marketing of eyeglasses in the Middle East — 171

Fig. 27: Foyagian optician's store, trade card in French and Ottoman Turkish, here at Bahçe Kapı, old Istanbul, but also at Yüksek Kaldırım, Galata, Istanbul, selling "eyeglasses of all sorts", as well as optical and chirurgical instruments (hence the somewhat morbid skeleton), rubber articles, etc. (courtesy: Ara the Rat, Los Angeles).

large demand in Vienna, which is ascribed to its many educational institutions. At the same time, the report claims that prejudices against wearing glasses in the Russian countryside delayed their diffusion. Yet France, and probably Germany as well, had already enacted laws which required that corrective glasses had to be prescribed by an oculist, here in the sense of an optician.

As to Istanbul, the local American vice-consul there, Oscar Heizer, reported that eyeglasses were sold by the few established opticians, but also stocked by department stores, jewelers, and fancy goods stores.[85] More remarkable was the observation that the cheaper ones were hawked by itinerant dealers "from whom the poorer classes of the population buy a glass that will enable them to see at a distance or read print easily, quite irrespective of what their peculiar visual weakness may be".[86] Some more well-to-do people in Salonika – says the local consular agent

---

85 However, we did not encounter eyeglasses among the articles sold by the Orosdi-Back chain.
86 *Optical-goods Trade*, pp. 63–64.

there – could afford to travel to Vienna to get their glasses. In any case, as we know, traveling to Europe to make all sorts of purchases had become a fashion among the richest. And those of Aden and Oman in need of eyeglasses ordered them from Bombay (India was then already producing cheaper models). Most local consular agents reported that the sale of glasses, both expensive and cheap models, were an urban affair or, as the Consul Ravndal in Beirut remarked, *fellahin* are illiterate, and anyway do not practice eye care.[87]

As far as French dealers in Istanbul were concerned, one name, Verdoux, stands out. Not surprisingly, its location was the Grande Rue de Pera (today's İstiklal Caddesi), near the *Tünel* underground funicular. Verdoux was known not only as an optician but also as an electrician, as well as a commercial agent for all sorts of technical instruments and supplies.[88] Several others, watchmakers, jewelers, and pharmacists dealt in eyeglasses too.[89]

At the time, the glasses sold by Verdoux were, typically for that store, mostly French makes.[90] We know of at least one prominent manufacturer, Lamy in Morez (in the French Jura, where watches were a distinct specialization) who expanded business into the Mediterranean countries: Lamy recorded in 1868 still a mere 21 clients ("debiteurs") in the Ottoman Empire and Egypt. Apart from Malta, which is prominently mentioned in their records, we see a steep rise in trading partners in Ottoman lands, including Egypt, from 21 in 1887 to 243 in 1914. Among the number of Ottoman trade partners, 66 were in Beirut, 50 in Alexandria, and 27 in Istanbul.[91] Obviously, there were also competitors from Italy and other countries who exported to the same region.

One way to follow the diffusion of eyeglasses is to look at specific advertisements for glasses, as well those placed by opticians and ophthalmologists. We found in 1877, Lawrence & Mayo, two British Jewish opticians, who had established themselves simultaneously in Cairo and Kolkata. In Cairo, their store was located on the street level beneath the Shepheard's Hotel, an almost unparalleled upper

---

87 *Optical-goods Trade.*, pp. 72–73.
88 *Annuaire Orientale*, (1891), pp. 452, 459, 557, 628, 634, and 1110. Also Ernest Giraud, *La France à Constantinople* (Constantinople 1907), p. 9 and p. 21. See further Nahide Isik Demirakın, "The City as a Reflecting Mirror: Being an Urbanite in the 19th Century Ottoman Empire", (PhD Thesis, Bilkent University, Ankara 2015), p.132.
89 The *Annuaire Orientale du Commerce* of 1891 (Istanbul 1891) mentions a dozen opticians, the majority with foreign names, and almost all established in the Grande Rue de Pera or Yüksek Kaldırım in nearby Galata. We counted eight *oculists*. The profession of optician was licensed by law only in 1940, Istanbul Ticaret Odası, *Türkiye Optik Sektörü*, pp. 21–22.
90 Kocaman, p.244, mentions the Muneaux, Videpied & Cie brand, a Saint-Simonean manufacturer in Paris with progressive roots (founded in 1848, relocated to Morez 1877).
91 "Des Lunettiers Moreziens à l'Échelle du Monde: Les Fils d'Aimé Lamy (1889–1914)", p. 10.

class spot for that city.[92] We suppose that this prime location could cause some threshold fear, and anyway the perusal of an even illustrated magazine was a first hurdle to be taken.

Yet we ought to give some thought to Lawrence & Mayo's regular advertisements in the illustrated magazine *al-Lata'if al-Musawwara*, inviting would-be candidates for the civil service to step in for an eye test.[93] It gives us a hint as to literacy and upward mobility in the ranks of employment. But not more than a hint, because we would like to know more about education, literacy and need for glasses. For instance – though a very different situation –, although Ottoman authorities around that time preferred literate gendarmes, many were not, which is, no doubt, the reason why an elderly serviceman in Izmir, as an exception, is described as wearing heavy eyeglasses (Fig. 28).[94]

Information from elite biased sources, must be contrasted with an invaluable observation by Ahmed Rasim, a renowned Turkish journalist and historian, to the effect that around 1911–1912, the supply of eyeglasses was not only more diversified in the big cities, in his case in Istanbul, but prices (and supposedly qualities) were graded too: While Verdoux would sell a pair for 30 *kuruş*, adding that a doctor's prescription automatically doubled the price, the same would cost 25 in lower down-town Galata, 15 in Bahçe Kapı, in the older commercial part of the city, or by the (Galata) Bridge.[95] And there, one could buy a pair from a peddler for 8 *kuruş*.[96]

---

92 Its device was "Helping visionaries see better since 1877". In India, Lawrence & Mayo were later taken over by a certain L.H. Aliston, and thereafter by the Mendonca (Mendonsa) family, and remain active today as a chain of opticians with 44 stores, see *Business Standard*, 3 and 6 June 2013.
93 *Al-Lata'if al-Musawwara*, regularly during 1921–1923, sometimes with a drawing of a *pince-nez*. In the United Kingdom, the Education Act of 1907 had made eye examinations obligatory for pupils.
94 Nadir Özbek, "Policing the Countryside", *International Journal of Middle Eastern Studies*, vol. 40 (2008), p. 55.
95 We have not undertaken an exhaustive search for advertisements, but a few specimens from *Yeni Mecmua* of 1923, kindly supplied by Fruma Zachs, show one by a merchant named Ismail Hakki who also sold cameras, binoculars, watches, and jewelry (at Sultan Hamam near the Yeni Cami), and by a pharmacist Evliya Zadeh Nureddin (Bahçekapı). Both for metal-framed oval eyeglasses as such depicted and called *fenni gözlük*, professional glasses. A later case is Tevfik Aydın Saat, a renowned watch dealer, who pioneered the sale of eyeglasses as well in Trabzon sometime after 1889s, and moved to Istanbul in 1940, *Yüzyıllık Hikayeler* (Istanbul 2015), p. 180 (see there an advertisement with watches and eyeglasses of 1920).
96 Ahmed Rasim, *Şehir Mektupları* (Istanbul: Oğlak Yayıncılık 2005), p. 34–35, a source which I would probably not have found without Kocaman's article. For itinerant urban peddlers (*seyyar satıcılar*) see Ahmet Kal'a, *İstanbul Esnaf Birlikleri ve Nizamları 1* (Istanbul, İstanbul Araştırmaları Merkezi 1998) but none of the numerous illustrations shows one selling eyeglasses. Muhammad Jamal al-Din al-Qasimi (et al.), *Qamus al-Sina'at al-Shamiyya* (Paris: Mouton 1960 in two volumes), p. 94, writing on Damascus around 1900, devoted only few lines to the occupation of the *haddar*,

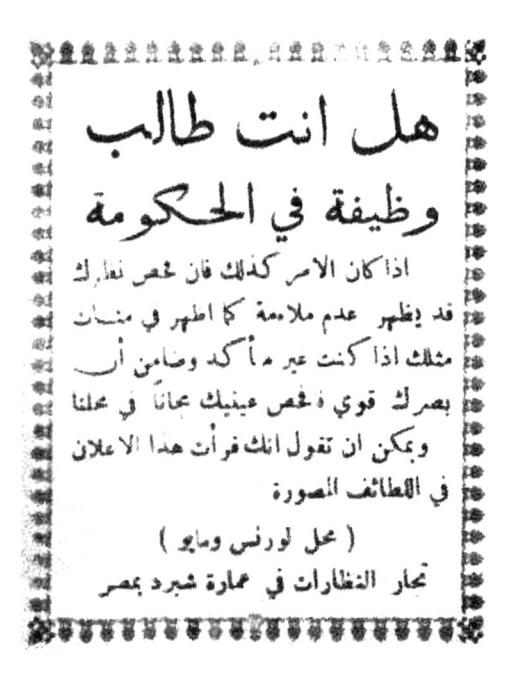

```
هل انت طالب
وظيفة في الحكومة
اذا كان الامر كذلك فان خص نظرك
قد يظهر عدم ملاءمة كما اظهر في مئات
مثلك اذا كنت غير مؤكد وضامن أ...
بصرك قوي فاخص عينيك مجانًا في محلنا
ويمكن ان تقول انك قرأت هذا الاعلان
في اللطائف المصورة
( محل لورنس ومايو )
تجار النظارات في عمارة شبرد بمصر
```

Fig. 28: "Do you seek a government job?" If you do, not convinced that your vision is bright and strong, then come in to have your eyes examined free of charge." Text of a frequent advertisement by Laurens and Mayo opticians' store at the Shepheard's building in Cairo (*al-Lata'if al-Musawwara*, 18 Dec 1922).

Apart from the grading of the urban quarters, this is an invaluable indication because itinerant peddling of modern objects in street markets often escapes our view owing to lack of sources, and our usual focus on big merchants. Unfortunately, nothing is known about the kind of cheaper makes on the market, and their places of manufacture.

Similarly, the English traveler, Douglas Sladen, observed in 1910 that in the Muski district in Cairo, many stalls sold spectacles which were "almost as much part of the costume as watches and chains", which may seem a bit of an over-statement. But it does confirm that peddlers in several cities supplied part of the demand for glasses.[97] He also mentioned that a Turkish [sic] merchant in the Khiya-

---

who usually sells small merchandise in villages (in Egypt also books as we know from Sayyid Qutb's autobiography).

97 Cf. Reynolds, p. 142–143 on *mutayawwalin* selling glasses in the 1940s in Cairo. Stalls with sunglasses have today remained common.

miyya bazaar area kept change and spectacles, probably his own, in his snuff box.[98]

5.7 Photographic evidence

While pictorial evidence on wearing glasses was, by and large, lacking in Muslim lands – leaving us with a relative gap in our sources – some 19th century European Orientalist painters, in fact outsiders, have portrayed "aged" scribes with eyeglasses.[99] This, too, indicates that by then reading glasses were already commonly used at least by literate professionals.

The advent of photography may enable us to fill the gap – of course. only from the mid-19th century onward when that invention reached the Middle East. Going through early photographic albums depicting prominent persons, some are seen to be wearing eyeglasses. It stands to reason that these were worn permanently, on prescription, not as reading glasses. A famous studio photograph of Midhat Pasha (1822–1883), the great Ottoman reformer, shows him with oval, metal-framed eyeglasses (Fig. 29). Though not specifically relating to that modern fixture, Stephen Sheehi suggests that it was "an *imago* of the Ottoman ego-ideal", hung in state offices, and printed in journals, newspapers and encyclopedias.[100] Wearing glasses could nevertheless, at the time, still have been exceptional: Among a total of 96 deputies in the first parliament of 1876, maybe only three can be counted as such.[101] Nancy Micklewright has meticulously analyzed the portrait of an anonymous senior Ottoman civil servant, however without noticing his rimless – then up-to-date style – glasses. On the other hand, she speculates that the portrait was possibly commissioned to emphasize the arrival of the then innovative candlestick

---

[98] Douglas Sladen, *Oriental Cairo, the City of the 'Arabian Nights'* (Philadelphia: Lippincott 1911), pp. 68, 70, and 90–91.

[99] One image of an *arzuhalci*, a public scribe by the Maltese painter and traveler, Amadeo Preziosi, (1816–1882) has often been copied, see the cover of Yuval Ben-Bassat, *Petitioning the Sultan, Protests and Justice in Late Ottoman Palestine* (London: Tauris 2013). Also p. 52, an unidentified engraving published in 1844. On the internet several more such "picturesque" portraits of bespectacled scribes can be found by Antonio de Dominicis, Fausto Zonaro, and Martinus Rørbye.

[100] Stephen Sheehi, *The Arab Imago, A Social History of Portrait Photography 1860–1910* (Princeton and Oxford: Princeton University Press 2016) pp. 153–156 (the portrait itself, taken around the late 1850s, possibly in France, is on p.155).

[101] For their photographs see T. Cengiz Göncü, *Belgeler ve Fotoğraflarla Meclis-i Mebûsân (1870–1920)* (Istanbul: 2010), pp. 114–115. Midhat's photo is on p. 72. For two more examples see Celaleddin Arif Bey (1875–1930), an Ottoman and later Turkish politician from Erzurum, on p. 198, and Dr. Besim Ömer Pasha, a politician and later famous gynecologist, on p. 206.

telephone on the picture (Fig. 30).[102] Ibrahim Hakki Pasha, Grand Vizier from 1910–1911, is also photographed in official uniform with a *pince-nez*.[103] The last reigning Sultan Mehmed Vahideddin (1918–1922) wore one, too.

Apart from the speculation that photographers using a flash may have wanted to avoid disturbing reflection, we might hypothesize several perfectly human explanations for this relative rareness: people simply did not want to be depicted with glasses. After all, also in Europe, men still used monocles and ladies discreetly used lorgnettes. But such prejudices disappeared in the course of time. On the contrary, the desire to wear glasses became a marker of modernity, elegance, and fashion.[104]

5.8 Ophthalmology

It would seem that medical advances did not directly contribute to the spread of eyeglasses because ophthalmology remained focused on eye diseases. This had been the case ever since the western scientists who had accompanied Napoleon at the end of the 18th century had been mostly interested in specific health aberrations among the Egyptian population, e.g., widespread trachoma, inflammation of the eye, and the effects of syphilis.[105] Eye diseases in Egypt remained rampant, caused by neglect, poor hygiene, and sand, often leading to blindness, but it took at least a century before public and charitable institutions took over from army hospitals.[106] From the mid-century, young students from Turkey, Egypt,

---

102 Nancy Micklewright, "Alternative Histories of Photography in the Ottoman Middle East" in Ali Behdad and Luke Gartlan (eds.), *Photography's Orientalism: New Essays on Colonial Representation*, (Los Angeles: Getty Research Institute 2013). In her earlier "An Ottoman Portrait", *International Journal of Middle East Studies*, vol.40/3 (2008), pp. 372–373, she had not yet paid attention to the glasses. We suppose that the photograph dates from around 1910 (till 1908, telephones had been held up, see the typewriter narrative in Chapter Four).
103 Carter Vaughn Findley, *Ottoman Civil Officialdom* (Princeton: Princeton University Press 1989), p.198.
104 This is the later (1970s) confession of a Kuwaiti boy scout, Saif Abbas Abdullah Dehrab who borrowed his brother's glasses, see Elizabeth Warnock Fernea (ed.), *Remembering Childhood in the Middle East, Memoirs from a Country of Change* (Austin, University of Texas Press 2002), p. 185
105 Marianne Wagemans and O. Paul Van Bijsterveld, "The French Egyptian Campaign and its Effects on Ophthalmology," *Documenta Ophthalmologica*, vol. 68/1–2 (1988): pp. 135–144.
106 Hirschberg, vol. 10, p. 338, quotes the Irish physician, Richard Madden, to the effect that none of the *fallahin* recruited into the army in the 1820s had two healthy eyes.

Fig. 29: Portrait of the famous reformer Midhat Pasa (ca. 1877) (Felix Nadar Archives, Paris, public domain)

and elsewhere were sent to Germany, France, or Austria, to specialize in ophthalmology.[107]

Mid-nineteenth century saw a tremendous ophthalmological advancement in Western Europe, to the extent that it became one of the first genuine medical specializations. This was due also to the introduction of new diagnostic devices, which were the result of advances in physics: e.g., the ophthalmoscope (1851), and also the Snellen chart (1862) to measure vision.[108] Innovative equipment enabled more precise diagnoses and prescriptions, which shifted their dispensing from the market to the clinic. Thus far, studies on the transfer of new medical knowledge from Eu-

---

[107] Serge Jagailloux, *La Médicalisation de l'Égypte au XIXe Siècle (1798–1918)* (Paris: Synthèse 1986), p. 65, mentions Mustapha al-Subki, Husayn ʿAwf and Ibrahim Dasuqi.

[108] The standard chart was developed by the Dutch ophthalmologist, Henri Snellen, but needed an adaptation for illiterate populations. Indeed, such a chart was introduced by E. Landolt only in 1899, and Arabic charts even later, see M.H.M. Emarah, "Arabic Eye Types for the Determination of Visual Acuity", *British Journal of Ophthalmology* 52 (1968), pp. 489–491; S. al-Khatabi and A.O. Danuton, "Arabic Visual Acuity Chart for Vision Examinations", *Ophthalmic and Physiological Optics* 14/3 (1994), pp. 314–316.

Fig. 30: Portrait of an unknown high official (ca. 1910) (permission: Getty Research Institution, Los Angeles).

rope to the Middle East have paid little attention to the new technical devices which were eclipsing the erstwhile supremacy of traditional Middle Eastern practice.

Around 1900, in Damascus, al-Qasimi's exhaustive compendium of traditional occupations, which was biased against (most) western innovations, still highlighted traditional practices (e.g., the ancient *kahhal* eye healer associated with black *kuhl* (kohl), an application recommended by the Prophet Muhammad). Qasimi often, deliberately, it seems, ignored western professions.[109] However, western innovations in the field of ophthalmology were known to a select public. This is attested, for instance, by a textbook *Kitab Sihhat al-'Ayn* (The Book of Eye Health) which appeared, with official sanction, in Beirut in 1890. Its author, Dr. Shakir al-Khuri, surveyed at length the state of the science of ophthalmology with regard to diseases and aberrations, including the use of spectacles (called *'uwaynāt* from *'ayn*, for

---

[109] Qasimi, *Qamus*, pp.385–386, biased as he is against modern occupations, mentions the *kahhal* as a traditional healer of eye diseases, ignoring eyeglasses. All over the region such eye-practitioners could be found in bazaars and coffee houses; some were itinerant.

eye).[110] No need to explain their use, as the author himself is depicted on an inner page of his book, wearing rimless glasses. This indicates that, by then and there, the prescription of corrective spectacles had entered the medical profession.

Overall, it might be that ophthalmologists, both foreign and local, were more interested in treating eye diseases than in fitting optical lenses to patients. A prominent Turkish specialist, Esat Isik, trained in Paris, who established a specialized clinic in 1899 in Istanbul, deserved praise for having developed a new type of ophthalmoscope but this, too, was a device to diagnose eye diseases.[111]

At the same time, it is interesting to investigate the impact of certain European specialists who had established themselves in Egypt or Turkey. Take the work of the German ophthalmologist, Dr. Edwin van Millingen, who had come to Istanbul in the 1870s and published several statistical reports on eye afflictions. Remarkably for his main professional activity, Van Millingen also performed innovative cataract operations with glass implants.[112] In 1880–1881, however, he did examine cases of myopia. As we have seen, he ascribed the rareness of this deficiency among the Ottoman population to cultural factors, with school-going Greeks amounting to about a quarter of his sample.[113]

Another learned western eye specialist in the Middle East was the German-Jewish Max Meyerhof in Cairo, who published over 300 articles on local eye diseases, relying extensively on the history of Arab medicine. The titles of his few articles on eye corrections would indicate that he considered myopia as a hereditary disease in Egypt, but in 1910 he wrote that very few Egyptians who should or who

---

[110] Shakir al-Khuri (spelled as Chaker Khouri), *Kitab Sihhat al-'Ayn* (Beirut: al-Matba'a al-'Umumiyya al-Kathulikiyya, Beirut 1890), pp. 170–216 on spectacles. Al-Khuri was a graduate of the Medical School in Cairo as well as licensed by the Higher Medical School in Istanbul.
[111] *Bilim Tarihi*, "The Essad Ophthalmoscope", http://www.bilimtarihi.org/OBA/2007-08_9-1-2.htm.
[112] Julius Hirschberg, *History of Ophthalmology* (Bonn: Wayenborgh Verlag 1991), vol. 7, part 2. Innovative, as Islamic ophthalmology knew how to perform cataract operations, albeit with poor results often ending in blindness.
[113] Clearly, this sample (95) half of which consisted of foreigners, reflected the practice of his private clinic (*Augenheilanstalt*), Edwin van Millingen, *Bericht über die Jahre 1880 und 1881* (Salzburg: Selbstverlag 1883)), p,.20. See also his earlier *Tri-annual Report of 5703 Cases of Eye Diseases Seen and Treated in Private Practice at Constantinople in 1877, 1878 & 1879* and also a *Tabular Analysis of 1.118 Cases Treated at the Imperial Naval Hospital*. On the other hand, van Millingen is quoted sweepingly saying that "among the Mohammedans, who smoke excessively but do not drink, he never observed scotoma" (degenerative visual acuity), as attested by Emil Gruening from New York, *Transactions of the American Ophthalmological Society*, vol. 13/2 (1913), p. 519.

could be helped with eyeglasses, actually turned to a doctor to have them prescribed.[114]

During the nineteenth century, many foreign charitable eye clinics had been established all over the Middle East, in particular in Egypt and in Turkey, but also in Palestine, e.g., St Johns in Jerusalem, the Scottish Mission in Tiberias, and others at the beginning of the 20th century such as Hadassa in Jerusalem. Even campaigns were held to combat eye diseases, such as in the 1930s at the annual Nabi Rubin festival in Palestine. But we do not know of eyeglasses which were dispensed.

5.9 Iconic figures

We are aware that the acquisition of spectacles without a pressing medical need, either tinted lenses or gold-rimmed frames, might have become a sign of elegance or status among intellectuals, giving them a modern appearance.[115] From one popular account we quote a statement to the effect that in Istanbul, *monocles* – a feature of aristocracy – which had first been introduced by Germans, and more so with the arrival in the early 1920s of White Russian refugees who were seen working in odd jobs, became a symbol hinting at their erstwhile status.[116] In any case, the phenomenon of wearing glasses for the sake of outward appearances was known not only in the Middle East.

Before continuing our discussion of the gradual diffusion of glasses from the late nineteenth century onward, we must dwell on a few iconic eyewear users in Egypt. The first one is Ya'qub Sanu', an activist journalist, playwright, and satirist, also known as the publisher of a journal entitled *Abou Naddara*, literally "the man

---

114 For a short biography and his extensive academic bibliography see Joseph Schacht. "Max Meyerhof", *Osiris* 9 (1950), pp. 6–32. He worked in Egypt intermittently from 1900 to 1945, and upon Hitler's ascendancy even acquired Egyptian citizenship.
115 In China, spectacles were sometimes acquired to show wealth or erudition, to gain scholarly status and respect, rather than to help vision, Dikötter, "Objects and Agency", pp. 164–165. Or as *Optical-goods Trade* 1911, 77, observed: "Members of literary and official classes have worn them for centuries, recently joined by merchants and others, "sometimes as aids to vision, but more frequently for ornament".
116 Jak Deleon, *The White Russians of Istanbul* (Istanbul: Remzi Kitabevi Publications 1995), p. 34. On the other hand, *monocles* were ridiculed in satirical papers, see Palmira Brummett, *Image and Imperialism in the Ottoman Press, 1908–1911* (Albany: SUNY Press), pp. 133 (Mehmed Caved, the Minister of Finance), p. 143 (a cartoon of Deputies), and p. 201.

with the glasses", which made its first appearance in Cairo in 1878.[117] His attacks on the Khedive Isma'il would soon lead to his exile in Paris where he undauntedly continued to publish the journal. In Egypt it then circulated clandestinely, under a dozen changing titles to prevent confiscation.[118] Much has been written on Sanu', but what interests us here is obviously the meaning of the eyeglasses in the title of the paper. It stands to reason that a pair of glasses was chosen to signify Sanu''s intention to take a closer, sharper look at the events in Egypt (Fig. 31).[119] As far as we can ascertain, the drawing of an oblong model of eyeglasses in the headline refers to a model which had been introduced in England around 1837 and became a recognized commodity for the rest of the nineteenth century.[120] One or two extant photographs of unknown date do show Sanu' himself wearing such glasses.

Some later issues of Sanu''s paper carried the title *Abou Naddara Zarka'*, the Man with the Blue Glasses. Though the model which we mentioned, was made of blue steel wire, the color refers to the lenses themselves. This asks for some further elucidation. In the 1660s, the famous French traveler, Jean de Thévenot, had already observed in Malta that *kaights* (*qa'id*s, commanders, or knights?) there, wore green spectacles, supposedly against the sunlight.[121] Blue or green tinted lenses, as a matter of fact, had been recommended to relieve eye problems by James Ayscough, the English optician of the mid-eighteenth century, who had ear-

---

[117] The periodical carried in the title the French trans-literation *Abou Naddara* while the conventional scholarly transcription would be Abū Naẓẓāra (Zarqā'). We use here the transcription from the title. See further Ami Ayalon, *The Press in the Arab Middle East* (Oxford: Oxford UP 1995), p. 146 who discusses its wide circulation.
[118] Irene L Gendzier, *The Practical Visions of Ya'qub Sanu'* (Cambridge Mass: Harvard UP 1966); Keren Zdafee, *Cartooning for Modern Egypt* (Leiden: Brill 2019), pp. 6–7, 56–57 *et passim*.
[119] It has also been suggested that Sanu' ultimately chose this title after consulting Jamal al-Din al Afghani and Muhammad 'Abduh, neither of whom wore spectacles, see Doaa Adel Mahmoud Kandil, "Abu Naddara, the Forerunner of the Egyptian Satirical Press", *Journal of the Association of Arab Universities for Tourism and Hospitality*, vol. 13/1 (2016), p. 11: "He revealed his intentions to look into the shady zones of the Egyptian society in one of its issues that was later issued in Paris saying that at the beginning of each month, he used to sit at his observatory in Paris, put on his glasses and direct them towards Egypt and watch people from there and write down their news".
[120] See a description in the British Optical Association Museum, https://www.college-optometrists.org/the-british-optical-association-museum (last accessed 6 March 2023). Wire drawing was developed since 1837, probably blued steel was an innovation as well. The illustrations in the papers could have been his own drawings.
[121] Jean de Thévenot, *The Travels of Monsieur de Thévenot into the Levant*, transl. part I (London 1687), p. 6. Blue glasses may have become in fashion even earlier, e.g., in Spain, see Bohne, p. 188. By the way, the lenses of the earliest eyeglasses were greenish as well.

ABOU-NADDARA

ORGANE DE LA JEUNESSE D'EGYPTE

5ᵗ Année   Numéro 1

RÉDACTEUR EN CHEF : James SANUA, 48 Avenue de Clichy à PARIS.

Fig. 31: Title page of the Abou Naddara satirical magazine, here vol. 5/1 (1881), always with the glasses of the publisher (Paris, BULAC)

lier been among the "inventors" of side arms.[122] A French officer in the English campaign of 1807 in Egypt used dark spectacles (and eyewash) to protect himself against acute infections which were common at the time.[123] It seems that Ibrahim Adham Pasha, the government inspector of schools in Egypt had begun wearing such green- or blue-colored glasses around the 1840s, and – more noteworthy – that by the 1860s the Khedive also wore them on inspection tours.[124] Furthermore, in the 1880s in Istanbul, even Sultan Abdülhamid bought at least two pairs of blue tinted glasses from Verdoux.[125] In fact, blue, green, or yellow, and even smoked crystal glasses as protection against the strong sunlight, may be considered as precursors of our present-day sunglasses.[126]

---

[122] Colored glasses were known since 1380 or 1420–30, says Ilardi, p. 66, but a random search on the internet indicates that green spectacles were rare enough to draw attention; see the obituary for a certain John Pinkerton in Paris (d.1826), *The Annual Biography and Obituary for the Year 1827*, vol. 11 (London: Longman 1827); or *Bentley's Miscellany*, vol. 16, a magazine edited by Charles Dickens and others (London 1844), vol. 16, p. 249 on an eccentric Irishman, Dr. Fogy.

[123] Max Meyerhof," A Short History of Ophthalmia During the Egyptian Campaigns of 1798–1807", *The British Journal of Ophthalmology* (March 1932), p. 141.

[124] Timothy Mitchell, *Colonizing Egypt* (Berkeley 1988), p. 26. On Ibrahim Adham, see also p. 69

[125] Kocaman, pp. 249–250. The court became a big client of Verdoux, see pp. 253, 254, 255.

[126] *Optical-goods Trade* 1911, pp. 88–89.

Anecdotally, the title *Abou Naddara Zarqa* had allegedly occurred to Sanu' when donkey-boys addressed him as "You, with the blue glasses, hire my donkey!"[127] We do not know whether his use of a colored type of glasses originated in a genuine medical prescription.[128] Irene Gendzier, one of the earliest biographers of Sanu', has hinted that as a child he had in fact contracted an eye disease. Rather than Timothy Mitchell's interpretation of "an invisible gaze, the ... ability to see without being seen", we take it as another marker of the European impact.[129]

Given Sanu''s satire, was there anything funny or negative about eyeglasses?[130] One can find 15th and 16th century images in Europe of people with absurd large spectacles which hint at senility or foolishness.[131] But here we encounter them as a sign of education, intelligence, and sharp judgement. Take an advertisement by Nestlé in Turkey to promote their baby milk, depicting a dark man with a fez and – unexplained – eyeglasses.[132] One would extend this question to a famous Egyptian caricature, more than half a century later, that of al-Misri Efendi. This hybrid representation of an "average" Egyptian was "invented" around 1930 by the newspaper *Ruz al-Yusuf* (Fig. 32).[133] A particularly short, round man with a touch of the dandy, a western suit, a necktie but, on the other hand, with a fez and a rosary. And besides – most conspicuous for our argument –, with an oversized pair of glasses, with heavy horn-rimmed lenses, as if to emphasize his ability

---

127 Matti Moosa, *The Origins of Modern Arabic Fiction* (Boulder, 2nd ed. 1997), p. 42.
128 Gendzier, p. 139.
129 Mitchell, p. 26.
130 Cf. Brummett, pp.133–134. But the Americans, too, turned the Japanese general Tojo with his round tortoiseshell glasses into a (negative) caricature, Ian Buruma, *Inventing Japan 1853–1964* (London: Weidenfeld and Nicolson 2003), p. 147. In Palestine/ Israel, slang expressions such as *abu-arba'* or *mishkefofer* (spectacled) were pejorative.
131 For a reference to subsequent German editions of Sebastian Brant's satirical *Narrenschiff* (from 1494) with representations of a stereotyped bespectacled "book fool" (in some editions by Albrecht Dürer), see Manguel, pp. 296–299; for later see Jean-Claude Margolin, "Des Lunettes et des Hommes ou la Satire des Mal-Voyants au XVIe Siècle, *Annales. Economies, Sociétés, Civilisations.* 30/ 2–3 (1975). pp. 375–393. Also, Dreyfus, p. 98.
132 Yavuz Köse, *Westlicher Konsum am Bosporus, Warenhäuser, Nestlé & Co im Späten Osmanischen Reich (1855–1923)* (München: Oldenburg 2010), p. 407.
133 Zdafee, pp. 123–136 *et passim*. Al-Misri Effendi was apparently "invented" by the Armenian-Egyptian illustrator Saruhian, though there are other opinions too. In any case, competing invented figures, e.g. Mishmish Effendi did not wear glasses. Neither were the iconic *Juha* or *Ibn al-Balad* characters depicted with glasses. For Ibn al-Balad see Sawsan El-Messiri, *Ibn al-Balad, a Concept of Egyptian Identity* (Leiden: Brill 1978), and Ahmad Amin, *Qamus al-'Adat wal-Taqalid wal-Ta 'abiral-Misriyya* (Cairo: al-Majlis al-A'la lil-Thaqafa, ed.1999), pp. 75–77. However, in Turkey as well, the satirical journal *Lala* (1910–1911) had a "Chic Bey", a dandy with fez, cane, and glasses, see Brummett, pp. 408–409.

to read.¹³⁴ The latter attribute was then still above average in Egypt. For that reason, the more down to earth stereotype of Ibn al-Balad (literally son of the country or the place) would somewhat later become more fitting.

Fig. 32: al-Misri Effendi, cartoon, front page of *Ruz al-Yusuf*, 12 June 1926.

Leafing through the Egyptian illustrated weekly *al-Lata'if al-Musawwara* in the 1920s– we have not systematically investigated Ottoman parallels but they can probably be found as well – one can find cartoons, or we would today say series of comics, as another iconic source which proves that eyeglasses had become a regular fixture. When not representing real persons, e.g., Prime Minister Muhammad Tawfiq Nasim Pasha who indeed did wear glasses, such images depicted archetypical *effendi* (schooled) politicians, doctors, journalists, booksellers and the like.¹³⁵

---

134 Israel Gershoni and James P. Jankowski, *Redefining the Egyptian Nation 1930–1945 (Cambridge: Cambridge University Press 1995)*, pp. 8–10.
135 On Nasim Pasha, see the issue of 26 March 1923. For an effendi getting annoyed by the newly installed telephone service, see On Barak, *On Time, Technology and Temporarily in Modern Egypt* (Berkeley: University of California Press 2013), p. 222, from *al-Fukaha* of 31 October 1928.

One thinks of two additional iconic Egyptian figures, far from being caricatures, firstly the blind writer and minister of education, Taha Husayn. His handicap did not impede his career, but he was always depicted with deep black glasses as if to prove the point. Even on posthumous postage stamps. Similarly, Umm Kulthum was often portrayed wearing elegant sunglasses. It was said that this was meant as a protection against strong stage lighting, which is hardly convincing because, usually, she did not wear them while performing. It might have been a means to lend her respectability, instead of a face veil, like the contemporary habit of women in the Gulf or in Iran wearing sunglasses.[136]

5.10 Some conclusions

From the turn of the century, many prominent persons wearing glasses can be seen in the illustrated press, in Egypt as well as Turkey and elsewhere in the region, including the Khedive 'Abbas Hilmi, and the nationalist leaders, Muhammad Farid and Muhammad Husayn Haykal. By the 1920s or the 1930s, the diffusion of (permanently worn) eyeglasses began to reach the state of "Confirmation" according to Everett Rogers' model, at least in the higher and middle classes in urban centers. One way to follow the progressive diffusion of eyeglasses would be to check the illustrated press, early photographic albums, and biographical collections with portraits: Have a look at a picture of King Fuad's cabinet in 1927: of twelve ministers, two wear glasses. As a western impact, the advent of movies with bespectacled heroes may – as elsewhere – also have contributed to the acceptance of eyeglasses.

Le Mondain Égyptien (edition 1939), an annual Who's Who for the elites, contains nearly 600 headshot portraits of notables; by our superficial count, 7.4% of the Egyptians are seen wearing eyeglasses, as against 6.3% of the foreigners.[137] A check on a Palestinian compendium, al-Shakhsiyyat al-Filastiniyya (1948, thus relatively late), shows 142 portraits, of which fourteen people with glasses (among them three physicians, two merchants, two journalists, one teacher, one (religious) shaykh, and one blind person).

---

[136] Cf. Thorstein Botz-Bornstein, *Veils, Nudity, and Tattoos: The New Feminine Aesthetics* (Lanham: Lexington 2015). Sunglasses, however, are a topic to themselves and are beyond our discussion here.

[137] E.J. Blattner (ed.), *Le Mondain Égyptien* (Cairo), was an annual. Admittedly, this is only an indication, not a sound statistical criterion. Most people wore metal framed spectacles, with only a couple of them (princes and high-placed foreigners) wearing a *monocle* or a *pince-nez*.

Eyeglasses were no longer solely a sign of modernity, as was maybe still the case before the 1940s, or a proof of literacy, intelligence and cleverness, but a normative need.[138] Though we would like to have more statistical proofs to support our assumption, it seems that the countries of the Middle East had narrowed the gap: We presume that the comparatively slow mass diffusion of eyeglasses in the Middle East had not been caused by lack of awareness, shortages on the market, or a prohibitive price, but by the fact that printing and literacy had come late, and that their local manufacturing came even later.

---

138 Nancy Reynolds describes a saleswoman who wears modern clothing and eyeglasses and a watch, p. 265. As to spectacles lending a clever appearance, we may draw attention to (later) comics. The popular youth magazine *Majid* (Abu Dhabbi, 1980s) featured a girl Zakiyya with glasses ostensibly to emphasize her cleverness, Allan Douglas and Fedwa Malti-Douglas, *Arab Comics, Politics of an Emerging Mass Culture* (Bloomington: Indiana UP 1994), pp. 155–156.

6 Did the piano have a chance in the Middle East?

This chapter explores the piano as another small technology, maybe somewhat irreverently, as a mere "tool", an instrument to make music.[1] Yet its success in Europe and America not only pertained to performing manifold forms of music but also to teaching, composing, sustaining song and dance, or accompanying theatrical and gymnastic activities. Its increasing sophistication, moreover, encouraged virtuosity and – aided by music notation (which was in fact another technology) –, created a growing repertoire of its own, albeit mostly – but not exclusively – remaining within the orbit of Western urban forms of music.

Though the piano became familiar all over the world, its diffusion depended not only on affordability and space, but also on its contextual utility and adaptability, be it in the public sphere or domestically. Far from becoming ubiquitous in the sense of full integration in all cultures, the piano therefore has different narratives in different countries.[2] Sometimes called the "Queen of instruments" for its imagined fascination and grace, in the Middle East at least, "she" failed to get the allegiance of most music-makers and lovers.

Thus, in spite of its broad geographical diffusion, the piano remained anchored in European bourgeois ideals. This is exactly the reason why – beyond the continent of its birth –, it became a status symbol, and consequently a requirement for refined education, in particular, for higher and upper middle-class girls.[3] It did,

---

[1] A pristine version of this chapter was given as a lecture on 3 May 2015 at the staff seminar of the Department of Middle Eastern History, University of Haifa.
[2] Consider China, Japan, and India, each of which has – to some extent – appropriated the piano. In China (which maintains a pentatonic musical scale), after the Cultural Revolution, the piano struck roots in its emerging middle class. Mao Ze Dong's widow, Jiang Qing, liked the piano, and saved it, though paradoxically she disliked western music, calling it "a coffin with the rambling bones of the bourgeoisie". See Richard Kraus, *Pianos and Politics in China* (New York 1989), p. vii. Also, Veronica Gaspar, "History of a Cultural Conquest: The Piano in Japan", *Acta Asiatica Varsoviensa*, no. 27 (2014), pp. 1–15. Apart from this, China today is the world's largest piano producer, followed by Japan which exports quality instruments all over the world (e.g., Yamaha, Kawai brands). No place in the Middle East produces pianos.
[3] See various social histories, e.g. Arthur Loesser, *Men, Women and Pianos* (London: Dover 1954), and James Parakilas, *Piano Roles, Three Hundred Years with the Piano* (New Haven: Yale UP 1999. See also paintings by Renoir and by others. In terms of gender, keyboard skill was often supposed to be a female ability, see Laura Vorachek, "'The Instrument of the Century': the Piano as an Icon of Female Sexuality in the Nineteenth Century", in *George Eliot-George Henry Lewes Studies*, vol. 38/39 (2000), pp. 26–43; on the later association with typewriting, see Friedrich Kittler, *Gramophone, Film, Typewriter* (Stanford: Stanford UP transl. 1999), pp. 194–195. The piano differs from the violin

https://doi.org/10.1515/9783110777222-007

to some extent, trickle down the social ladder, from the ruling courts and from the most affluent, to aspiring middle classes with their schools and institutions, but its costs and requirements of space, hampered the acquisition.

As to Europe, not a few outstanding sociologists have given us insights into its contextual significance and class distinction, e.g., Theodore Adorno, who was an accomplished pianist himself, and Pierre Bourdieu who placed the piano on top – and the accordion low – on his "cultural capital" scale of the 1960s.[4] But such scholarly studies on the piano hardly touch upon non-European contexts. In the countries of the Middle East, the piano turned out to be nearly incompatible with regional musical traditions.[5] As such, it proves that not all Western innovations enjoyed broad acceptance among the populations of the Middle East, and we will see why.

Historians are habitually more interested in successes than in failures. We think, nevertheless, that the piano's social history in the Middle East deserves a closer look.[6] It was not a great success but also not a total failure or a "non-story". This chapter proposes to highlight some notable junctures of acceptance, mainly in the region's upper classes, – those which aspired to look and behave like their counterparts in Europe. The cultural tensions involved in that process are well-known. Thereby we focus on major urban hubs such as Istanbul, Cairo and Alexandria, and Beirut – with their similarities and differences –, taking additional glimpses at other urban centers in Palestine and in Iran.[7] But it is at the

---

in its class setting, being more expensive, and needing more space. Violins, on the other hand, were more mobile, a reason why they have often been associated with migrating populations such as Jews and Roma.

4 Pierre Bourdieu, *La Distinction* (Paris: Minuit 1979), pp. 140–141.

5 Occasional anti-music trends in Islamic environments are irrelevant here (as, for instance, the prohibition to depict human beings which was not consistently heeded everywhere either).

6 The Italian word *piano* as a foreign word did enter Turkish, Arabic, as well as Farsi. To this effect Vedat Kosal. *Western Classical Music in the Ottoman Empire* (Istanbul: Stock Exchange 1999) [with thanks for the publisher's complimentary copy] quotes on p. 21 a poem by Cenap Şahabeddin (b. 1870), but we assume that it was adopted earlier. For the Arabic (*biyanu or biyan*) see Mohammed Sawaie, "Rifa'a Rafi al-Tahtawi and his Contribution to the Lexical Development of Modern Literary Arabic", *IJMES* 32/3 (2000), p. 402 n.1. Hebrew adopted a different term, *psanter* from *psanterin* after the Bible, book Daniel Ch. 3, but this was equally a foreign word, derived from *psalterion* which probably was a harp. The language re-newer Eliezer Ben Yehuda had in vain proposed *makoshit*, something like "knocker".

7 For several references on Iran, I owe thanks to Willem Floor. Our common interest in this topic goes many years back. There are many parallels between the Ottoman lands and Iran (the incompatibility of the music itself, the high costs of pianos, and difficulties of transport, the initial role of diplomats and missionaries), and also some differences (a somewhat slower diffusion). Wilem Floor is presently writing a history of the piano in Iran.

same time the narrative of persistent, but unsuccessful, attempts to make the very same instrument suitable to indigenous music.

6.1 A short history of the piano

The countries of the Middle East have a long musical tradition of their own, in a spectrum encompassing rural folkloristic, or ritual-religious, to urban art arrangements and songs. It has sometimes erroneously been assumed that the piano was a later heir to popular Middle Eastern instruments such as the *qanun* and the *santur*, but this is unfounded because the latter were not keyboard instruments.[8] Those traditional instruments were played with mallets or by plucking the strings. The difference with a mechanical piano keyboard remains fundamental.[9] Middle Eastern musical traditions could do well with the existing array of instruments, and had no immediate need for the piano, at least not in the way in which other European or American innovative small technologies struck roots.

The *pianoforte* was the outcome of a gradual European technological evolution, with the clavichord and dulcimer or zither as antecedents. It may be seen as the reflection of an urge to improve and innovate both the instrumental technology and the music itself. Usually, the year 1700 and the name of Bartolomeo Cristofori of Padua, later of Firenze, serve as critical landmarks of the "invention" (rather the development) of the modern piano. Its hammers enabled the playing of soft and strong tones, and thus led to the term *pianoforte*. In the course of the 18th century, a chain of piano builders such as Gottfried Silbermann, Johann Andreas Stein, Johannes Zumpe, and in the 19th century Sebastien Érard added sophisticated double escapements.

In a process of interaction, innovative mechanical changes in the piano impacted the refinement of European music. J.S. Bach' s *Well Tempered Clavier* (1720s) is still considered as a landmark of standardization of the keyboard and tuning. Mozart composed 27 concertos mainly for himself or, initially, for his students, but they have also survived as normative compositions. The "interplay" between manufacturers, composers, and artists, and the technical maturing of the instrument generated ever more sophisticated pianos, but also new compositions with higher demands on musicality and virtuosity. Larger concert halls, orchestral performances, recitals and other individual performances, as well as printed

---

[8] The *qanun* is a plucked Arab string instrument, the *santur* is s hammered sort of dulcimer.
[9] The 'ud (or oud), a usually short-necked pear-shaped lute did come from the Orient. In the west, *al-'ud* 'developed into "lute".

scores, changed the music scene and brought forth great celebrities such as Frederic Chopin, Franz Liszt, Clara and Robert Schumann, Johannes Brahms, and Sergei Rachmaninoff.

Novel manufacturing methods of the Industrial Revolution, such as a steel frame and carbon steel wires, softer and more effective hammers, and other new materials, as well as the extension of the keyboard (with the black and white colors) from 54 to 88 keys, turned the grand piano into a sophisticated and unparalleled instrument. Grand pianos could have 2.500 different parts, more than any other musical instrument.[10] Some of the still famous brands of today emerged in the 19th century; in fact, three of the greatest, Steinway, Blüthner, and Bechstein were launched in the same year, 1853.

Very important for the diffusion of modern pianos was the early 19th century development of upright pianos by a number of Englishmen, in particular Robert Wornum (around 1826). Apart from the revolution in vertical stringing, later even diagonal, and a related novel hammer action, it caused the beginning of a breakthrough in social diffusion, without changing the music itself. Though hardly affordable by the lowest classes, such more compact models, now came within reach of upper middle classes, as well as schools and institutions. At certain places in Europe and America, in the course of time, working classes could afford it too.

To synchronize our discussion with comparable new European small technologies, examples of which we have discussed in the preceding chapters – all products of mass manufacturing –, we take the 19th century piano as our starting point. By the second half of that century, pianos – grands as well as uprights – became much in demand and were marketed by hundreds of manufacturers in France (rather than Italy), then also in Britain, Austria, and Germany, and subsequently on a very large scale in the USA.[11] From an individual artisan craft, piano-making became a fully-fledged industry.[12] The output, worldwide, was impressive. It is estimated that it increased from 50.000 a year in the 1860s to 650.000 in 1910, grands as well as uprights, and production continued to rise steeply until about the 1920s.[13]

---

[10] Compare this to the saxophone which was invented in the early 1840s with a few hundred parts.

[11] Steinway (Steinweg) started in Germany but became a leading manufacturer in the USA after 1860.

[12] Sonja Petersen, "Piano Manufacturing Between Craft and Industry: Advertising and Image Cultivation of Early 20th century German Piano Production", *ICON, Journal of the International Committee for the History of Technology*, vol. 17 (2011), pp. 12–30.

[13] Cyril Ehrlich, *The Piano, a History* (Oxford., rev ed. 1990), p. 108 et passim.

6.2 *Maqam*s and notation

I have written about several novel small technologies which made their appearance in the Middle East, of which typewriters were ingeniously adapted to serve local vernacular scripts. Yet there is no exact parallel with the conventional piano. The reason is clear at first "hearing". With their usual seven and a half octaves, conventional pianos cannot render *maqam*s (except to imitate one or two), which are the tone scales typical of Middle Eastern musical traditions.[14] Be they – rather broadly speaking –, Arab, Turkish or Persian, in all their subtle differences, their characteristic intervals, intonation, or ornamentation and nuances in tone, each with their own name.[15] *Maqam*s, whether vocal or produced by string instruments, are really melody types or modulations, with subtle regional, social, and ethnic differences.[16]

Still, one perceived the 17-tone system per octave as usually tracing back to Arab and Persian theoreticians from the 9th-13th centuries, in particular Safi al-Din al-Urmawi. As we shall see, a scale division of 24 tones is possible too. In this respect, the term "quartertone" (so-called from the 19th century onwards) is misleading, because it suggests an equal mathematical division (temperament) of tones and intervals.[17] In fact, also the European scale system could then differ

---

14 *Encyclopaedia of Islam*, 2nd edition, s.v. *Maqam*. Theoretical development by Arab, Turkish and Persian scholars goes back to the 9th-13th centuries, and disproves the assumption that Middle Eastern scales lack systematic order. Notes have Arabic or Persian names, e.g. *sīkā* for E half-flat, or *'awj* for B half-flat. The regional-ethnical differences between the *maqam*s, especially in the Persian tradition, might have significance for the experiments with the piano, as we shall see below.
15 Ruth Katz, "The Lachmann Problem", an Unsung Chapter in Comparative Musicology (Jerusalem: Magnes Press 2003), pp. 333. Dr. Robert Lachmann, a German scholar who emigrated to Palestine in 1935 and set up the core of a musicological department at the Hebrew University, gave a series of talks on "Oriental Music" for the Palestinian Broadcasting Station Radio in 1936/7, the first of which typically opens with a contrasting comparison on pianos, pp. 329 ff.
16 Of the various Turkish *maqam*s, though in general close to their Arab parallels, some could be harmonized to more diatonic pitches, see Feza Tansuğ, "Ottoman Elites and their Music: Music Making in the Nineteenth Century Istanbul", *Musicology Today*, vol. 39/3 (2019), pp. 206–207.
17 The prominent sociologist, Max Weber, who primarily sought to prove the universality of his theories, in a lesser-known book *The Rational and Social Foundations of Music* (1921 posthumous, 1958), discussed music and harmony. Not surprisingly, he stressed the "rationality" of the well-tempered scale of Western music, and of related instruments such as the violin, organ, and piano. Though not ignoring the Arab scale system (*maqam*s), he interpreted it as a lack of order, and hence connected it to the near impossibility of notation. Claiming to explain the universal success of the modern piano, he reduced it to western educated elites, thereby ignoring its Central European and Italian roots, Weber interestingly added the climatic conditions of Northern Europe with its closed homes, as a factor enhancing its diffusion.

from place to place and became fixed only over several centuries. The musicologist Robert Adelson, drew my attention to the fact that in 17th century Italy, some harpsichords (such as the Transuntino of 1606) had 31 notes per octave, including quartertones, to allow singers and some string instrumentalists more spontaneous expression.[18]

Actually, there have been many attempts to adopt a suitable, "orderly" (imagine: westernized), method of notation.[19] One example had been the 17–18th century Moldavian prince Kantemiroğlu (or Dimitri Cantemir). Somewhat later, an Armenian church musician Hampartsum designed a system of symbols (somewhat similar to stenographic script) which remained in use until well into the 19th century. Then, around 1840, the multi-talented Lebanese scholar Mikhail Mishaqa, (app. 1840), designed a quartertone system with 24 equal intervals.[20] But not everyone would agree: The highly estimated Father Xavier Collangettes (a professor of physics at St. Joseph University in Beirut) published in 1904 an article in which he basically defended the *maqam* tone system, which according to him had existed from the 13th century.[21] Such an equal division of intervals caused profound disagreements among experts in Lebanon and Egypt, and some have argued that this was one of the reasons which impeded the acceptance of pianos.[22]

---

[18] Private correspondence, 8 April 2021.
[19] It would seem that a visiting German priest, Salomon Schweiger, had in 1576 been the first to notate a piece played by the Janissary band. The first Ottoman textbooks with notation methods would appear only toward the end of the 19th century, see Recep Uslu, "Western Music Theory Works Among the Ottomans, 19th Century", Turkish Music Portal, http://www.turkishmusicportal.org/en/articles/western-music-theory-works-among-the-ottomans-19th-century (last accessed 6 March 2023).
[20] Mikhail Mishaqa had converted from the Greek-Orthodox church to Protestantism and had studied in France. Mishaqa's treatise was translated by the well-known missionary, Eli Smith, and published in the first issue of the *Journal of the American Oriental Society* 1849, pp. 171–217. Nowhere did it refer to the possibility of transcribing quartertone music for the piano. See further Shireen Maalouf, "Mikha'il Mishaqa: Virtual Founder of the Twenty-Four Equal Quartertone Scale", *Journal of the American Oriental Society*, vol. 123/4 (2003), pp. 835–840; and Fruma Zachs, "Mikhail Mishaqa: The First Historian of Modern Syria", *British Journal of Middle Eastern Studies*, vol 28/1 (2001), pp. 67–87. No doubt, his Western formation played a role.
[21] Christian Poché, "Vers une Musique Libanaise de 1850 a 1950", *Les Cahiers de l'Oronte*, no.7 (1960) (Beirut), pp. 80–81. In this illuminating article (pp, 77–100) the order of pages was unfortunately hashed up.
[22] For a thorough discussion see Anas Ghrab, "The Western Study of Intervals in 'Arabic Music' from the Eighteenth century to the Cairo Congress", *The World of Music*, vol. 47/3 (2005), pp. 55–79. Alternative (western) divisions and terms such as "three-quarter-tones", "third-tones", "sixth-tones", or "micro-tones" are less relevant to our discussion here.

6.3 *Alla Turca* and other orientalisms

On the other hand, in Europe, Mozart's famous *rondo alla turca* (KV 331, 1778 or 1783) had become well-known. But as a musical genre, *alla turca* was never meant as a musical rapprochement between Europe and the Middle East, let alone to advance the piano in the Ottoman regions. It remained an imagined representation of the Orient.[23]

The genre was allegedly born in the wake of the military confrontations (whether 1554 and 1683), when Ottoman Sultans had made conciliatory gestures by sending a military band to European rulers. It may have begun with a Janissary band accompanying the new Ottoman envoy to Vienna in 1665, and continued with various Polish, Russian, and Prussian rulers in the 18th century who hosted such a band at their courts.[24] Paradoxically, this *mehterhane* of the Janissary army corps was known to precede the troops into battle. What made an impression in Europe was not only its loud music but also its rhythm. The innovation did not lie so much in its trumpets or horns, but in its drums, cymbals, bells, triangles and other jingling, ringing, and beating instruments, previously unheard of in European military bands.

Ironically, the Janissary corps was eliminated in 1826 (the so-called "Auspicious Event", an overthrow after which modern army reforms began). By then the European *alla turca* genre was already in decline, succeeded by other themes of aesthetic interest in Middle Eastern music, as well as in further "exotic" countries and peoples.[25]

There was one temporary *alla turca* addition to the sound of pianos which deserves mention. While our present-day pianos usually have three pedals, the models which appeared at the end of the 18th and the beginning of the 19th centuries were fitted with at least one more, to bring about different sound effects, a so-called *Janissary stop*. When put down, such pedals activated a tinkling bell and

---

23 The *alla turca* genre embraced a much wider musical sphere such as orchestral, opera, and military music. Consider Gluck; Haydn's military symphony no. 100 (1793–1794); Mozart's Turkish march in the opera *Entführung aus dem Serail* (1782); and Beethoven, *Ruinen von Athen* (1812) and his ninth symphony (1824). The latter may have used field work notations from the Middle East but more systematic interest in Middle Eastern music by composers and opera-writers developed only later (Saint-Saëns for instance). Cf. the various articles in Jonathan Bellman (ed.), *The Exotic in Western Music* (Boston: NEU Press 1998).
24 Michael Pirker, "Janissary Music" in *Grove Music Online* (2001).
25 A revived version of the *mehterhane* is performed for tourists in Istanbul (and sometimes abroad), complete with allegedly authentic clothing and instruments.

an imitation brass drum.[26] There were even music scores with the pertinent notation for their use. This particular instrumental embellishment, mainly a Viennese fancy, was not regularly used, and disappeared quite soon. Carl Czerny (d. 1857) who lives on as a music pedagogue, rather than as a composer – his finger exercises are not fondly remembered by most piano pupils – claimed that the added pedals could easily be dispensed with.[27] As a remnant of this adaptation, in the 1930s, in Egypt, there were still pianos sold with an added pedal which produced a timbre reminiscent of the *qanun*.[28] Even so, it could not make the piano popular: After all, there was no resemblance with *maqam*s, and European music remained within the realm of polyphony.

6.4 When did modern pianos begin to reach the Middle East?

With regard to almost all novel western consumer products one might ask who really "needed" them in the Middle East. Pianos could have become a generally accepted instrument only if they were able to perform or generate indigenous melodies. The immediate answer is therefore that pianos were not really needed, but were imposed on elites, and only gradually desired by (certain) higher classes for social motives. In fact, where the latter strove to resemble Europeans, they acted equally under foreign cultural impact, albeit not necessarily by coercion.

Pianos first arrived as presents and exchanges, and were meant as representations of singularity, to prove a country's superiority. Such traditional diplomatic exchanges remind us of the elaborate mechanical clocks which reached the Topkapı Palace. They had limited practical use in an essentially different time system which lasted to the beginning of the 20th century and ended up as display objects only.

Eastern courts had a long indigenous tradition of music-making, though not necessarily of experimenting with new instruments.[29] Nevertheless, even before

---

26 On "Turkish music pedals" for bells, cymbals, drums, and the like see David Rowland, *A History of Pianoforte Pedaling* (Cambridge: Cambridge UP 1993), pp. 19, and 154–155. For the new upright Giraffe models, see Edwin M. Good, *Giraffes, Black Dragons, and Other Pianos, a Technological History from Cristofori to the Modern Concert Grand* (Stanford: Stanford UP, 2nd. ed. 2001), pp. 136–137, p. 182, and p. 267.
27 Rowland, p. 155, quoting Czerny's opinion of 1838–9 on "childish toys".
28 Salwa Castelo-Branco, "Western Music, Colonialism, Cosmopolitanism, and Modernity in Egypt", in Virginia Danielson et.al. (eds), *The Garland Encyclopedia of World Music*, vol. 6: The Middle East (New York and London: Routledge 2002), p. 610.
29 Ursula Reinhard, "Turkey an Overview", in *Garland Encyclopedia*, vol. 6, p. 771.

pianos had assumed the capabilities of the nineteenth century, some earlier instruments had arrived in the royal courts of the Middle East. The French King François I (d. 1547) sent an entire orchestra to the Sultan, but his associates apparently feared that it would demoralize the troops, and their instruments were burnt, and the players sent back.[30] A similar story concerns Queen Elizabeth I's gift in 1599 of an organ (with a clock and a few self-playing tunes) to Sultan Murad III. [31] It was allegedly destroyed by his successor Ahmed I owing to religious objections. One of the gifts of the King for the Sultan, which an Ottoman delegation to France, led by Mehmet Said Eff. in 1741, carried back was "a great organ" but we do not know whether it survived.[32]

When Napoleon occupied Egypt in 1798, one of the musicians in his "scientific delegation", Henri-Jean Rigel, a pianist and composer, most likely brought a piano along, but it was not sufficiently impressive to draw the attention even of the critical historian, Jabarti.[33] One parallel noteworthy case concerns a five-octave piano which arrived in Iran in 1806 as a gift by Napoleon to Fath Ali Shah of Persia, but there were no pianists around. The court *santur* player, Surur al-Mulk, tried in vain to play it.[34]

Three decades later, an idealist French group of Saint-Simonians arrived in Egypt. The aspirations of its members were in the field of civil engineering and medicine, and ultimately failed in their idealist mission. Yet several of its members may also be remembered in the field of music. Their occasional concerts were said to have been acclaimed by an undefined audience, but this did not generate a lasting effect. The young composer, Félicien David, had come along with a portable iron piano of five-and-a-half octaves, specially designed for the mission by the Chavanne firm in Toulouse. It is said that during the two years he stayed in Cairo

---

[30] However, Ömer Eğecioğlu, *History of Istanbul*, vol.7, https://istanbultarihi.ist/enc (accessed 6 March 2023), claims that under French impact in the sixteenth century new rhythms, called *frenkcin* and *frengifer* were introduced into Turkish music.
[31] John Mole, *The Diary of Thomas Dallam, 1599, London to Constantinople and Adventures on the Way* (London: Fortune 2012). The young builder Thomas Dallam played it himself at the Court, but care had to be taken that he would not sit too close to the Sultan with his back turned toward him. See also Evren Kutlay, "A Historical Case of Anglo-Ottoman Musical Interactions: The English Autopiano of Sultan Abdulhamid II", *European History Quarterly*, vol. 49/3 (2019), pp. 388–389. Part of her book *Osmanlı'nın Avrupalı Müzisyenleri* (Istanbul Kapı Yayınları 2010) can be found in translation on the internet.
[32] Fatma Müge Göçek, *East Encounters West: France and the Ottoman Empire in the Eighteenth Century* (New York: OUP 1987), p. 144.
[33] El-Shawan, p.609.
[34] It became no more than a decorative object in the Palace, "Iranian Piano", https://en.wikipedia.org/wiki/Iranian_Piano (accessed 6 March 2023).

(1833–1835) he earned a living by giving piano lessons. The French music magazine, *Le Ménestrel*, has it that Muhammad 'Ali invited him to teach his women, but when he arrived at the Citadel, the eunuchs requested to demonstrate to them how they should instruct the women. Which David refused. Then, in 1835, David fled the local cholera epidemic to Beirut, and subsequently returned to France. Having left behind his piano, some of his companions took it along to Odessa, where it served the Dutch consul's family for some time, and finally returned to Paris, where it disappeared after the Prussian invasion of 1871.[35]

The most remarkable piano narrative of the nineteenth century evolved at the Ottoman Sultans' Court, – somewhat paradoxically as an outgrowth of the pervasive military reforms.[36] Following the abrogation of the Janissary corps in 1826, and with it of the *mehter* band, a novel Imperial orchestra (*muzika-i humayun* as the word *muzıqa* then meant) was founded.[37] Basically, it remained within the military sphere under a commander with the rank of *pasha*, later *miralay*, but it grew to become a new royal service embracing 500 musicians. Later it was reduced to 350 and lastly to 120 members, while being split up for different sorts of events. A brass band would play military marches, a genre which remained very popular at the court, but besides, new ensembles of harem ladies were formed, and small orchestras which played in operas and at other performances, be it at the theatre of the Dolmabahçe or Çirağan Palaces, or at the well-known Naum Theatre outside. Indeed, operas were another very popular trend of European music at the time and may have sustained the piano.[38]

The instructors were mostly recruited from Italy.[39] The most famous musician hired was Giuseppe Donizetti Pasha, the elder brother of the well-known opera composer, Gaetano Donizetti (1828–1856). Engaged by Sultan Mahmud II, he stayed

---

35 Ralph P. Locke, *Music, Musicians, and the Saint-Simonians*, (Chicago and London: Chicago UP 1986), pp. 173, 190, 202, 220, and 225. The piano had suffered badly in Egypt from the climate, as well as a fall on arrival in Alexandria, p. 220.
36 Three elaborate articles in *Le Monde Musical*, vol. 40, April 1929, vol. 41 January 1930, and vol. 42, March 1931, and many later accounts in particular in Turkey itself.
37 The word *muziqa* (*musiqa*), of Greek origin, in Arabic and Turkish was usually associated with western (not primarily religious) music. Terms of the 19th century such as *muziqar* for a formally educated musician indicate a difference from an *alati*, of lesser stature, who learnt to play by hearing only.
38 "Anything that was composed for the military bands (in European forms) had also a piano transcription", Evren Kutlay (mail 13 Jan. 2021) explaining that the piano was "incredibly popular".
39 An extensive list can be found in A. Orkun Gündoğdu, "Osmanlı/Türk Müzik Kültüründe Avrupa Müziği'nin Yaygınlaşması Süreci ve Levanten Müzikçiler" (Thesis, Başkent Üniversitesi, Ankara 2016), esp. pp.114–120 *et passim*.

in Istanbul from 1828 till his death in 1856 (under Sultan Abdülmecid).[40] Donizetti was said to have ordered "many" pianos from Vienna for the ladies of the court, though an unnamed skeptic observer added "I do not know how they are going to play, since no one so far has succeeded in going anywhere near them".[41] The correspondent of the French magazine *Le Ménestrel* wondered how he ordered pianos from Vienna while there were only two piano teachers in Istanbul.[42] But this must have changed progressively after 1836.

Donizetti was succeeded by more Italians, Bartolomeo Pisani, and then by Callisto Guatelli (app. 1856–1899/1900) who went on to serve Sultan Abdülaziz. Soon Pasha Guatelli directed, taught many local students, and also composed (again, marches, some imitating Janissary sounds). With a firm in Naples, he even developed new instruments for the military band, e.g., a small bombardino tromba. After Guatelli came Dussap Pasha, and the Spanish-French musician d'Arenda, and then a Prussian-German organist (and pianist), Paul Lange. Only in the 1920s, did a Turkish musician take over. Evidently, one of the assignments of all the recruited foreigners was giving piano lessons.[43]

The upshot was that the Ottoman Court began to appreciate European music. However, traditional music was not discarded. This symbiosis was similar to many other cultural, technological, and political imitations, including western dress, furniture, and habits. In a way, the Ottoman Sultans competed with other royal courts elsewhere, and pianos became very much part of it. Celebrated foreign performers came to Istanbul to give concerts at the Palace, as well as at embassies and elite salons. Celebrity Franz Liszt's visit in 1847 is particularly remembered, not in

---

[40] Donizetti was preceded by a Frenchman, Manguel. Several Italian composers (e.g. Baldassare Galuppi, Tomasso Traetta, Giovanni Paisiello) had served in the 18th century in a similar capacity the Court of Czarina Catharina II. Sultan Mahmud liked European music. The first Ottoman delegation of 158 students who had been sent to Europe included also one studying the piano, Amnon Shiloah, *Music in the World of Islam: A Socio-Cultural Study* (Detroit: Wayne State University Press 1995), p. 91, also p. 105. Once, Mahmud invited a whole orchestra from Vienna, *Le Ménestrel*, 1 May 1836.
[41] Quoted from *Le Ménestrel*, 18 December 1836, by Emre Araci, "Giuseppe Donizetti at the Ottoman Court: A Levantine Life", *The Musical Times*, vol. 143, no 1880 (Autumn 2002). Donizetti also composed, mainly marches, and allegedly also the first Ottoman national anthem, for which he devised a new scale notation system.
[42] 18 December 1836.
[43] Italo Selvelli, son of Italian parents, who usually conducted operas, gave piano lessons at the Court as well. Several later Ottoman musicians began their carrier at the *muzika-i humayun*, see Recep Uslu, "Western Music Theory Works Among the Ottomans", http://www.turkishmusicportal.org/en/articles/western-music-theory-works-among-the-ottomans-19th-century (accessed 4 August 2022).

the least because an impostor by the name of Listman almost stole the show.[44] The French piano maker Érard even sent a grand for the occasion.[45]

Mahmud II (1785–1839, Sultan from 1808) himself played no western instrument, and still held on to Turkish music. Besides he composed a total of 26 known compositions, an ability which was considered as a proof of mastery. But from Sultan Abdülmecid (1839–1861) onwards, all Ottoman sultans learnt to play the piano; at least two or three of them are said to have excelled in it. They composed (mainly) military marches. Typically for the double-track musical culture at the court, Sultan Abdülaziz (1861–1876) was capable of playing the 'ud and the ney (flute), and also composed in both Eastern and Western styles. Sultan Murad V (reigning briefly in 1876) played piano as well as violin, and composed no less than 488 pieces for piano and other instruments (today preserved in the University Library of Istanbul). Sultan Mehmed Reşad V, albeit closer to traditional Mevlevi sufi music, and his successors Mehmed V and Abdülmecid II, all played the piano. [46] The latter was also an excellent painter and left a striking picture of a Beethoven concert at the Bağlarbaşı summer palace (Fig. 33).[47]

Music lessons for women (mostly by women) had long been a tradition at the court, and musical events were frequent, as vividly described by Leyla Saz, who as a young girl had been part of the entourage of the Çırağan palace in Abdülmecid's times. She had learned to play the piano from a certain Therese Roma, in Kadiköy, a district of Istanbul outside the palace, but later taught herself others at the Court.[48] "There were pianos pretty much everywhere in the Serail", she writes, "often of inlayed carved wood and decorated with mother-of-pearl."[49] She, too, composed western-style music along with Turkish music. Her great-grandson re-

---

44 Ömer Eğecioğlu, "The Liszt–Listmann Incident", *Studia Musicologica*, vol. 49/3–4 (September 2008), pp. 275–293. The home of music dealer, Comendinger in Pera (Beyoğlu), where Liszt stayed, still bears a plaque to commemorate his visit. We mention also visits of Leopold de Meyer and Camille Saint-Saëns, but not all the celebrities were pianists: Henri Vieuxtemps was a violinist and Paris Alvars a harpist, to note but two. Liszt also had a famous Ottoman pupil Faik bey (Franz) della Sudda.
45 *Le Ménestrel*, 13 June 1848.
46 Kosal, *passim*.
47 "Sarayda (or Haremde) Beethoven", 1915, Painting in Istanbul Resim ve Heykal Müzesi, showing the Sultan's spouse and son each playing the violin, and his niece Hatice ("Ophelia") at the piano
48 Leyla (Saz) Hanımefendi, *The Imperial Harem of the Sultans, Daily Life at the Çırağan Palace during the 19th Century* (Istanbul: Hil Yayın 1955), pp. 58, 134–135, 187, 271, 279, 280, 283, *et passim*. See also Kathryn Woodard, "Music in the Imperial Harem and the life of Composer Leyla Saz (1850–1936)", *Sonic Crossroads* (2011). Woodard also analyzes her songs and marches, with their inclination to imitate Turkish *maqams*.
49 Saz, pp. 41 and 58.

Fig. 33: A painting of 1915 by Abdülmecid II, the last Caliph (1922–1924), entitled "Beethoven in the Harem", depicting his Circassian wife Şehsuvar Kadınefendi playing violin, princess Hatice (daughter of Sultan Murad V, also known as Lady Ophelia) playing piano, and his son Ömer Faruk playing cello, while two other women are listening, one possibly his third wife Mehisti, at his summer palace in Bağlarbaşı (from the İstanbul State Art and Sculpture Museum, public domain).

members that she sometimes preferred the harmonium, once a gift from her father, as it could produce quartertones.[50]

Most remarkable is Sultan Abdülhamid II (1876–1909), until recent decades much maligned as a reactionary ruler. He is remembered as an admirer of western music by his daughter, Ayşe Osmanoğlu, in her memoirs. He himself played the piano and the violin.[51] As we have seen, the Court had previously acquired various

---

50 Ibid., Appendix to her book, p.287.
51 Ayşe Osmanoğlu, *Babam Sultan Abdülhamid (Hâtıralarım)* (Istanbul,1984), esp. pp. 119–120 where he rigorously follows the piano skills of his daughter. Among his teachers were Guatelli,,Düssap Aleksan, and Lombardi. This propensity is also discussed by Douglas Scott Brookes, *The Concubine, the Princess, and the Teacher; Voices from the Ottoman Harem* (Austin: Univ. of Texas Press 2008), pp. 62, 65, 103, 114, and 151. See also Vedat Kosal p. 42. A pianist himself, the late Kosal has recorded compositions by Sultans Abdülaziz and Murad, along with some by Donizetti and Guatelli Pashas (disk *Ottoman Court Music* 2003).

pianos but now added at least a high quality Érard grand, and one or more Bechsteins, as well as a British Kastner self-playing piano in 1907 with a selection of piano and opera tracks (none of which survived his exile in Salonika).[52] The Dolmabahçe Palace today still has a grand piano on show for the visitors. Allegedly, there was a total of twelve high-quality pianos in the palace.[53] Besides, Abdülhamid, in his enthusiasm for the piano, seems to have donated a German grand with his *tughra* engraved on it to one of the princes at Bait al-Din in Lebanon.[54] At least as remarkable is a story that the Sultan had pianos made locally by a carpenter from Kastamonu, one sample of which he allegedly sent as a gift to Kaiser Wilhelm II.[55]

In Egypt since the ascendancy of Muhammad 'Ali (1805–1848), the ruling court went through a similar phase of Nizam-i Jadid military reforms, but there, the instructors were mostly Ottomans, with one German and several Italians. Egypt, still being a formal part of the Empire, had its five military music schools following the example of Istanbul, though apparently not with regard to the piano.[56]

In 1869 Empress Eugénie donated an upright piano (manufactured in Stuttgart) to Khedive Isma'il for his royal yacht *al-Mahrusa* which headed the opening of the Suez Canal. She even played on it herself for the occasion. Both the piano and the yacht still survive. Princess Nimetullah, a daughter of Isma'il, was taught to play by an English governess, and thereafter by the conductor of the royal band, a German-Huguenot.[57]

At the beginning of the twentieth century, more Egyptian princesses and princes were taught the piano. Prince Faruq, to become king in 1936, was taught by the Italian maestro Ettore Puglisi, (a fact which, according to the renowned musicologist, Salwa al-Shawan, led to his lack of interest in Arab music).[58] Prince Hasan

---

52 Kutlay, p. 397 and p. 411. We also read about a self-playing piano in Iran in 1908, William Penn Cresson, *Persia: The Awakening East* (Philadelphia: Lippincott 1908), p. 87.
53 *Daily News* 26 May 2016. But some instruments may have been at other palaces such as Çiragan.
54 The instrument was built by Kaps in Dresden between 1901 and 1902, see Jan Altaner, "Kaps – a Grand Piano from Dresden in the Chouf Mountains", Goethe Institut, Lebanon, https://www.goethe.de/ins/lb/en/kul/sup/spu/20980887 (last accessed 31 March 2023). Intriguingly, if it was presented to one of the Ottoman *mutasarrif*s residing at Bayt al-Din, Naum Pasha was a Greek Orthodox and Muzaffir Pasha a Roman Catholic.
55 Kutlay, p. 413.
56 James Heyworth-Dunne, *Introduction to the History of Education in Modern Egypt* (London: Luzac 1939), pp. 134–135, and 219.
57 Fanny Davis, *The Ottoman Lady, A Social History from 1718 to 1918* (New York: Greenwood 1986), p. 54.
58 Salwa El-Shawan Castelo-Branco, a prominent (Christian) ethnomusicologist based in Lisbon is the daughter of the Egyptian composer 'Aziz al-Shawan. She started playing the piano from the age

'Aziz Hasan and princess Nevine 'Abbas Halim received piano lessons at the Tiegerman conservatory about which more below.[59] Maybe unexpectedly, even the Hashimite court in Amman seems to have owned at least one piano, which ended up in private hands in Israel.[60] Similarly, in Qajar Iran, new music schools were opened, for which western, mostly Italian, experts were hired. There it was the Dar al-Fonun which opened the new era. A French military musician, Alfred Jean Baptiste Lemaire, served from 1867–1907 as head of the music division. He is credited with the composition of the first Persian national anthem, but also transcribed Persian songs for the piano which were somewhat later performed in Paris.[61] Possibly unique for Iran, it worked in two directions. The prominent *santur* player, Sarvar al-Mulk, seems to have been literally instrumental in altering several keys on one of the newly imported pianos in accordance with the Persian *dastgah* mode. Princess Ismat al-Dawla and prince Azud al-Dawla learnt to play Persian music on it.[62] Another well-known *santur* player at the court, Muhamad Sadiq Khan, developed a new Persian style for the piano whereby the right hand mainly did the improvisations, while the left hand doubled the melody line. Tremolos could in that way give somewhat the impression of the strokes on a *tar* or *santur*.[63]

---

of five, and then went on with Ettore Puglisi, a pupil of Ferroccio Busoni. Puglisi, born in Palermo, had established himself in Cairo in the early 1930s as a private piano teacher and then taught at the National Conservatory till 1978, A.Stoichiţă, "Trois Continents, Une Passion. Entretien Avec Salwa El-Shawan Castelo-Branco", *Cahiers d'Ethnomusicologie*, vol. 26 (2013), pp. 241–254.

59 Hasan Aziz Hasan, *In the House of Muhammad Ali: A Family Album 1805–1952* (Cairo, AUCP 2000), pp. 106–108; Samuel Nkruma, "Follies of Fate", *al-Ahram Weekly* 24/30.12.2009 interviewed Princess Nevine, granddaughter of Muhammad 'Ali, who had been taught by Madame Marianos (of Greek descent). Ara Guzelimian of the Juilliard School mentions "an ornate white piano" now at the Cairo Opera House which reputedly once belonged to Faruq", quoted in Laura Robson," A Civilizing Mission? Music and the Cosmopolitan in Edward Said", *Mashriq & Mahjar* 2/1 (2014), pp. 112–113 and 115–117.

60 Information from our former student C.G., who related that Emir Abdallah had donated a Schimmel baby grand to his chief veterinarian, Haim Appelboim, upon his retirement from the British army (it is not clear whether it had become redundant at the Palace); it later landed with a family in Benyamina.

61 On his merits and a book, he wrote on Persian music, see "Iranian Piano", *Wikipedia*; https://en.wikipedia.org/wiki/Iranian_Piano; and Hormoz Farhat, "Piano in Persian Music", https://referenceworks.brillonline.com/browse/encyclopaedia-iranica-online/alphaRange/Ph%20-%20Pn/Phttps://iranicaonline.

62 C.J. Wills, *Persia as It Is* (London: Sampson Low 1886), p. 311 relates that the Shah's third son was taught to play the piano. He further mentions a young French woman who had arrived to give lessons. The Iranian Prince Salar al-Din had a piano at home (information from W. Floor).

63 "Iranian Piano." *Wikipedia*, https://en.wikipedia.org/wiki/Iranian_Piano.

The long reign of Shah Nasr al-Din (1848–1896) stands out, like that of Sultan 'Abdülhamid's in Ottoman Turkey. Evidently, he had been taught by one of the piano teachers who had established themselves in Tehran. The diary of an anonymous lady in his household describes him playing the piano first thing every morning.[64] There also is a tile at the Golestan Palace which shows the Shah listening to a piano performance in a house concert.[65] But piano entertaining also slowly entered certain public theatres, which had traditionally combined drama with music, albeit in intermissions.[66]

6.5 Missionaries, colonial officers, and diplomats

A specific category of "piano promotors" were Protestant – mostly American – missionaries who in their zeal to win over new converts, used their musical instruments to accompany religious hymns.[67] Henry Jessup (1856–1910 in Lebanon), one of the most famous missionaries, considered it one of his aims to educate Arabs singing "our tunes" correctly, that is by western standards of harmony, "Pianos have become quite common, and the Oriental taste is becoming gradually inclined to European musical standards".[68] Evidence from Lebanon, and later also from neighboring countries, suggests a gradual transition from organs and harmoniums to pianos. These were imported mostly by Protestant missionaries from the

---

64 Piotr Bachtin, "The Royal Harem of Nasser al-Din Shah Qajar (r. 1848–1896): The Literary Portrayal of Women's Lives by Taj al-Saltana an Anonymous 'Lady from Kerman', *Middle Eastern Studies* 51/6 (2015), pp. 993–994 and 1006n. She claims that at the time (app. 1892–1894) there were five or six pianos in Tehran, "and hardly anyone could play it".
65 Jennifer Scarce, *Domestic Culture in the Middle East* (Edinburgh 1996) p. 54, the said tile was made from a photograph and installed on a frieze in 1887.
66 Willem Floor, *The History of Theater in Iran* (Washington DC: Mage 2005), p. 235, quoting Kuhestani-Nezhad (2002).
67 Western missionaries in different parts of the world are said to have introduced science, medical knowledge, and technological innovations to advance their proselytizing campaigns; the clavichord, and later the piano was one means, see Uta Zeuge-Buberl, *The Mission of the American Board in Syria, Implications of a Transcultural Dialogue* (Stuttgart: Steiner 2017), p. 92. On the importance of music, especially organs for religious worship, see Deanna Ferree Womack, *Protestants, Gender, and the Arab Renaissance in Late Ottoman Syria* (Edinburgh: Edinburgh UP 200), pp. 165, 182–183, 210.
68 Henry Jessup, *Fifty-Three Years in Syria* (New York: Revell c.1910), vol. I, p. 251. See also Shmuel Moreh, *Modern Arabic Poetry: 1800–1970: the Development of its Forms and Themes* (Leiden: Brill 1976), p. 24–25 and p.114 even sees an impact of the piano on traditional forms of Arabic poetry.

United States, but also from Britain and Germany.[69] This included the training of pupils, and even of piano tuners.[70] It conforms with the views of the prominent ethnomusicologist Ali Jihad Racy (of Lebanese descent) who attests to the impact of American missionaries and their hymnals, but in particular to an increasing interest in western instruments after the First World War. [71] Though the process needs more profound investigation, we argue that what began as a Protestant perception of the piano as an edifying instrument, then became part of a wider musical taste for western classical music, but by no means a general trend.

As we will see, Lebanon became an important chain in our narrative, which – like its early press and intellectual lead – had an impact on Egypt as well. Undeniably, the Christian influence on piano education – through school curricula, and not primarily aimed at religious conversion –, could be seen in Palestinian cities too.[72] Though rooted in the European or American Christian tradition, not all piano education or performance had straightforward missionary goals.[73] One of many examples is Bir Zeit which began as a Protestant Christian girls' school and later became a college (and today a university); piano lessons there were given by Salvador 'Arnita, about whom more later.[74] It extended even further: In

---

[69] Julia Hauser, *German Religious Women in Late Ottoman Beirut, Competing Missions* (Leiden: Brill 2015), p. 209.
[70] Jessup, vol. I, p. 252; vol. II, p. 566 and 699.
[71] Ali Jihad Racy, "Words and Music in Beirut: A Study of Attitudes", *Ethnomusicology*, vol. 30 (1986), p. 415. We suggest that the process had already begun before World War I.
[72] Rachel Beckless Willson, pp. 144–150 on bishop Blyth's spouse and daughters. Also, Mona Hajjar Halaby, "School Days in Mandate Jerusalem at the Dames de Sion", Jerusalem *Quarterly* 31 (07), p. 50 and p. 55. This also applied to a certain extent also to Iran, see the daughter of a Qajar aristocrat going to the American Bethel School which had a piano for singing hymns., Camron Michael Amin, *The Making of Modern Iranian Women, Gender, State Policy, and Popular Culture, 1865–1946* (Gainesville, University Press of Florida 2005), p. 153.
[73] Also a prominent German Protestant philologist and folklorist such as Gustaf Dalman who worked in Jerusalem in the first decades of the 20th century, and among many other studies published a collection of Palestinian songs, had a piano at his home in Jerusalem where social meetings took place, see Moslih Kanaaneh et al. (eds.) *Palestinian Music and Song Expression, and Resistance since 1900* (Bloomington & Indianapolis: Indiana University Press 2013), p. 31.
[74] Rima Tarazi, "The Palestinian National Song, a Personal Testimony", *This Week in Palestine*, 27 June 2010; Saleem Zoughi, "Salvador Arnita", *ibid.*, January 2021. 'Arnita, b.1914, had started his career as a church organist in Jerusalem and Alexandria, pursued studies in Rome, taught and composed in Jerusalem, led the YMCA orchestra in Jerusalem, fled in 1948 to Lebanon where he joined the AUB music department. In fact, the Arab division of the Palestine Broadcasting Corporation (from 1937), had employed Yusuf al-Batruni, Salvator 'Arnita, and Augustine Lama (all of whom had started as church organists), see Issa I. Boulos, "The Palestinian Music-Making Experience in the West Bank, 1920s to 1959: Nationalism, Colonialism, and Identity", PhD Thesis, Leiden University 2020, p. 149, p. 171 and pp. 252–253. Many examples of Palestinians, generally of Christian

Egypt during World War I, the nineteen YMCA centers provided pianos, and other recreational amenities for the British troops.[75]

Consular and diplomatic agents, like the foreign powers themselves, inserted the piano into their (more secular) *mission civilisatrice*. Memoirs on Palestine stand out. In 1856, in Haifa, the Dutch vice-consul "Signor V." (apparently an Italian) invited the local notables to listen to what was said to be the first piano ever introduced there. We know this amusing story from Mary Rogers, sister of the British Vice-Consul, who was requested to play it, and astounded the audience by performing "their national Ottoman anthem" from noted scores. This exceptional audience consisted of some twenty Muslim notables, In her memoirs she even added pompously "This was the dawn of a new era in the history of the little European colony of Haifa".[76] Moritz Busch, the author of a German travel account, noted in 1859 that in Jerusalem five pianos could be counted, of which only two of good quality.[77] Around the same time, the French consul in Jerusalem, Paul-Emile Botta, also had a piano to entertain his local guests.[78] We know of several British, French, Prussian, and Russian representatives elsewhere in the Middle East, who had brought along pianos to their places of posting to impress local populations with their merits, or for their own pleasure.[79]

---

origin, who composed piano music, are discussed by Yuval Shaked, "On Contemporary Music", https://www.searchnewmusic.org/shaked.pdf (last accessed 6 March 2023)

75 Lanver Mak, *The British Community in Occupied Cairo, 1882–1922* (PhD Thesis, SOAS, London 2001), p. 205. However, ill-behaved Australian soldiers on at least one occasion threw a piano out of the window, p. 202.

76 Mary Eliza Rogers, *Domestic Life in Palestine* "(London: Bell and Daldy 1862), pp. 351–352. Rogers tried to persuade an impressed Muslim notable to buy a piano for his wife, and declared herself prepared to teach her, but the guest implied insolently that local women would be not capable of learning to play it. The Vice Consul might have been Vincent Marcopoli.

77 Moritz Busch, *Eine Wallfahrt nach Jerusalem, Bilder ohne Heiligenscheine* (Grunow, 3rd. ed. 1881), p. 308. Thanks to Arndt Engelhardt who drew my attention to this source.

78 Rachel Beckles Willson, *Orientalism and Musical Mission, Palestine and the West* (Cambridge: Cambridge UP 2013), pp. 79–80, which she calls "a novelty" as described by the Swiss clergyman Titus Tobler, (1865?) who found Beethoven somewhat odd in the Jerusalem environment.

79 In Jerusalem, the Prussian consul Rosen, who was married to the daughter of a famous pianist Moscheles, possessed a piano. His wife "delighted" the Pasha with a recital, according to Mrs. Finn, the wife of the British consul, who herself also performed on it, see Arnold Blumberg (ed.), *A View from Jerusalem 1859–1858* (Cranbury NJ: AUP 1980), pp. 96 and 216. But see also Elizabeth A. Finn *Reminiscences of Mrs. Finn* (London: Marshall Morgan 1929), pp. 97 and 171. When they gave a reception for the visiting guest, and the Consul-General Colonel Rose, played himself," the Pasha was as delighted as a child to watch." For Egypt see Archie Hunter, *Power and Passion in Egypt: Life of Sir Eldon Gorst 1861–1911* (London: Tauris 2007), p. 147. He possessed a Steinway.

6.5 Missionaries, colonial officers, and diplomats — 205

For a typical colonial officer such as Ronald Storrs, who served in Cairo and subsequently as Governor of Jerusalem (1917–1926), a piano was an indispensable proof of civilization. In Cairo he complained about "so little really good music", which he resolved with quite disciplined house concerts, attended by visiting French, Austrian, and Italian ladies who played on a donated Blüthner baby grand. He even used the term "piano famine both in Egypt and in Palestine". Soon after arriving in Jerusalem, he specially invited two pianists from Jaffa for an "elite civil musical party", but he ultimately failed to establish a mixed music school.[80] However, showing off a piano could equally be a commercial interest: a French consul in Jerusalem asked Paris for a Pleyel grand for his cultural events, in order not to depend on a hired German Blüthner.[81] By which we do not overlook that foreigners could also have a strong desire to import a piano for their own pleasure.[82] This seems true even for the Ottoman diplomat, Ahmad Vefik, coming to Tehran as early as 1852.[83]

In the region, it often meant that problems of transport and repairing damages had to be overcome, a challenge which existed even in Europe, but all the more so in the Middle East. Foreign observers noted that porters even carried pianos on their own backs.[84] Coastal cities did not pose the most arduous problem, except sometimes as such a heavy colossus had to be hoisted ashore, but inland Anatolia,

---

[80] Ronald Storrs, *Orientations* (London: Nicholson, ed. 1945), p. 97 where he prides himself on strictly clearing away the tea service before the concerts began, having barking dogs chased away and traffic diverted. On Jerusalem, see p. 316, the music school was finally handed over to the Jewish community for lack of Muslim and Christian interest. On pianos see further pp. 27, 89, 99, 129. Even so, Storrs is remembered for his concerts by one of the famous local oud players, see Salim Tamari and Isam Nassar (eds.), *The Storyteller of Jerusalem, the Life and Times of Wasif Jawhariyyeh, 1904–1948* (Jerusalem: Institute for Palestine Studies 2014), p. 126.
[81] Dominique Trimbur, "Fortune and Misfortune of a Consul of France in Jerusalem: Amédée Outrey, 1938–1941", *Bulletin du Centre de Recherche Français Jerusalem*, vol 1998/2), p. 136.
[82] See James Robert Kneip, *A.S. Griboeder: His Life and Works as a Russian Diplomat, 1817–1829* (Ohio State U., PhD Thesis 1976), pp. 15, 64, and 74 on getting his piano to Tabriz, and acquiring another in Tiflis. Also, Pierson, superintendent of the Telegraph Department had apparently taken a piano to Iran, C.J. Wills, *In the Land of the Lion and the Sun, Modern Persia* (London: Ward, Lock & Co 1891), p. 71.
[83] Lady Sheil, *Glimpses of Life and Manners in Persia* (London: Murray 1856) pp. 282–283. She was the wife of the British minister. From her diary it is not completely clear whether Ahmad Vefik, grandson of a Greek convert, who had done part of his studies in France, himself played or somebody else from his delegation.
[84] Donald Stuart, *The Struggle for Persia* (London: Methuen 1902), p. 250. Also, Storrs, p. 97, where his piano was "duly carried up the staircase on the back of one Egyptian porter". Douglas W. Sladen too was surprised by individual Cairene porters carrying pianos, *Oriental Cairo, The City of the Arabian Nights* (1911, repr. New Haven 2022), p. 41.

Syria, and Iran with their mountains and deserts did. One reads with amazement about the trouble and determination to get grand pianos to their destination on camelback over long distances (in not a few cases with damage).[85] James Baker, the owner of the well-known department store in Istanbul, apparently had a potential piano market in mind when in 1888 he conceived of a piano which could be taken apart into four pieces for easier transport.[86]

6.6 Trickling down?

Thus, the courts, the higher classes, and certain Christian minorities had become familiar with pianos and piano music.[87] Certainly, Middle Eastern travelers to Europe and publicists had been aware of the importance of the piano in bourgeois circles there.[88] It is unlikely that the "ordinary" people knew what was, so to say, playing at the courts. History wherever often saw subjects or dependent populations striving to emulate their rulers, but here it was hardly the case. Music involves more internalizing than the mere imitation of clothing, furniture, or food.

---

[85] In 1810, in a caravan from the King of England to the Shah of Persia, porters allegedly carried a piano on their backs over a distance of 620 miles from Bushire to Teheran (too heavy for camels, and no carts!). Most gifts arrived with damage, James Morier, *A Second Journey Through Persia, Armenia and Asia Minor* (London: Longman 1818), p.197. Moritz von Kotzebue, *Narrative of a Journey into Persia in the Suite of the Imperial Russian Embassy in the Year 1817* (London: Longman 1819), p. 66 talks about a piano which arrived without injury from St. Petersburg in Tiflis. However, he adds that at a performance, the Georgians "first laughed, then yawned and fell asleep". Another such story can be found in a typescript of the Presbyterian Labaree family of missionaries but all also amateur musicians, who, in 1904, had a hard time getting their piano overland to Urmia (and into their house); it arrived with some damage, "A Chronicle of the Labaree Family, 1860–1920", p. 114 (with thanks to Robert Labaree for sending me a copy). It was in 1920 destroyed by local insurgents. Another Presbyterian, Dr. Frame, based in Rasht, had purchased his piano in Baku, where porters had difficulties loading it into a boat, R.E. Hoffman, "A Medical Missionary's Journey to Persia in War Time", *Cleveland Medical Journal*, vol. 15 (1916), p. 586. Cf. Sophie Roberts, *The Lost Pianos of Siberia* (NY: Grove Press 2020), or Bauer, p. 135 on getting pianos by mules over the Andes.
[86] Ersin Altin," Baker Mağazalari: Göstere Göstere Tüketmek", *Manifold* (https://manifold.press/baker-magazalari-gostere-gostere-tuketmek), quoting from Victoria and Albert Museum, *From East to West: Textiles from GP and J Baker* (1984), p. 12. [With thanks to the author for the reference].
[87] Merih Erol, "The 'Musical Question' and the Educated Elite of Greek Orthodox Society in Late Nineteenth-Century Constantinople", *Journal of Modern Greek Studies*, vol 32 (201), pp. 133–164. The author distinguishes between the upper layers and the less educated of the community, while the clergy was critical.
[88] An early example for Iran is Mirza Saleh Shirazi in 1815, Camron Michael Amin, p. 52.

Some impact of the West on the musical scene cannot be denied, but the question is how far it became diffused. In Istanbul, from around the 1860s, more or less independently from the court, an Armenian theatre director and opera composer, Dikran Tchouhadjian (Chukhajian), gave private piano lessons. In Parisian fashion, typical high-ranking bureaucrats in Istanbul and Cairo would occasionally take their wives and daughters to the theater, adding that the latter often knew how to play the piano.[89]

Later in the nineteenth century, high bourgeois circles, western-oriented minorities, and – naturally – foreign residential elites in Middle Eastern cities, became interested in acquiring pianos. Pianos as status symbols gradually made their appearance in some of the literary salons in the main cities, though we do not have the particulars on who played what compositions. In Istanbul, one such regularly held salon was hosted by the Armenian feminist, Srpouhi Vahanian, whose French husband, Paul Dussap, a pianist, served as a court musician.[90] Other famous examples were those of Maryana Marrash (Aleppo), Mary Ajami (Damascus), Princess Nazli Fadil (Cairo), and the pioneer feminist, Huda al-Sha'rawi (Cairo). [91] Charity concerts, organized by embassies, as well as foreign or westernized residents, also featured the piano.[92]

In imitation of Europe, piano playing was becoming a requirement for well-educated girls, and respectable upper-class spouses.[93] However, anti-feminist

---

[89] Adam Mestyan, 'A Garden with Mellow Fruits of Refinement', Music Theatre and Cultural Politics in Cairo and Istanbul 1867–1921 (PhD Thesis, Central European University, Budapest, 2011), pp. 69, 73, 168, and 295. The all-round musician Medeni Aziz Efendi (1842–1895), who played Turkish and European instruments (piano included), is said to have taught the granddaughters of Mahmud II and other girls.

[90] As "S. Ex Dussap Pacha" he figures among the nine local French chant and piano teachers, Ernest Giraud, La France à Constantinople (Constantinople: Imprimerie Française 1907), p. 43.

[91] Sufuri [pseudonym]: al-Muqtataf, June 1928), p. 686. On the salons, see Joseph T. Zeidan, Arab Women Novelists, The Formative Years and Beyond (New York, SUNY, 1995), pp. 5–52. Also, Nada Anid, Les Très Riches Heures d'Antoine Naufal, Librairie à Beyrouth (Paris, Calman-Levy 2012), p. 44–45, 92. See further the rare images of home interiors on the internet (Arab Image Foundation). As to Egypt, Huda al-Sha'rawi writes that her husband, from whom she later divorced, disliked piano playing, Harem Years: The Memoirs of an Egyptian Feminist (transl. Magot Badran, New York: SUNY 1986), p. 60.

[92] Malte Fuhrmann, The Port Cities of the Eastern Mediterranean (Cambridge: Cambridge UP 2021), pp. 114–115.

[93] For an excellent survey of the piano in the Middle East see Frederic Lagrange, Musiciens et Poètes en Egypte au Temps de la Nahda (Thesis, Paris 1994), on girls see pp. 96, 300, and 312. See also Fuhrmann, pp. 276–279, 281.

men could sometimes oppose it.[94] Not a few (auto-) biographical works by women have in recent years drawn the attention of scholarly researchers, and several of them mention the piano lessons which had been part of their education.[95] We even hear a few of them complaining about the discipline it required.[96] As musical instruments were mostly associated with women, the gramophone, upon entering private homes, was, for a long time, also considered as a domestic musical instrument.[97]

One comes across related references in novels as well. The young girl Asma, a protagonist in Salim al-Bustani's novel of 1873, tremendously disliked her compulsory piano lessons (a theme well-known elsewhere too). However, it would appear that Adèle, Salim's wife, was a keen pianist, who had imported a piano, one of the first in Lebanon, which remained for years in the historic Bayt Bustani, and still survives today in family hands.[98]

---

94 Fuhrmann, p. 285, cites "middle class" criticism against teaching girls piano, dance, and French, as being skin-deep only to impress. Abigail Jacobson quotes from the diary of a gentleman: "I don't want somebody who can play the piano but doesn't know how to handle housework", "Negotiating Ottomanism in Times of War: Jerusalem in World War I through the Eyes of a Local Muslim Resident, *International Journal of Middle Eastern Studies*, vol. 40/1 (2008), p. 78. Similar disapproval could be found earlier in *al-Lata'if al-Musawwara* (1888), according to Mona Russell, "Competing, Overlapping and Contradictory Agendas: Egyptian Education on under the British Occupation, 1882–1922", *Comparative Studies of South Asia, Africa and the Middle East*, vol. 21/1–2 (2001), p. 56.
95 Elizabeth Warnock Fernea (ed.), *Remembering Childhood in the Middle East* (Austin: U of Texas Press 2002): p. 74 on Hoda al-Naamani, a poet who went to a French school in Damascus and later lived in Cairo; p. 253, on the younger Randa Abou-Bakr, b. 1966, a lecturer of English at Cairo University with a passion for music. Beth Baron, *Egypt as a Woman, Nationalism, Gender and Politics* (Berkeley: University of California Press 2005), p. 192 mentions the feminist writer Labiba, Ahmad, who was the daughter of a medical doctor; Samia Mehrez (ed.) *The Literary Atlas of Cairo: One Hundred Years on the Streets of the City* (Cairo: AUCP 2010), p. 86 where Yasser Abdelalif remembers Madame Nabile Habáshi playing the black piano at the French Lycée. For Lebanon after the First World War, see, for instance, Fay Afaf Kanafani, *Nadia, Captive of Hope, Memoir of an Arab Woman* (London: Routledge 1998), pp. 40, 44, 58 et passim.
96 Mona N. Mikhail, *Seen and Heard, a Century of Arab Women in Literature and Culture* (Northampton Olive Branch Press 2004), pp. 7 and 9; on Randi Abou-Bakr in Egypt see Fernea, pp. 350–354; But boys too played, see Edward Said below.
97 This may also be concluded from a few divorce cases in Beirut, in 1910, in which the female partner received the gramophone, it being identified as an instrument of pleasure, for music had no specified value, while the male would take the household appliances, see Toufoul Abou-Hodeib, "Taste and Class in Late Ottoman Beirut, IJMES, vol. 43/3 (2011), p.485–486. The famous Baidaphone label (founded by a Lebanese family) occasionally issued records with piano accompaniment, Lagrange, pp. 180–184.
98 "Lost Levantine Houses of Beirut, Beirut Arts Club, Mar Mikhael (Beit Boustani)", http://www.levantineheritage.com/lost-houses-beirut.html (accessed 14.7.2020). Since the house was sold in 2003, the piano is now a in a cousin's apartment in the Ashrafiyya quarter.

Cultural magazines of the time – intensely observing the West – acted as additional channels. A prominent Ottoman women's magazine such as *Hanımlara Mahsûs Gazete* sometimes devoted articles to the benefit of the piano for girls' education, and occasionally even published sheet music and lyrics of famous songs.[99] The magazine explained that women playing the piano upheld the happiness of the family and would please their husbands coming home tired. The piano was said to be more elegant than the violin, but the *qanun* or *'ud* could serve as substitutes. Manuals on the ideal (upper or upper-middle class) outlay of the home, and household duties, of the prevalent *tadbir al-manzil* (household management) genre, generally recommended having a piano.[100]

In the course of the following decades, this came to apply not only to the main urban centers but also to provincial cities. One example from the available literature was the township Tirebolu on the Black Sea in the early years of the Republic, where pianos (and gramophones) became the elite's symbols of modernity and of girls' education.[101] Probably this was true for more peripheral places in the region.

---

[99] Ersin Altın, *Rationalizing Everyday Life in Late Nineteenth Century Istanbul c. 1900* (dissertation, New Jersey Institute of Technology, 2013) pp. 132–134; Ayşe Zeren Enis, *Everyday Life of Ottoman Muslim Women: Hanımlar Mahsûs Gazete (Newspaper for Ladies) (1895–1908)* (Istanbul: Libra Kitap 2013), pp. 407–411. Nahide Işık Demirakın, "The city as a Reflecting Mirror: Being an Urbanite in the 19th Century Ottoman Empire" (PhD Thesis, İhsan Doğramaci Bilkent University 2015), p. 94, a young woman in a novel taught the piano, and French and English, to become a worthy wife. Also Ayşe Nursenal, "A Reading in the Late 19th Century Istanbul, Public Life and Space Through the Tanzimat Novels", (MSc. Thesis, Izmir Institute of Technology 210), p. 84, and Fabian Steininger, "Morality, Emotions, and Political Community in the Late Ottoman Empire (1878–1908)", (PhD Thesis, Free University Berlin 2017), pp. 235–238. Cf. Haris Exertzoglou, "The Cultural Uses of Consumption: Negotiating Class, Gender, and Nation in the Ottoman Urban Centers during the 19th Century, *International Journal of Middle Eastern Studies*, vol. 35/1 (2003), p. 81 (to a great extent focusing on the Greek-Orthodox middle class).

[100] Mona Russell, "Modernity, National Identity, and Consumerism: Visions of the Egyptian Home, 1805–1922", in Relli Shechter (ed.), *Transitions in Domestic Consumption and Family Life in the Modern Middle East* (New York 2003), pp. 45, and p. 48 quoting from Francis Mikha'il's advice manuals and articles. However, in "Competing, Overlapping and Contradictory Agendas: Egyptian Education under the British Occupation, 1882–1922", *Comparative Studies of South Asia, Africa and the Middle East*, vol. 21/1–1 (2001), p. 56, she mentions a contemporary concern that practical homemaking skills should get priority. See for Turkey also *Nevsal-i Afiyet* (Yearbook of Health), quoted by Ersin Altın, *Rationalizing*, p. 109. We also found the mandolin as an instrument for women. A somewhat later example (1925) are the drawings in the Iranian magazine *Khalq* contrasting Iranian and European families (the latter with a piano), Camron Michael Amin, pp. 76–77.

[101] Hale Yılmaz, *Becoming Turkish: Nationalist Reforms and Cultural Negotiations in Early Republican Turkey, 1923–1945* (Syracuse: Syracuse UP 2013), pp. 108–110, and p.115 where a person buys a Pleyel for his daughters from a bank director in Trabzon who was transferred elsewhere.

Even in the Egyptian countryside, pianos could be found in the mansions of village shaykhs who were apt to adopt urban manners.[102]

We have already referred to diplomats and missionaries in Palestine. As elsewhere, gradually, higher classes there acquired pianos. A British (non-Jewish) supporter of Zionism, Lawrence Oliphant, somewhat cynically describes opulent residences of Syrian-Christian families having "a 300-dollar piano, on which the lady never plays".[103] Piano ownership definitely applied to high class Christian milieus where girls were really taught how to play. The well-known educator Khalil Totah (a Quaker) owned a piano; he and his wife even cooperated with the Iraqi poet Ma'aruf al-Rusafi in setting his poems to music.[104] The prominent Greek-Orthodox Sakakini family, who were lovers of classical music acquired an expensive German Blüthner grand.[105] More unusual was the case of Husam al-Din Jarallah, a leading *'alim* and later Jordanian Chief-Qadi, who sent his daughter Sa'ida to the Catholic Schmidt's College and the Dames de Sion school where she learned to play the piano.[106] Undoubtedly, more members of the Palestinian upper classes had pianos at their homes.

Palestine is rather a special case due to its Jewish immigrant population bringing along their pianos from Europe. An anecdote (possibly made up) tells us how in 1909 the piano of Shoshanna Bluwstein, sister of the famed poetess Rachel, was unloaded in the harbor of Jaffa where the Ottoman customs officer allegedly (but a bit unlikely) asked her "Who needs a piano here in Palestine? This is not Europe or Paris; belly dancing is not possible on piano music."[107] In the 1930s, not a few ref-

---

102 Mona Abaza, *The Cotton Plantation Remembered, an Egyptians' Family Story* (Cairo: American University in Cairo Press 2013). The large salon of the mansion in Kafr Tanbul in the Delta featured a piano, and both young girls in the family could play it. This confirms Gabriel Baer's description, "The Village Shaykh, 1800–1950 in his *Studies in the Social History of Modern Egypt* (Chicago: University of Chicago Press 1969), p. 51.
103 Lawrence Oliphant, *Haifa or Life in the Holy Land* (Jerusalem 1887), pp.115–116 (thanks to Dick Bruggeman for the reference).
104 Rachel Beckles Willson, "Doing More than Representing Music" in Joshua S. Walden (ed.) *Representation in Eastern Music* (Cambridge: Cambridge UP 2013), p. 262. Rusafi lived only for a short time in Jerusalem.
105 Itamar Radai, "The Rise and Fall of the Palestinian-Arab Middle Class, *Journal of Contemporary History*, vol 51/3 (2016), p. 494. See also Adam Raz, Looting of Arab Property During Israel's War of Independence [in Hebrew] (Jerusalem: Carmel/Aqevot 2020), quoting from Sakakini's diary, p.64
106 Suzanne Schneider, "Religious Education and Political Activism in Mandate Palestine" (PhD Thesis; Columbia University 2014), p. 201.
107 Amos Bar, *HaMeshoreret MiKineret, Sippura shel Rahel* [The Poetess from Kinneret, The Story of Rachel] (Jerusalem: Ben Zvi and Am Oved 1993), p. 31. (Thanks to Yuval Ben-Bassat for the reference).

ugees from Germany managed to take their pianos along (in 1935 alone as many as 372). Radio broadcasting (from 1936), as elsewhere, and concert life, contributed their share too, and not only among the Jewish population.

6.7 Further diffusion

It is nearly impossible to assess the wider diffusion of pianos in the Middle East. Even if it remained limited to the upper layers of the population it was not totally insignificant. Local commercial handbooks (often in French), foreign trade statistics, and consular reports, not in spite of, but because of, their common bias toward the elite, may give us some indication.

Allegedly, pianos began to be imported commercially into the Ottoman Empire after the Crimean War (with many other new European commodities), probably for the courts but also for other upper-class families. This is corroborated by the emergence in the latter part of the 19th century of specialized stores in the eminent urban centers which offered various well-known European brands. In 1863, the Grande Rue de Pera (İstiklal Caddesi of today) had already four stores which sold pianos.[108] Among the most prominent importers was Comendinger, the music instrument dealer and publisher originally from Leipzig, whose store regularly advertised in foreign-language papers, magazines, and almanacs (Fig. 34).[109] The firm, meaningfully styled Fournisseur de S.E le Sultan, offered prestigious brands such as Bechstein, Érard, Pleyel, as well as the lesser-known Kaps from Dresden. In Cairo and Alexandria, outstanding names of piano dealers were Calderon, Papasian, and Bodenstein. Their names were conspicuous enough as markers of the foreign residential elite; we can add others such as Belefantis, Bertero, and Poliakine, Granato, as well as Boulous and Chidiac (possibly a Lebanese Christian).[110] Izmir in 1889 had at least seven piano teachers (all with Levantine names),

---

[108] Mestyan, p. 295 quoting the annual Cervati and Sargoglu, *L'Indicateur Constantinopolitain* of 1863. Paul Cervati, a member of the publisher's family, was himself an accomplished pianist and composer. Kutlay, p. 396, counted 11 piano sellers in Istanbul in 1868 and 41 in 1913. Also, the number of indispensable piano tuners grew and reached 11 between 1903 and 1909.
[109] Comendinger, "Maison Fondée en 1859" carrying the predicate "Fournisseur de SM. Le Sultan" with a *tughra* (the Sultanic monogram), sold and also published itself sheet music, and maintained "a musical library". Sheet music (probably many marches but also polkas, waltzes, and quadrilles), was an important medium for piano playing, and deserves closer investigation. Comendinger also published piano works by local composers, e.g, Edgar Manas, an Armenian, co-author of the later Turkish national anthem.
[110] Orosdi-Back in Istanbul, Izmir etc. sold western musical instruments, see Kupferschmidt, *Orosdi-Back Saga*, pp. 40 and 65 but apparently not pianos.

Fig. 34: Regular advertisement from the *Annuaire Oriental du Commerce* (here 1889–90) by the Comendinger music store in Istanbul, priding itself to be purveyor of the Sultanic court. Among the reknowned piano brands they represented were Érard, Kaps, Bechstein, Pleyel and some others. (permission: Salt Research Yearbooks, Istanbul).

and according to another source of 1900, fifteen shops there were dealing with pianos.[111] A similar guide for Cairo, as late as 1951, lists eleven piano dealers, and five in Alexandria, and one each even in peripheral places such as Bani-Suwaif, Fayum, and Minya.[112]

At least one Italian piano maker, Balatti, had established himself in Istanbul, apparently then still in conjunction with Comendinger. It would seem that his pianos were considered relatively expensive.[113] We know the names of a few private piano teachers, almost all female, and piano tuners (males, mostly Italians). A bit later, the *Hanımlara Mahsûs Gazete* and similar magazines regularly advertised piano teachers, with foreign names or diplomas.[114] For a long time, teaching the piano was mainly consigned to foreigners, often Italians and Frenchmen, or to local Armenians or Jews. By 1901, one could find also least two Italian piano makers and five tuners in the Ottoman capital.[115]

Upper class Cairo followed its formal capital Istanbul, albeit with its own aspirations to adopt western culture. There too existed a connection between the elite's love for the opera, and a demand for piano lessons, especially for the daughters of the "better educated Egyptians".[116] The small interested Egyptian audience at the time, meaning Cairo and Alexandria, could easily attend performances by resident or visiting European musicians. A few of them were musicologists or composers and pianists who tried to connect to Arab or Egyptian music. But, with all their intentions, once more, these were imitations, only hinting at the oriental music, and not a real bridge.

---

111 Alex Baltazzi, review of *Izmir Rumlarının Müziği 1900–1922*. Internet March 2013
112 Max Fischer (ed.) *The Egyptian Directory* (Cairo: 1951), p. 702 where one finds three piano tuners in Cairo and one in Alexandria. El-Shawan, p. 610, even counts twenty [western] music stores in Cairo. Berque, pp. 332–333 observed that one shop on 'Imad al-Din street, had been sequestered in 1914 – possibly Austrian – , and came into the hands of a Spaniard. El-Shawan p. 610. mentions twenty music stores in Cairo.
113 *Coup d'Oeil Général sur l'Exposition Nationale de Constantinople* (Istanbul 1863), p. 120. In the interesting chapter on instruments, mention is also made of an Italian tuner and piano repairer, Luigi Gambara, who exhibited an intriguing double stringed piano, his own design, made of local wood, with a pedal to shift sound effects, p. 122.
114 Ersin Altin, *Rationalizing*, p. 410. See also Burhan Çağlar, "Brief History of an English-Language Journal in the Ottoman Empire: *The Levant Herald* and *Constantinople Messenger* (1859–1878)", (MA Thesis University of Toronto 2017), p. 41. Edgar Whitaker, the owner, was an accomplished pianist himself. A later example, for 1925, is a Turkish piano teacher Qadriye Abdulreşid Hanım who advertised, emphasizing her education by Mayerman in Berlin in *Asar-i Nisvan* (1 Haziran 1341)
115 Daniel J. Grange, *L'Italie et la Méditerranée (1896–1911)*, (Rome 1994), p. 486.
116 Adam Mestyan, *Art and Empire: Khedive Ismail and the Foundation of the Cairo Opera House* (M.A. thesis, Central European University Budapest 2007), p. 82 mentions a certain Frederick Kitchener of the Opera, an Englishman, who from 1909 taught piano to pupils of "22 nationalities", p. 82.

There was no lack of performances of western classical music. The Czech, Josef Hüttel, served as the conductor of the large Egyptian Radio Symphonic Orchestra which comprised Egyptian as well as foreign-residential musicians (many Italians).[117] Not only the diffusion of gramophone records, but also radio broadcasts and the screening of Arab musical movies, in all countries, began to contribute their share to making western music more familiar, and to the dissemination of western music. This also lent more publicity to local pianists. One of them was Jenő Takács, a Hungarian musician, who held a piano teaching position at the Cairo Conservatory from 1927 to 1932. During these years he composed and performed several pieces drawn from Arab folk tunes, and lectured also on the radio, but his impact in Europe was greater than in Egypt.[118]

Iran, where according to one source, a French company started selling pianos in 1866, saw several musicians continuing experiments with the tuning of the standard piano. Morteza Khan-i Mahjubi, born in 1900 into a family of musicians, retuned the instrument to play in the so-called *dastgah* mode, in the way we have seen before in the Qajar court. To this end he also designed a new notation system, It is interesting to note that the family had a piano at home.[119]

6.8 Statistics

Joseph Grünzel, an Austrian trade researcher, writing in 1903 on the importation of western musical instruments into the Ottoman Empire, noted that pianos were among the fastest to strike roots. French manufacturers, who had still been dom-

---

117 In 1921 he came to Egypt, where he was conductor of the Alexandria Philharmonic (1929–1934) and head of the European music dept. of Cairo Radio (1934–44). On Hüttel (born in 1883) see further Zein Nassar, "A History of Music and Singing in Egyptian Radio and Television" in Michael Frishkopf (ed.) *Music and Media in the Arab World* (Cairo: American University in Cairo Press 2010), p. 69. Western classical concert life still needs more research.
118 Wolfgang Suppan, *Jenő Takács, Dokumente, Analysen, Kommentare* (Eisenstadt: Burgenländische Forschungen 1977); idem, "Jenő Takács – ein Arabischer Bartok", *Jahrbuch für Volksliedforschung* (1982–1983), pp. 297–306. Later he also worked in the Philippines (with thanks to the publishers for sending me a copy).
119 His contemporary Javad Maroufi (b.1912), also from a family of musicians, is also mentioned as a composer who tried to harmonize Persian *santur* music for a retuned piano, but more following western lines, Maryam Farshadfar, "Techniques of Piano Performance with Regard to Persian Music: a Study on the Traditional and the Innovated Styles, such as the Use of Fluid Piano or the Pianos Tuned Equal Temperament (A Special Study Project, Fall 2012), https://universityofmontreal. academia.edu/MaryamFarshadfar (last accessed 6 March 2023). See also "Iranian Piano", *Wikipedia*.

inant some thirty years before, had possibly been overtaken by German companies such as Blüthner and Bechstein.[120] Still, a grand piano would cost on the order of 800–1000 francs, as compared to a mere 2–50 Francs for a very cheap German or Austrian violin. Clearly, only few could afford them.

Considering 1910 as a top year for world production, attaining 600.000 pianos, with 370.000 in the USA alone, Ottoman imports appear to be negligible even as far as the ongoing aspirations of the elite are concerned. Very few historians who follow the cultural impact of globalization trends, mention the piano. The French scholar, Jacques Berque, was an exception, pointing out that Egypt imported pianos "in Louis XV or Louis XVI style… decorated with mother-of-pearl inlay, and carved moldings in mahogany", and – as we have mentioned – with a third pedal for a "mandolin effect". However, for a population of about 13 million, his statement of 3.232 instruments imported in 1923, seems unlikely.[121] In 1930–1931 this number would be 300, a more reasonable number.[122] Yet, rich families who had several daughters even bought a piano for each of them.[123]

It is impossible to assess, even by the mid-20th century, how many pianos had ever been acquired by wider populations in the Middle East. Not to forget, by schools, theatres, clubs, and the larger hotels and cafes which catered to foreign visitors, some of which had piano bars,[124] In the Cairo of 1910, it seems that piano-organs could be seen and heard in several cafes and bars, particularly the modern ones distinguished by electric lighting.[125] During the short-lived phase of the silent cinema, piano-music had accompanied movies. After 1917 Istanbul saw the influx of a number of refugee pianists from the Soviet Union.[126] None of the household surveys which begin to appear as late as the mid-20th century, as far

---

120 Josef Grünzel, *Bericht über die Wirtschaftlichen Verhältnisse des Osmanischen Reiches* (Vienna 1903): Imports of musical instruments in 1314 (1898–99) to the amount of 1.7 m. piaster included pianos, organs and harmoniums, violins, mandolins and guitars. Wind instruments (*Blasinstrumente*) from Austria, also France and Germany, mainly for military bands in the capital.
121 Jacques Berque, *Egypt, Imperialism and Revolution* (London: Faber, transl. 1972), pp. 332–333. Quoted also by Magda Baraka, *The Egyptian Upper Class Between Revolutions, 1919–1952* (Reading: Ithaca 1998), p. 163 who, however, fails to elaborate on this important aspect.
122 David Gurevich, *Foreign Trade of the Middle East…1930–1931* (Jerusalem: The Jewish Agency 1933), p. 218 lists for that year 300 imported by Egypt, 94 by Syria and Lebanon, 93 by Palestine, 99 by Iran, and 12 by Turkey (probably incorrect. Most instruments imported to Palestine came from Germany and Austria, while Iran imported most from Russia).
123 Suzy Eban, *A Sense of Purpose, Recollections* (London 2009), here quoted from the Hebrew edition, p.83, says that three pianos in her paternal home were no exception.
124 Mehrez, pp. 102–103, 203, 271.
125 Sladen, *Oriental Cairo*, pp. 63 and 68.
126 Jak Deleon, *The White Russians* (Istanbul: Remzi 1995), pp. 63–68, 85–86, 90–92, 103 [Thanks to Canan Balan for the reference].

as is known to us, mention pianos. One additional way to arrive at an assessment would be to look at pictures of families and households, but indoor photography before the Second World War was rare, for lack of flashlights.[127] A few, indeed, do show a piano, but this is far too biased to serve as proof.

6.9 An "oriental piano" for the Middle East?

Given the acquisition of conventional European pianos, while not always discarding authentic local music, the question arises whether anybody considered the option to adapt the instrument in such a way that it could faithfully render local music. Think of European small technologies again, many of which blew over to the countries of the Middle East.

In Europe, at least from the sixteenth century onward, explorations took place into the theory of microtonal music, generally in connection with standardizing tuning systems. [128] We assert that this was for a long time not known in the Middle East. Around World War I, experiments with practicable quartertone harmoniums and piano keyboards had come to fruition. One example is the German composer, Willy von Möllendorff, who presented a "bichromatic" 24-tone harmonium to the public in Berlin and Vienna in 1917. But a harmonium with sounds deriving from bellows and reeds, greatly differs from the hammer and string percussion mechanism of the piano.

Such European "inventors" of quartertone pianos did not aspire to an instrument which would faithfully encompass the range of Middle Eastern, or in fact Arab, music. Their experiments were aimed at the composing of new musical art forms, and never yielded a large repertoire, let alone one which was offered to Middle Eastern audiences.

Around 1918, the Russian avant-garde composer, Ivan Wyschnegradsky, had begun experimenting with a set of two differently tuned upright pianos. It would seem that Wyschnegradsky, too, was not interested in bridging the divide between European and Eastern music. Refused by the leading manufacturers, Gav-

---

127 See Erol Ahmed, *Photographic Trust, Photographic Truth in the Hamidian Period* (BA Thesis, University of Michigan 2009), pp. 36–37, undated photographs by Ali Sami, a court photographer, respectively of his wife and of an unknown man at upright pianos. Search also the collection of the Beirut Arab Image Foundation.
128 The British musicologist, Hugh Davies, published an excellent survey "Microtonal Instruments" in *Grove Music Online*. See also Franck Jedrzejewski, *Dictionnaire des Musiques Microtonales* (Paris; l'Harmatan 2003), on Habá, pp. 85–95, on quartertone pianos, pp. 199–201, on Wyschnegradsky, pp. 250–273.

eau and Érard, and in 1921 disappointed with Pleyel, he continued his search for a potential manufacturer of a suitable innovative instrument in Germany. There, in 1923, he met the Czech composer, Alois Habá, his peer in age, but originally educated as a violinist, and much more interested in folk music. He had begun composing in quartertones in 1920, at first for violin and other string instruments, but soon for piano as well. In 1923 Habá had engaged the German-Czech firm Förster to manufacture a quartertone piano, – a very sophisticated instrument with an intricate mechanism consisting of a double keyboard and double strings. Gerard Förster delivered a first grand model in 1924 to Habá, followed by another one to Wyschnegradsky in 1928.[129]

Alois Habá performed his quartertone compositions, and demonstrated the innovative piano in 1924 in Prague, Paris Sienna, Warsaw, and Berlin, together with his assistant, later composer, Karel Reiner. Berlin at the time was the scholarly hub of ethnomusicology, a new academic field which attracted several Central and East European composers who were interested in folklore.[130] He also gave demonstrations and lectures elsewhere in Europe. According to Habá himself, it was his first opera, "Matka", which aroused the interest of the Egyptian Mahmud al-Hifni who was at the time studying with the ethno-musicologists Curt Sachs and Erich von Hornbostel in Berlin. A decade later, this led to an invitation to Cairo.[131]

---

[129] Letter to this author by Wolfgang Förster, nephew of Gerard, 25th Nov. 2010 [with many thanks for the additional material]. As indicated, Wyschnegradsky had ordered a quartertone piano from the famous French manufacturer, Pleyel, which disappointed him. The piano was noted in the Egyptian press and appeared with a photograph in *al-Lata'if al-Musawwara*, see below. Wyschnegradsky's piano was exhibited at the Leipzig fair of 1928. Two more models were made for the Oriental Music Institute in Cairo in 1930 and for an unnamed client in Alexandria in 1931. They have since become museum objects: Habá's first is now in Prague, and Wyschnegradsky's in Basel. Förster also built one sixth-tone piano for Habá, as well as several lighter quartertone harmoniums for didactic purposes. In 1926 the Straube firm in Germany built such a harmonium for Mordecai Sandberg, a Romanian, later Palestinian, composer and physician who set the Psalms to quartertone music. On the latter see Roman Brotbeck, "Völkerverbindende Tondifferenzierung: Mordecai Sandberg – ein verkannter Pionier der Mikrotonalität", *Neue Zeitschrift für Musik*, vol. 152/4 (1991), pp. 38–44.
[130] Israel J. Katz, *Henri George Farmer and the First International Congress of Arab Music (Cairo 1932)* (Leiden: Brill 2015), pp. 176–177. Reiner would survive the Theresienstadt concentration camp where he played an important role in its music scene.
[131] Alois Habá, *Mein Weg zur Viertel- und Sechsteltonmusik* (Düsseldorf: Orpheus, Gesellschaft zur Förderung der Systematischen Musikwissenschaft 1971) in which memoirs he elaborates on his career and the planning of a quartertone piano. On Cairo, see there, p. 55, but strangely, he does not elaborate on his time in Egypt. His quartertone compositions were criticized by Nazis and Communists alike, but Habá to some extent resumed his activity after World War II. In 1956 he was invited to lecture in Lebanon, but he admits that his quartertone music did not catch on there either, p. 63.

Across the ocean, in 1928, the American composer, Charles Ives, commissioned a quartertone piano with a single two-layer keyboard, to be built by the Chickering company. Ives had composed a few pieces for it, and the instrument was to be played by the renowned German-born pianist, Hans Barth, who had settled in New York in 1908. Barth himself was also a composer of quartertone concertos.[132] However, Ives's health declined, and he could not continue his composing. This episode was therefore not very fruitful either.

In the Middle East, the music scene was not static, but evolved from a completely different vantage point. By the end of the 19th and the beginning of the 20th centuries, vivid discussions, especially in new specialized music magazines, related to the advancement of musical composition (basically meaning westernization) or, as some claimed, on the supposed retardation of Arab music. Modernists among them put the blame to the absence of harmony and polyphony, and the lack of notation and equal temperament. Some would even plead for the need of new instruments, blaming the limited capabilities of the 'ud which hindered the evolution of the musical creativeness. In fact, in those years, there were songs arranged for piano.[133]

From among the countries in the region, it was Lebanon, or rather specific segments of its population, which might have been exposed to the earliest experiments with an "Eastern piano". Not surprisingly, almost all musicians who took part in this process had a Christian – often a Protestant – background, as well as some French educational connection. The construction of the first local quartertone piano is ascribed to a Lebanese musician named Najib Nahad in 1912. This, at least, is the opinion of the Aleppo-born ethno-musicologist Christian Poché who, however, tells us no more about him.[134] In Nahad's footsteps followed his more fa-

---

[132] Charles Ives, *Selected Correspondence of Charles Ives* (Oakland: Univ. of California Press 2007), pp. 151–152. See also Howard Boatwright, "Ives's Quarter-Tone Impressions", *Perspectives of New Music*, vol. 3 (1965), pp. 22–31, and Peter Thoegersen, "Charles Ives's Use of Quartertones: Are They Structural or Expressive?', https://www.academia.edu/35604414/Charles_Ivess_Use_of_Quartertones_Are_They_Structural_or_Expressive See also his "Quartertones Are Not Out of Tune: You are!", https://www.academia.edu/2562791/Quartertones_Are_Not_Out_of_Tune_You_Are (accessed 4 March 2023). Barth was born in Leipzig and inspired by Feruccio Busoni who was equally interested in new scales. In 1930, Barth performed his own quartertone piano concerto in Philadelphia.
[133] Tess Judith Popper, "Musical Writings from the Modern Arab 'Renaissance' in Nineteenth and Twentieth Century Syria and Egypt", PhD Thesis University of California, Santa Barbara 2019).
[134] Poché, "Vers une Musique Libanaise", p. 81 only mentions that this particular piano had 17 keys based on the scale of Safi al-Din mentioned before. See further Lagrange, pp. 184–185. The Poche family (originally without Gallicized accent), Protestants from Bohemia, were a prominent commercial family in Aleppo. It is said that his grandfather Joseph Poche and Leon Orosdi, son of Adolf who had fled from Hungary, had played together in an amateur string quartet, see Mafalda

mous fellow countryman Wadi' Sabra (1876–1952), a Protestant by birth.[135] His initial music education had been accompanied by the American missionary Henry Jessup. After piano and organ studies at the Paris Conservatory, and conducting episodes in Istanbul, he cooperated in 1919 with Gustave Lyon of the Pleyel firm in Paris to construct a piano with 36 sixth-tones.[136] Sabra would later declare that he had worked on it for twenty years. He also published sheet music for the piano, mainly Oriental marches, and Arabic poems, as well as several books and articles on the Arab tone scale. In fact, he argued that Europe, to a large extent, owed its harmony system to medieval Arab musical theory. Sabra is today mainly remembered as the composer of the Lebanese national anthem and the founder of the National Lebanese Conservatory (1930), rather than for his Oriental piano.[137]

The invention of an "Oriental piano" in Lebanon – if we can call it an invention because it was rather a development –, might or might not have been directly linked to the experiments which were taking place in Paris and Berlin. In any case, the trends going on in Lebanon drew attention in Egypt as well. There, certain musicians were attempting to anchor its musical heritage in a national culture and education system, and some had argued that the unavailability of a suitable piano had delayed the development of "Oriental" music, later called Arab or Egyptian music.

The year 1922 saw a few important try-outs in which the outstanding singer and composer, Sayyid Darwish, was involved.[138] In fact, he had recognized the possibilities of the (western) piano, and even had cautiously made use of it in his musical theater pieces, a genre which was becoming increasingly popular in Egypt at the time. In April of that year, Najib (Naguib) Nahhas, a lawyer of Lebanese extraction, who served as advocate in the Mixed Courts in Alexandria, demonstrated a

---

Ade, *Picknick mit den Paschas, Aleppo und die Levantinische Handelsfirma Fratelli Poche (1853–1880)* (Beirut: Orient-Institut 2013), p. 59.
135 Among other publications, see Wadi' Sabra, often spelled Wadia or Wadih, *Exposé d'un Nouveau Système Perfectioné de Partage 12 Demi-Tons de l'Octave* (Beyrouth 1940). See further Ilan Pappe, whose *The Modern Middle East* (London: Routledge 2005) is one of the few comprehensive social history textbooks on the region with a chapter on music, pp.163–181, on Beirut see p. 165.
136 Pleyel was at the time headed by Gustave Lyon who was a world-renown expert on acoustics. The dialogue between (mainly) Lebanese musicians and European piano manufacturers needs closer study.
137 See https://www.nytimes.com/1930/02/24/archives/hans-barth-gives-a-unique-recital-performance-on-quartertone-piano.html *NYT* 1930, also *al-Hilal*. For our days, see several didactic items on quartertones on YouTube.
138 As described by his son, Hasan Darwish, *Min Ajal Abi Sayyid Darwish* (Cairo: al-Hay'a al-Misriyya al-'Amma 1990), pp. 362–373.

quartertone piano of his own design at the Club Syrien d'Alexandrie.[139] This also pointed to a Lebanese connection.

In September 1922, the magazine *al-Lata'if al-Musawwara* showed a picture of Emile Eff. al-'Ariyan (Arian) with his "Eastern (or Oriental) piano" (*biyanu sharqi*), manufactured in Paris, which was said to reproduce both Western and Eastern tunes, in this case 24 quartertones per octave (Fig. 35).[140] 'Ariyan, the designer, an engineer of Lebanese descent, demonstrated it on 24 November at a meeting in Cairo which was attended by some prominent musicians, including Sayyid Darwish. The instrument was symbolically endorsed for use as "beneficial" (*saliha*) by those present.[141]

Not everyone, of course, agreed. In the following years, a vivid debate went on, starting from prominent Lebanese and Egyptian musicologists, and continuing in Egyptian musical magazines such as *Rawdat al-Balabil* and *al-Midmar*.[142] 'Ariyan himself defended his adoption of the 24-tone scale in the Francophone *Bulletin de l'Institut d'Egypte* of 1924).[143]

Alberto Hemsi, a prominent Jewish composer from Alexandria, also stepped into the debate. Born in Turgutlu (Western Anatolia) he received his musical education in Izmir and Milan. However, a career as a pianist was precluded after being wounded in Italian military service. Between 1928 and 1957 he was teaching, composing, and publishing in Alexandria, where he specialized in Sephardi folk music. In 1928 Hemsi wrote that an Oriental Piano, with due respect, would remain a hybrid instrument, unwanted, and lacking a suitable repertoire. He wrote "... malgré tout, *les milliers de pianos* se trouvant dans les pays de l'Islam n'ont pas encore donné la moindre petite mélodie susceptible d'avoir une place digne dans le domaine polyphonique" [*my italization*]. Even Habá's compositions, he concluded, were "difficult to play, and impossible to listen to". In short, he argued that there was little point in developing such an instrument if nobody could play it properly or compose for it. In consequence, he recommended the development

---

139 A prototype is still on exhibit at the new Cairo opera.
140 *Al-Lata'if al.Musawwara* , 11 September 1922
141 Hasan Darwish, *Min Ajal Abi*, pp. 359–373.
142 To which one can add *al-Radyu*, and the press in general, which proves that music remained an important cultural concern; this was the case also in Turkey and to some extent probably in other Middle Eastern countries as well.
143 Emile Aryan, "Preuve Irréfutable de la Division de l'Échelle Musicale Orientale en 24 Quarts de Ton", *Bulletin de l'Institut de l'Egypte*, vol. 6 (1924), pp. 159–167.

of local music education along two parallel lines, the Eastern and European traditions.[144]

6.10 The Arab Music Congress of Cairo in 1932

It is not surprising that the long-standing debate on the so-called Oriental Piano became one of several aspects of the Arab Music Congress which convened in Cairo for three weeks in 1932. As a landmark, the Congress was a convergence of cultural and political tensions of the time, either reform and westernization, or Egyptian, Arab, and Eastern identity and hegemony. It was certainly one of King Fuad's cultural ambitions, intended to give Egypt a regional lead in this musical field and to re-classify "Oriental" as "Arab" music. In 1929 Fuad had already established an Institute of Arab Music which reflected trends which could tentatively be called reformist.[145] In view of the critical aspirations in different Egyptian musical circles, one could see the program of the Congress as a reform agenda.[146] The initiative to hold an international conference was shared by Mahmud al-Hifni, the above-mentioned Egyptian who had studied musicology in Berlin. The Congress was attended by prominent musicians and musicologists from Egypt, the Arab East and the Maghrib, Turkey and Iran, as well as by first-rank European musicologists and composers. It stands out as one of the most important scientific East-West dialogues in the history of Arab music research. Among the foreigners were Bela Bartók, Paul Hindemith, and Alois Habá.[147] The Hungarian Jenő Takácz also attended. Teaching piano at one of Cairo's conservatories since 1927, the latter had to his name several pieces for piano in quartertones, inspired by indigenous melodies.[148]

---

[144] Alberto Hemsi, *La Musique Orientale en Egypte, Études et Polémiques* (Alexandria, Edition Orientale de Musique 1929), esp. pp. 11–16, the quotes are from p. 12 and 13. See also David Maslowski, "Les Modèles Culturels des Juifs d'Égypte de la Fin de la Domination Ottomane (1882) jusqu'à la Révolution des Officiers Libres (1952)", M.A.Thesis Paris I). who ascribes the founding of the Philharmonic Orchestra of Alexandria to Hemsi, as well as of the conservatory which taught both Oriental and Western classical music. He also founded the Édition Orientale publishing house.
[145] Philippe Vigreux, "Centralité de la Musique Égyptienne", *Egypte/Monde Arabe*, vol. 7/1991.
[146] Anne Elise Thomas, "Intervention and Reform of Arab Music in 1932 and Beyond", Conference on Music in the World of Islam, *Assilah* (8–13 August 2007).
[147] Linda Fathalla, "Instruments à Cordes et à Clavier dans les Recommendations du Congrès du Caire", in CEDEJ, *Musique Arabe, Le Congrès du Caire de 1932* (Cairo 1992), pp. 99–103 with relevant figures. The CEDEJ volume is the most comprehensive scholarly analysis on the Congress. See also, Lagrange, pp. 192–193.
[148] Wolfgang Suppan, *Jenő Takácz, Dokumente, Analysen, Kommentare* (Eisenstadt, Burgenländische Forschungen, vol. 66, 1977) [with thanks for sending me a copy], and idem, "Jenő Takácz –

However impressive the attendance of the Congress, it was also full of controversy. A closer reading of the debates in the Egyptian press has laid bare internal tensions and discords, as well as its failure to reach consensus on major issues.[149] And one of these concerned the piano, as we shall see.

Firstly, scholarly researchers have found an alleged bias in the selection of the participants, ignoring, for instance, *ālātī* (traditional) instrumentalists. Secondly, this also implied the relative deprecation of a country such as Lebanon. And thirdly, most Egyptian participants expected an unequivocal endorsement of their perception that Arab music had to catch up with western standards of harmony, polyphony, and what they considered as musical progress. European musicologists, on the other hand, turned out to be conservatives in the sense that they were anxious to preserve Arab music as was, and to prevent any "ravaging effects" of the West on it (as Bartók once called it).[150] These European priorities are equally reflected in the selection of recordings made at the Congress which are still considered as one of its achievements. True, using advanced technologies, western musicologists such as the German Curt Sachs and the Hungarian composer Bela Bartók (with his strong ethno-musicological interests), had their way, pushing strongly for the preservation of traditional or folk music.

Regarding one of the principal questions, namely the standardization of Oriental scales and modes, the relevant sub-commission of the Congress, led by the above-mentioned Lebanese Père Collangettes, did not reach a consensus and, even more interestingly, some of the participants were vacillating.[151] The French delegate, Carra de Vaux, argued that this was an impossible task in view of the too many regional variations in pitches.[152] For instance, the exact notation of the so-called *sīkā* (E half-flat), led to a heated debate on the entire *maqam* scale system. Consequently, the proposal to adopt the tempered 24-note quartertone

---

Ein Arabischer Bartók", *Jahrbuch für Volksliedforschung*, vol. 2/28 (1982/3), pp. 297–306. Alois Habá, Bela Bartók, and Charles Ives all composed some quartertone music but were less directly inspired by Arab music. Takács later worked for some time in the Philippines. None of his works necessitated a quartertone piano.

149 Ali Jihad Racy, "Historical World Views of Early Musicologists: An East-West Encounter in Cairo, 1932" in Stephen Blum et al, (eds.), *Ethnomusicology and Modern Music History* (Urbana and Chicago: Univ. of Illinois Press 1993), pp. 69–91. And see also his "Musicologues Comparatistes Européens et Musique Égyptienne au Congrès du Caire" in CEDEJ, *Musique Arabe*, pp. 109–122.

150 "La musique de divertissement européenne fait sentir ici aussi son effect ravageur", CEDEJ, *Musique Arabe*, p. 95 and 97n.

151 Vigreux (paragraph 44) writes that some considered a musical evolution culminating in European harmony as "a sort of passport to musical universality".

152 But in correspondence with Najib Nahhas he may have been prepared to yield, and the same seems true for Collangettes, CEDEJ, *Musique Arabe*, p. 232.

scale, as had been suggested by the Lebanese Mikhail Mishaqa three of four generations earlier, was not fully endorsed. While Egyptian representatives such as Emile 'Ariyan and Najib Nahhas, both of whom had a stake in quartertone pianos of their own design, supported the 24-note scale, Wadi' Sabra, the director of the Beirut Conservatory, who led the Lebanese delegation, and who had brought along a different proto-type, protested. The authoritative musicologist, Rauf Yekta Bey from Turkey, opposed the tempered scale, while some of the other prominent musicologists with different perceptions were undecided.[153]

The issue of the admissibility of the piano, which was in essence connected to the debate on scales, was debated by another sub-commission under Curt Sachs, which had to decide on accepting (western) instruments in Eastern or Arab music in general. This had implications for the debate on the so-called Oriental piano.

Fig. 35: The 'Ariyan quartertone piano of 1922 (*al-Lata'if al-Musawwara*, 11 September 1922).

---

153 Wadia Sabra, *Congrès de Musique Arabe du Caire, Étude Détaillée sur les Travaux des Commissions* (n.p, n.d), pp. 3–5.

At least five different quartertone pianos were tried out at the Congress.[154] Najib Nahhas, as we have seen, had demonstrated one prototype in Alexandria in 1922. Since then, he had worked on two improved models, now presenting a model built by the Gaveau and Herberger firms in Paris.[155] George Samman, a musician from Alexandria, came with a simpler model, built by a German firm called Schmitt, which apparently did not encompass the entire range of Arab quartertones.[156] Moreover, Nahhas accused him of plagiarism. Emile 'Ariyan presented his model, which had already been heard in 1922. Wadi' Sabra (heading the Lebanese delegation) presented a quartertone piano which was equipped with "lobomatic" hammers and could allegedly play any Oriental *maqam* with its 36 keys (or 90 commas) per octave. He told the participants of the Congress that over twenty years earlier he had visited Paris (apparently Pleyel) several times to accomplish this model but that he had built it himself. However, it was said in the press that the development of his model was in fact to be largely ascribed to Bishara Ferzeïn (Firzan), a piano professor at the Beirut Conservatory. Worse, during one of the evenings of the Congress, a performance at the French Lycée led to a debacle, when two upcoming singers of the day failed to harmonize their singing with the instrument; singing was the ultimate proof of the correct pitch. One was Umm Kulthum, then 34 years old; the other, 'Abd al-Wahhab, then 30 years old. A fifth model quartertone piano demonstrated there was the one designed by Alois Habá and built by Förster in 1926. It had two rows of keys (Figs. 36 and 37, see also Fig. 13).[157] Unfortunately, no recordings are available, though there seems to be a short movie (which we have not seen) on Nahhas in which he plays the piano.

The members of this commission were mostly German scholars who were reserved on the matter, at least as long as the conventional piano had not been convincingly adapted to quartertones. On the other hand, one of the Egyptians, Muhammad Fathi Bey, tabled a motion to the effect that Oriental music played on a piano was not only a matter of scales, but also of form and rhythm, and that

---

154 CEDEJ, *Musique Arabe*, pp. 347–354 and highly interesting photographs of the various propotypes. The total number of pianos presented at the congress is confusing as some were imported and exported without proper customs registration, or maybe also owing to French-German rivalries. See also Mark Thorn, "Conflating Instruments and Music: 'The Piano Controversy' in Cairo", *Thorn Pricks*, 10 November 2006, http://thornpricks.blogspot.com (accessed 4 August 2022).
155 One still survives at the new Opera building in Cairo.
156 *Al-Lata'if al-Musawwara*, 4 April 1932, carries a story on the piano "invented" by George Shaman.
157 CEDEJ, *Musique Arabe*, pp. 347–354. There might have been two more models there. Förster manufactured very few of them (1924–1928?) and sold even fewer to the Arab world. But there was no commercial incentive to produce such pianos on a larger scale. Nevertheless, at least one was acquired by the Arab Music Institute in Egypt.

Figs. 36 and 37: A Főrster quartertone piano of 1926 in the National Music Museum of Prague (photo: Naama and Zwi Kupferschmidt, by permission of the museum).

the audience would over time get used to it. This pending question was then also referred to the plenum of the Congress.[158]

It is worth noting that opinions on the piano were not strictly divided according to European-Middle Eastern lines.[159] In the final plenary session, Muhammad Fathi made one more attempt to press his view that the piano was like other "scientific inventions" which transcended culture and nationality. In fact, he argued that in the Middle Ages, the East had lent its instruments to the West. On his side were Alois Habá, Fortunato Cantoni, the Italian conductor of the Cairo opera house, and various Arab delegates such as Najib Nahhas. However, the opposite draft proposal to the effect that the piano was absolutely unsuitable to Arab music, was narrowly defeated by a majority of one. It had been submitted by the German delegates, among them the ethno-musicologists Erich Von Hornbostel and Curt Sachs and the composer Paul Hindemith and was supported by the Lebanese Wadi' Sabra and the Turk Cemil Mesut. Henri Farmer, the prominent British orientalist historian of Eastern music, noted in his diary: "Better to admit the quartertone piano, than be forced to lose the identity of Arabian music by being circumvented by the European half-tone piano."[160] The admissibility of the piano remained in a diplomatic deadlock as concluded in the *Recueil* of the Congress.[161]

The Cairo Congress, of course, lacked authority to put a ban on the piano. In the following decade, the prominent Egyptian singer and composer, 'Abd al-Wahhab, occasionally included the piano in his works, as well as the accordion, along with some other "western" instruments such as the double-bass in a few of his cinema songs, albeit as modest accompaniments, – not as major musical leads. He employed a Greek pianist.[162] 'Abd al-Wahhab even imitated the tango and the waltz.

---

158 Sabra, *Congrès*, p. 12.
159 Robert Lachmann attending the Congress still as a German delegate, even speaks about "agitated discussions", Ruth Katz, *The Lachmann Problem*, pp. 315–316, also pp. 331–333.
160 Israel Katz, *Farmer*, pp. 170–171 and p. 182n.where he explains:" only the Pythagorean wholetone, fourth, fifth and seventh were *just* intervals, whereas the third and sixth were not". See also Ruth Katz, "The Lachmann Problem", p. 119. Elsewhere Farmer writes on Habá's demonstration of the Förster quartertone piano:" His instrument was tuned to the modern tempered scale, and his compositions were – to me – wholly acceptable in his own harmonic system. Where is all this going to lead us? I can only say, as Muslims themselves would express it, – 'Allah alone knows'", Henri George Farmer, *Studies in Oriental Music, reprints edited by Eckhard Neubauer* (Frankfurt a.M.: Goethe University 1997), p. 342.
161 *Recueil*, pp. 690–696.
162 Nabil Salim Azzam, *Muhammad 'Abd al-Wahhab in Modern Egyptian Music* (PhD Thesis, University of California 1990), pp. 137–138, and pp. 260–261; Walter Armbrust, *Mass Culture and Modernism in Egypt* (Cambridge: Cambridge University Press 1996), pp. 104, and 131, and 237n. (quoting A.J. Racy). But some movies of the 1930s already frowned upon aristocrats frequenting piano bars.

Also, Asmahan, sister of the famous singer Farid al-Atrash, in at least one famous song, used a piano as rhythmic accompaniment.[163] In a later phase, quartertone accordions and clarinets also made their appearance.

Though not accepting the conventional piano, nor any quartertone prototype of it, the sub-commission on western instruments, also rejected the cello, considering it too monotonous (*sonore*), but it did endorse the violin. The western 16th-century type of violin was not much older than the piano and had probably made its appearance in the Ottoman regions not much earlier than the piano. It would seem that western violins began to rival the *kemençe* (from Persian *kamanča* or *kaman*) sometimes referred to as the Eastern violin), and the *rababa* or *rebeb*, in the course of the nineteenth century. Surprisingly little has been written on this transition. There is evidence of western violins being introduced by Balkan minorities and Roma (gypsies) who came to perform in the Ottoman center.[164] A local Aleppan family, Sawwa, with three generations from 1830s onward, became famous as violinists and performed also in Egypt.[165] We indicated that violin lessons were given also at the Ottoman Court. In Morocco, by the way, the western violin was also introduced but retained the position on the knee (in a sort of *da gamba* style).

The explanation is clear: Violins could be tuned to Middle Eastern music with its quartertones and *maqam*s, whereas pianos could not. At the beginning of the 20th century, violins, and even cellos and basses, regularly entered *taht*s (ensembles, later: *firqa*s), the modern "Arab orchestras" which accompanied singers. From the 1930s, Umm Kulthum's orchestra, for instance, was partly based on traditional instruments such as the *'ud, qanun,* and *nay*; it would gradually include more and more violins, an accordion, and even a cello, but never a piano.[166] An innovation which was also reflected in the westernized dress of the players.

6.11 Piano education

The Cairo Congress devoted much attention to the introduction of a modern curriculum of music education in Egyptian schools. In fact, it would seem that not

---

See also Viola Shafik, *Arab Cinema, History and Cultural Identity* (Cairo: American University in Cairo Press 2000), p. 134.
163 "Dakhalat Mara fi-Junayna", composed by Midhat 'Asim who was also a pianist and modernist advisor of the Egyptian Broadcasting Station.
164 Also in Morocco, Ruth Katz, *Lachmann*, p. 310.
165 Legrange, pp. 73, 85, 92, 124, 129, 133, and 149.
166 Legrange, p. 201. The *nay* is an end-blown flute.

only Egypt. but several other Arab countries as well, in principle, would accept the 24-tone system as standard.[167] But for the time being, in the 1930s, this did not include the piano. Here, one may argue that the piano was hardly anywhere in the world an instrument taught to all schoolchildren.

It meant that piano teaching remained largely within the realm of the urban elites. This could be observed in Istanbul, where individual teachers advertised their lessons in foreign-language periodicals, while several private conservatories sprung up as well. One was headed, for a short while, by the German Paul Lange who had in 1880 assumed positions as church organist and piano teacher. Toward World War I he also directed some ensembles of the Court. Such private conservatories, mainly teaching the piano, were duplicated elsewhere as well, and in the majority were staffed by foreigners.

Interwar Cairo had a plethora of such institutions and educators: Joseph Berggrün, Joseph Richter, Joseph Szulc, Ettore Puglisi, Vincenzo Caro, Rachel Salib, and Hans Hickmann.[168] For Alexandria we mention Piero Guarino and Guiseppe Galasso.[169] The *Conservatoire Berggrün*, founded in 1921 by a Russian pianist of that name, is often mentioned, especially since it was taken over and elevated by the more successful Ignace Tiegerman. The founder himself left for Argentina and subsequently for the USA.

Ignace Tiegerman, an outstanding piano performer and teacher, has often been mentioned as an extraordinary phenomenon: a Polish-Jewish pianist who in 1931 came to live in Egypt for its dry climate, and remained there till his death in 1968. Had it not been for his asthma, it is said that he could easily have made a splendid international career. In 1933 he took over the Berggrün *conservatoire*, which attracted the Egyptian upper class and foreign-residential pupils.[170] It is said that over the years nearly a thousand talented piano pupils were

---

167 Thomas, pp. 5–6.
168 Laura Robson,"A Civilizing Mission? pp. 107–129. The German ethno-musicologist Hans Hickmann, strongly interested in Pharaonic music, but also an organist and pianist, lived in Egypt from 1933–1957 conducted his conservatory, Musica Viva. See an obituary in *Ethnomusicology* 13/2 (1969) pp. 316–319. Later there was also a Claudio Monteverdi Conservatory. Caro and Salib were mentioned to me by Sonia Gergis (private correspondence, with thanks).
169 Marta Petricioli, *Oltre il Mito, l'Egitto degli Italiani (1917–1947)* (Milan: Mondadori 2007), p. 272.
170 Among the staff was Leila Birbary Wynn with a Lebanese background, who became married to Wilton Wynn, a legendary correspondent of AP. For her memories "Edward Said's Piano Teacher Recalls Ignace Tiegerman in Cairo" see *Arbiter* no. 25. November 2019. But violin lessons were given as well.

taught there, which is an impressive estimate.[171] Even more remarkable is that several of them such as Ignaz Friedmann, Henri Barda, and Selim Sednaoui would become internationally renowned pianists in their own right.[172]

From the fact that Egypt produced several well-known piano performers, one concludes that there was a (limited) milieu in which they could thrive. Apart from this, there was also a local Egyptian piano repertoire composed or arranged by composers trained in western music, e.g Mathilda and Sophie 'Abd al-Massih, Abu-Bakr Khayrat, and Halim al-Dab' (El-Dabh).[173] Another, Yusuf Greis, started as a violinist.

6.12 Edward Said

In this respect, the story of Edward Said is characteristic. He remains one of Tiegerman's better-known pupils. Starting from the field of comparative literature, Said primarily gained fame for his book *Orientalism* (1978) and for later studies, in which he criticized Western misrepresentation and distortion of Middle Eastern culture. But, at the same time, he was also an eminent music reviewer. In fact, among his roughly twenty-five books and volumes, there are at least four on classical western music.[174]

---

171 Nevine Miller, whose father Husayn Sirri, a politician who served three terms as Prime Minister, was taught by "a visiting Polish professor of the Warsaw conservatory"; she remembers having played on Prince Faruq's upright piano, *The Ivory Cell* (Bloomington: Author House 213), pp. 2,17, and 18. Gamal Nkrumah, "Follies of Fate", *al-Ahram Weekly* , 24/30 December 2009, interviewed princess Nevine 'Abbas Halim who was not gifted enough, was not taught by Tiegerman himself but by a Greek lady, Mme Marianos.
172 Samir Raafat, "Ignace Tiegerman, Could He Have Dethroned Horowitz?", Egy.Com-Judaica {internet], originally published *Egyptian Mail*, 20 September 1997, and Allan Evans, an audio restoration pioneer," Ignace Tiegerman: The Lost Legend of Cairo", *Arbiterrecords* (on internet 1999). Tiegerman also taught the princes Hasan 'Aziz Hasan (grandson of Khedive Ismai'il) and the young prince Faruk. See further Samir Raafat, "Pianissimo: Henri Barda, an Egyptian-born Virtuoso", *Egyptian Mail*, 1 April 1995 and Egy.com.on the web, as well as "'Cairo Confidential' – Une Rencontre avec le Pianiste Henri Barda", *Concerts de Monsieur Croche*, accessed 11 November 2019.
173 El-Shawan, p. 610.
174 Edward Said, *Musical Elaborations* (London: Chatto and Windus 1991); *Out of Place, a Memoir* (London; Granta Books 1999); *Reflections on Exile & Other Literary and Cultural Essays* (New York: 2000), The British social historian, John M. Mackenzie, a critic of Edward Said, has drawn up an extensive bibliography in his *Orientalism, History, Theory, and the Arts* (Manchester and London: Manchester University Press 1995), pp. 138–175.

The music he so fondly talks about is the classical European repertoire which he absorbed from a very young age onward, first in Jerusalem and then in Cairo.[175] Indeed, his parental home belonged to the bourgeois echelon we have described. Cairo's Opera House, established by the Khedive Isma'il in 1869, was a cultural linchpin for the milieu in which he grew up, and his parents also loved opera and opera recordings. It is maybe not too far-fetched to fancy that this was one – or, maybe par excellence the – medium which later made him pursue the theme of Orientalism. After all, a large part of the opera repertoire of the 18th to 20th centuries is permeated with dramatized, and often distorted, images of the Orient. In a television interview he once declared that he wanted to investigate the connection between operas and Orientalism, but nothing as thorough as his study of literature came of it. Said did publish an essay on Verdi's *Aida*, which he dissected to its Imperialist and Egyptologist bones, but he was generally milder on Mozart. Supposedly this was due to the latter's Masonic inclination as perceived in his *Magic Flute* and his *Abduction from the Seraglio*.[176] Or, as one researcher remarked: "And this, much more than the modish habits of 'Turkish' music drew Mozart sympathetically eastwards."[177] But to the best of our knowledge, Said never systematically addressed the topic of *alla turca* music, and what musical style could be more "Orientalist" than that?

It might be argued that for Said, Oriental and Western music remained in two different orbits.[178] And he may not be the only one. In a rare reference, he admitted that at least in his childhood, he had little appreciation for Umm Kulthum. In an interview in 2000, he related that his mother took him to a performance of the famous singer, when he was aged only ten:

> It was a dreadful experience for me .... There did not seem any order to it. The musicians would wander on stage ... her songs would go on for forty to forty-five minutes .... the tone was mournful, melancholic. I did not understand the words. So, I very early on rejected it and began to focus exclusively on Western music, for which I hungered more and more.[179]

---

175 Laura Robson's perceptive analysis (with some bias against the "distinctly Zionist" character of the classical music scene in Jerusalem under the British Mandate) points to the elite's cosmopolitan milieu in which Said grew up, be it Jerusalem or Cairo. She also mentions the family's Protestant Christian affiliation which connects to the above passages on Lebanon.
176 Edward Said, "The Imperial Spectacle" in *Grand Street*, vol.6/2 (1987), pp. 82–104.
177 *Culture and Imperialism*, p. 118.
178 On the other hand, there is also the East-West Divan Orchestra which he founded with Daniel Barenboim, and a classical conservatory in Ramallah. See Robson's article which is in large part focused on this project.
179 Quoted by Rokus de Groot, "Perspectives of Polyphony in Edwards Said's Writings", *Alif: Journal of Comparative Poetics*, no. 25 (2002), pp. 219–220; *Out of Place* (London 1999), p. 96–97.

So, what was Said's relation to the piano, one of the outstanding social markers of the milieu in which he was brought up? In his memoirs he mentions the piano lessons at home in Cairo, which proves that not only girls received them. As for many upper-class children, absolutely not only in the Middle East, but this was also a typical tortuous part of their education. In his autobiography he laments: "Music was, on the other hand, a dissatisfying and boring drill of piano exercises...I was chained to in mindless repetitions, that did not seem to improve my keyboard powers sufficiently...." And further on: "...Miss Cheridjian had replaced Leila Birbary, (her) weekly appearances, were unpleasant confrontations over my ability to follow her shouted instructions – count fa, fa, ti, ti, forte, piano, staccato, punctuated by loud slurps of the coffee and energetic chomps on the cake brought to her by Ahmed, our ironic head suffragi."[180]

It did not hinder Said from becoming an outstanding amateur pianist, and connoisseur of classical music. But he did not treat it in the same way as he did western literature. "My own enjoyment of pianism is pointed towards the past...It is private memory that is at the root of the pleasure we take in the piano."[181] Steering away from this apparent dissonance between aesthetics and ideology, he instead not only elaborated on public and private perceptions of music, but also contemplated on theoretical aspects such as lack of counterpoint and polyphony in Arab music.[182] As to Umm Kulthum, given Said's overall frame of reference, there seems to be a re-evaluation as he now praised her "feminine consciousness and domestic propriety".[183]

6.13 An acoustic "glass bottom"?

Still, we have to come back to explaining how the piano remained stuck in the upper and aspiring middle classes and did not trickle down. Affordability was, of course, one overriding, albeit differential, factor anywhere, but it was not the only one.[184]

---

180 *Out of Place, Memoir* (London: Granta Books 1999/2000), pp. 96 and p. 97. The term "head *suffragi* "(chief servant) definitely signifies a class designation. Said had started taking lessons at the Berggrün Conservatory with Mrs. Bourkser (probably the Russian wife of David Bourkser of the National Bank of Egypt) who was succeeded by Mrs. Leila Birbary, mentioned above.
181 *Reflections*, pp. 216–217.
182 De Groot, *passim.*
183 *Reflections*, p. 346, an article first published in *London Review of Books*, vol.12 no 17 (1990).
184 Visiting Russian social democrat activists were astonished in 1907 that families of skilled workers in London could afford a piano, and piano lessons for their daughters, Arthur P. Dudden

232 — 6 Did the piano have a chance in the Middle East?

To begin with, by the mid-twentieth century, the piano in Europe and America was losing much of its impact. There, it was affected by shifting class standings, competition from more popular and less expensive instruments, the gramophone, and the radio, limited domestic space, and a certain saturation of the market. Cyril Ehrlich, even went so far as to remark that in Europe and America "the household god was dethroned".[185]

As for the countries of the Middle East in the 1940s and 1950s, they went through pervasive political, social, and ideological changes too, albeit to varying degrees. Meanwhile, authentic art and popular music, with their *maqams* and other characteristics, held their own, and nationalism with its emphasis on authenticity, left the "foreign" piano – on the whole – hardly a chance to integrate into the dominant cultural trends.

The fact that western music, and the piano as one of its dominant instruments, did not become fully integrated into the popular music scene, cannot be considered as a proof that something "went wrong".[186] To describe the barrier – it would be tempting to use the metaphor of breaking a glass *bottom*, the reverse of a glass *ceiling*. It was not a matter of discrimination, nor always an ideological rejection, but an aesthetic or emotional impossibility. There remains a complex "musical divide", to various degrees, all over the countries of the Middle East.

The radical westernization process of Republican Turkey, as enforced in the 1920s under Atatürk, was ambiguous on the piano. It could not politically connect with the piano's antecedents at the Ottoman court and its place among related elites. With military men, bureaucrats, and intellectuals now in the lead, it was Ziya Gökalp, who developed a fully-fledged ideological program, which discredited Ottoman perceptions of the past, belittled even the authenticity of the *maqams*, and theoretically advocated instead, a harmonization of western and Turkish folk music. For some time, traditional music was even banned from the State Radio. On the other hand, typically, Atatürk's spouse, Latife Hanım, from a rich Izmir family and with a European education, was known to play the piano but, after their divorce, he apparently sold it.[187]

---

and Theodore H. von Laue, "The RSDLP and Joseph Fels: A Study in Intercultural Contact", *American Historical Review*, vol. 61/1 (!955), p. 24.
185 Ehrlich, *The Piano: A History* (Oxford, rev ed. 1990), p. 199.
186 Bernard Lewis, *What Went Wrong? The Clash Between Islam and Modernity in the Middle East* (NY: Harper 2003), pp. 127–129, 133–137, and 149. Lewis makes a point of different perceptions of time and measurement as if *maqams* lack order.
187 Aydın Ildem, manager of the Pera Palace Hotel, later of the Megara Palace Hotel, knowledgeable about piano restorations, mentioned three French pianos which had been in the Topkapı Pal-

The pianist Vedat Kosal deplored that much of the rich pluralism – as he sees it – which had marked musical life under the Ottomans had fallen into oblivion.[188] His efforts and those of others, more recently, have revived interest in the Ottoman heritage as described above. It would be more fitting to say that the piano has since shifted from one elite to another. Between 1923 and 1936 several conservatories were opened; and the National Conservatory in Ankara was directed by Cevat Dursunoğlu who had studied in Berlin. A music teachers' training institution, Darülelhan, became a conservatory, for some time led by Musa Süreyya, who also had studied piano in Berlin.[189] The German composer Paul Hindemith was asked to advise, and following his suggestion at least three prominent Jewish pianists found employment in Turkey: Ludwig Czaczkes, George Markowitz and Eduard Zuckmayer. The Muzika-i Humayun ultimately became the Symphonic Orchestra., and Turkey had composers in a western style, in particular the famous "Five", one of whom, Ahmet Adnan Saygun has several piano pieces to his name; in part they incorporated Turkish folk music. But while the nature of Turkish music remains a topic of local interest and research, it did not mean that Gökalp's ideals of a new collective blend were achieved. Also, with several Turkish pianists reaching international fame, it did not turn the piano into a popular instrument, or as one teacher remarked "Students learn to appreciate western classical music with their minds, but don't really feel it in their hearts".[190]

Rather, a new, spontaneous genre emerged in Turkey, the so-called *arabesque* music, combining Anatolian, Arab, and Kurdish melodies, and *maqam*s, which broke away from the prescribed hegemonic ideal.[191] Amidst ongoing debates on an ideal Turkish musical culture, the country's arguably most renowned pianist, Fazil Say, called *arabesque* "tantamount to high treason".[192] Short of adopting western music, or be it one type of music, at least among a majority of the population,

---

ace, one of which came into the possession of Atatürk who donated it to his friend Topal Osman, a (controversial) military figure, in 2021 accessed on https://megarahotel.com.
188 Kosal, pp. 84–85.
189 Some very good Steinway and Bösendorfer grands which had been donated to conservatories and radio stations are apparently neglected [Private mail from present-day well-known pianist D.E.O].
190 Matthew Brunwasser, "Western Classical Music Struggles to Find Audience in Turkey", *The World* [public radio program in US], 11 June 2013. See also Damla Bulut and Ferit Bulut, "The Problems of Piano Teachers in Fine Arts and Sports High Schools and the Solution Offers", *Procedia*, vol. 51 (2012), pp. 152–156.
191 The term *arabesque* was initially meant as denigrating, invented by pro-Western intellectuals.
192 "Arabesque Music is 'High Treason'", *Hürriyet Daily News*, 14 November 2012. Around the same time, a blasphemy case relating to beliefs on the Hereafter, was brought against Say, but his conviction was ultimately revoked.

traditional instruments such as the *'ud* , the *qanun*, the *santur*, the *nay* and the *tubul, darbuka* and other drums, continued to dominate.[193] A decade ago, it was reported that annually 400.000 *bağlama*s (the Turkish long-necked *'ud*) were sold, mostly made in Turkey, as against the import of 15.000 guitars, 50.000 violins, and "approximately" 10.000 pianos.[194]

In Egypt, the lifestyle of the royal Court soon belonged to the past (if it ever had inspired the lower strata) and the same happened to the erstwhile effendi classes and the foreign residential communities, who had been piano enthusiasts. One may ask what happened after Greeks, Italians, and Jews of various nationalities had been forced to leave the country and had to sell or leave behind their pianos. It seems that in Cairo there are still large storehouses with secondhand pianos for sale or for rent. The new state bourgeoisie under the Nasirist regime, in their search of nationalist authenticity, by and large, turned away from western cultural models. The same must have happened also in other countries.

The villa in Giza of the last pre-Revolution Minister of Education, the eminent writer Taha Husayn – nowadays the Ramatan Museum –, still contains his grand piano in the reception hall. It had been a matter of personal education and taste; we do not know how often it was used. In a way, the piano's standing was henceforward redefined from above, albeit limited. Dr. Tharwat 'Ukasha, a Free Officer, and Minister of Culture (who happened to be a piano lover) reorganized the Conservatory in 1959. One of the piano teachers there was Ettore Puglisi, who somehow symbolizes the continuation to the new regime, having educated prominent Egyptian pianists such as Olga Yassa, Sonia Gergis, and Rafi Armenian, all from well-to-do musical families, and other outstanding indigenous performers.[195] However, most of them are better known abroad as piano performers rather than in Egypt itself. Western concert life and broadcasts attracts only small audiences.[196]

---

193 For the violin, and some other instruments, see above. The guitar is a special case cf. Motti Regev, *'Ud ve-Gitara* (Tel Aviv: Beth Berl 1993) [in Hebrew].
194 *Hürriyet*, 7 April 2013, on the basis of an *Economist* survey which also referred to the existence of 1500 music stores (source not located). Also, the state embarked on a Piano for Every School campaign. Turkey then had a population of app. 77 m.
195 Kind information on Puglisi by Sonia Gergis to author, 15 February 2020.
196 Selim Sednaoui, "Western Classical Music in Umm Kulthum's Country" in Sherifa Zuhur (ed.) *Images of Enchantment, Visual and Performing Arts in the Middle East* (Cairo: AUPC 1998), pp. 123–134. The Higher Musical Committee was led by Abu al-'Uf another officer. The American pianist Gary Graffman who gave a few performances in 1979 (reportedly on several good but poorly maintained pianos) met "a contingent of Soviet teachers" (typically for that time) and about a two-thirds Egyptian audience in a environment of "no real concert life", "Visiting the Pyramids and Pianos of Egypt", 4 February 1979, https://www.nytimes.com/1979/02/04/archives/visiting-the-pyramids-and-pianos-of-egypt-visiting-egypt.html (accessed 4 August 2022).

A troubling but indicative case is Palestine: War circumstances proved that several prominent Palestinian families had possessed pianos but in fleeing Palestine in 1948 they had to leave them behind, and there are testimonies to the effect that they were appropriated by Jewish Israelis.[197]

Lebanon is again different, because it had a stronger piano culture than we have seen in neighboring countries, which is ascribed to its Christian populations and its relationship to Europe.[198] Not exceptionally, in 1929, the existing School of Music was transformed into a National Conservatory with the active involvement of Arkadie Konguel and his spouse, both Russian pianists.[199] More interesting are the Lebanese Rahbani brothers, who occasionally included a piano as accompaniment for the renowned singer, Fayrouz. This essentially differs from her equally famous counterpart, Umm Kulthum in Egypt.[200] Nevertheless, in Lebanon, there emerged scores of Oriental music written for pianos. Even so, it remained a limited experiment.[201]

6.14 Conclusions

Pianos, in spite of their qualities and different degrees of acceptance, historically and socially, have faded into the margins of musical scenes in the Middle East, and not only there. And yet, paradoxically, most Middle Eastern states maintain conservatories as essential educational markers of modernization. There, western classical music is taught, including the piano, albeit often in a two-track Eastern-Western approach. Most countries of the Middle East have brought forth performing

---

[197] Walid al-Khalidi, *Qabl al-Shatat* [Before Their Diaspora] (Beirut: Institute for Palestine Studies 1987), p. 140 has a photograph of the interior of a rich Arab house in Jaffa, app. 1935. See further Nathan Krystall, "The Fall of the New City", in Salim Tamari (ed.) Jerusalem 1948: The Arab Neighbourhoods and Their Fate in the War (Jerusalem: Institute of Jerusalem Studies 2002), p. 111; and Adam Raz, pp. 64, 69, 77–78, 80–81, 86–87, 95, 173 and 357.
[198] In Lebanon, this also applied to Sabri al-Sharif, who was born in Jaffa and had started his piano education at the Collège des Frères, and later established the band called 'Usbat al-Khamsa there, sometimes using the piano.
[199] Henri Gil-Marchex,"L'Institut Musical de Beyrouth", *Le Monde Musical*, April 1929, p. 138.
[200] Christopher Stone, *Popular Culture and Nationalism in Lebanon, the Fairouz and Rahbani Nation* (Abingdon: Routledge 2008) sees a limited impact of Sayyid Darwish's use of the piano on the Rahbanis, pp. 6, 50–51, 69, 171. Burkhalter, p. 156 says on the Rahbanis: "There are often just a few discreet hints at harmonics via the piano".
[201] Reuters reported on 22 March 2018 that a turbaned Shi'ite cleric, Sayed Husayn al-Husayni, was expelled from his seminary for posting a video of his piano playing, deemed "undignified" behavior.

pianists, usually educated by foreigners, or at institutions abroad.[202] This is also true for chamber and symphony orchestras and performances. Here and there, pianos serve as unexpected emblems of a sort of Occidentalism or modernism.[203]

Experiments with quartertone pianos continued for some time. Abdallah Chahine (Shahin), a Lebanese musician and founder of a renowned music store in Beirut, had a rapprochement between Eastern and Western music in mind while persisting for two decades with a new design. In 1954 he exhibited his novel, intricate quartertone piano with 119 notes on seven octaves and 280 chords, and with a middle pedal to shift scales. It was constructed according to his specifications at the piano factories of Hoffman in Vienna and Renner in Stuttgart. Having demonstrated his piano at public events, he recorded in 1962 some Oriental compositions, then played on yet another improved model.[204] A recent, very charming, graphic novel by his great-granddaughter, Zeina Abirached, has led to renewed interest in his piano but this is no more than a nostalgic act (Fig. 38).[205]

In 1974, the Syrian Wajiha 'Abd al-Haqq created a so-called *kithara* (literally something like "multiplier") with a double keyboard, – the second one for

---

[202] For a few decades it seemed that Russians took the dominant place of Italians. Several musicians were trained in Moscow. Also, the Syrian pianist and composer Ghazvan Ziriqli studied under two Russians at the Damascus Conservatory.

[203] For Syria see for instance: Jonathan Holt Shannon, *Among the Jasmine Trees, Music and Modernity in Contemporary Syria* (Middletown: Wesleyan UP 2006), p. 41, talks about a long waiting list for piano lessons at the Arab Music Institute in Aleppo, also p. 50. See also the "Piano Bar" and a few similar places in Damascus, Christa Salamandra, *A New Old Damascus, Authenticity and Distinction in Urban Syria* (Indiana UP 2004), pp. 2, 73–74, and 81.

[204] Chahine (1894–1975) began his career as a piano tuner and organist in the (Jesuit) St. Joseph's Church in Beirut, which has often served as a concert venue. His music business, founded in 1952, which specialized in western instruments, also produced records with the label Voix de l'Orient which promoted the singer Fayrouz. Since then, over three generations, the firm expanded into a well-known chain of stores, at least one carrying the name of one of his sons, Mozart Chahine. See Rita Eid, "Abdallah Chahine, When the Orient Meets the Occident Again", http://www.musimem.com/chahine-eng.htm (last accessed 6 March 2023). See further, Thomas Burkhalter. *Local Music Scenes and Globalization: Transnational Platforms in Beirut* (New York: Routledge 2013), pp. 152, 162, 171; and somewhat critically, Johnny Farraj and Sami Abu Shumays, *Inside Arabic Music: Arabic Maqam Performance and Theory in the 20th Century* (Oxford: Oxford UP 2019), p. 40–41. Exceptionally, Touffic Succar, a French-educated Lebanese composer and conservatory teacher, has composed several quartertone pieces for piano.

[205] *Le Piano Oriental* (Brussels; Casterman 2015) translated into English and German as well. From Abirached's tale it can be understood that Hoffman had hoped, in vain, for a hundred orders. Actually, she admits that synthesizers soon made it obsolete. A Greek French composer and pianist, Stéphane Tsapis, gives concerts in Paris and elsewhere on a Yamaha duplicate of Chahine's piano, and is recorded on disk.

Fig. 38: Cover of the graphic novel by the Lebanese artist Zeina Abirached, *Le Piano Oriental* (Brussels: Casterman 2015) on the quartertone piano invented by her grandfather 'Abdallah Chahine. Both depicted here, with the said piano (permission: Éditions Casterman).

micro-tones.[206] This might have been the instrument which Racy mentions as having heard in Baghdad in 1975.[207]

---

[206] Sheherazade Qassim Hassan, "Musical Instruments in the Arab World, in *Garland Encyclopedia*, vol. 6, p. 421.
[207] Ali Jihad Racy, "Historical World Views", p 91 n.10. Most recently, on the other hand, a group of Finish musicologists developed a new kind of quartertone piano with a triple-lined keyboard; its optical sensors are supposed to make playing (relatively) easier. Elisa Järvi, "The Keyboard Expands – Working on the New Quarter-Tone Piano" (PhD Thesis, Sibelius Academy, Helsinki

None of the quartertone pianos became successful anywhere, and no prototype ever reached the stage of commercial marketing. They were expensive, and difficult to tune and to play. And not really needed.

But then, one relatively persuasive solution to the adaptation of the piano to local music came up with advanced electronic synthesizers based on the time-tested keyboard: Where quartertone pianos had never been capable of producing satisfactorily the characteristic intervals and melodies of Middle Eastern music, synthesizers based on the piano keyboard now could, at least to one extent or another.[208] Called *urg* (an Arabic neologism from organ, but really more like a piano), they began to appear in the 1960s, soon followed by more advanced "Oriental keyboards" complete with software to produce a sheer endless variety of necessary *maqam*s and rhythms.[209] Like the prestigious piano manufacturers, there are nowadays several outstanding companies on the market.[210]

Such electronic keyboards made their entrée all over the countries of the Middle East, as well as in their diasporas. In Egypt, Magdi al-Husayni became wildly popular in the 1970s as a keyboard player. Popular bands habitually perform and record *sha'bi* (folksy) music at weddings and other celebrations, often combined with traditional instruments or electric guitars and drum kits.[211] This is equally true for their counterparts in Lebanon, Jordan, and other places.

2019). In the United Kingdom, Geoffrey Smith, following Mahjoubi, developed a so-called Fluid Piano (2003) with small sliding handles on each string to shift the tuning in order to play "international" music, including Persian melodies, Farshadfar, p. 14.

208 This is not entirely new; see similar conclusions by Racy, Ghrab and others.

209 Nicolas Puig, "Le Long Siècle de l'Avenue Muhammad 'Ali: d'Un Lieu et des Publics Musiciens", l'Égypte dans le Siècle 1901–2000, *Égypte Monde Arabe*, 2001, 1–2 (no. 4–5), p. 220, and same, "Egypt's Pop-Music Clashes and the 'World-Crossing', Destinies of Muhammad 'Ali Street Musicians" in Diana Singerman and Paul Ammar (eds.) *Cairo Cosmopolitan* (Cairo, AUCP 2006), p. 513–515; also, Floor, *Theatre in Iran*, p. 307. See further Anne K. Rasmussen, "Theory and Practice at the 'Arabic Org': Digital Technology in Contemporary Arab Music Performance', *Popular Music* vol. 15/3 (1996), pp. 345–365, based on field work mainly among Syrian and Lebanese Arab-Christian immigrant musicians in the USA (1986) contains important insights, such as the options of modulations and flexibility to tune the instrument beyond the Mishaqa 24-tone scale. Also, Shannon, p 14.

210 E.g., Korg, Yamaha, Casio, Ringway, Roland, Artesia, Kadence.

211 James R. Grippo, "What's Not on Egyptian Television and Radio! Locating the 'Popular' in Egyptian Sha'bi" in Frishkopf, *Music and Media*, pp. 141, 143, 148–149. Karin van Nieuwkerk, "Of Morals, Missions, and the Market: New Religiosity and 'Art with a Mission" in Egypt", in her ed., *Halal Soaps, and Revolutionary Theater: Artistic Developments in the Muslim World* (Austin: University of Texas Press 2011), p. 194, claiming that the synthesizer was even permitted as *halal* at strictly religious weddings and parties. Though loud beat music is frowned upon in

Clearly, the diffusion and integration of foreign musical instruments such as the piano could not make sense if they had no chance of being incorporated in the local music scene, or did not create new challenges for composers, arrangers, and performers. Instruments, players, and music itself, must act in harmony. Will electronic keyboards now make the difference?

Saudi Arabia, a few shopping malls have general music stores, some of which sell keyboards and even pianos.

Bibliography

Books and articles

Abaza, Mona. "Shopping Malls, Consumer Culture and the Reshaping of Public Space in Egypt." *Theory, Culture and Society* 18/5 (2001): 97–122.

Abaza, Mona. *Changing Consumer Cultures of Modern Egypt, Cairo's Urban Reshaping.* Leiden: Brill 2006.

Abdel-Latif, Abla M. "The Non-price Determinants of Export Success or Failure: The Egyptian Ready-Made Garment Industry, 1975–1989." *World Development* 21/10 (1993), pp. 1677–1684.

Abirached, Zeina. *Le Piano Oriental.* Brussels: Casterman, 2015.

Abou-Hodeib, Toufoul. "Taste and Class in Late Ottoman Beirut', *International Journal of Middle East Studies* 43:3 (2011): 475–492.

Abou-Hodeib, Toufoul. *A Taste for Home. The Modern Middle Class in Ottoman Beirut.* Stanford: Stanford University Press, 2017.

Adas, Michael. *Machines as the Measure of Men, Science, Ideology, and Technologies of Western Dominance.* Ithaca: Cornell University Press, 1989.

Ade, Mafalda. *Picknick mit den Paschas. Aleppo und die Levantinische Handelsfirma Fratelli Poche (1843–1880).* Beirut: Orient-Institut and Würzburg: Ergon Verlag, 2013.

Agstner, Rudolf, *Die Oesterreichisch-Ungarische Kolonie in Kairo vor dem ersten Weltkrieg, das Matrikelbuch des k. und k. Konsulates Kairo 1908–1914.* Cairo: Austrian Cultural Institute, 1994.

Agstner, Rudolf. "Das Wiener Kaufhausimperium 'S.Stein' im Osmanischen Reich", *Wiener Geschichtsblaetter* 59 (2004), pp .130–140.

Aharoni, Yisrael. *Zikhronot Zo'olog 'Ivri.* Tel Aviv: Ariel, 1946/7.

Ahmed, Leila. *A Border Passage.* New York: Farrar, Straus, and Giroux, 1999.

Akin, Nur. *19. Yüzyilin İkinci Yarısında Galata ve Pera.* Istanbul: Literatür Yayınları, 1998.

Ali, N. Mustafa M. "The Invisible Economy, Survival, and Empowerment, Five Cases from Abbara, Sudan." In R.A. Lobban (ed.), *Middle Eastern Women and the Invisible Economy.* Gainesville: University Press of Florida, 1998: 96–112..

Allen, Roger (ed.). *A Period of Time, A Study and Translation of Hadith 'Isa ibn Hisham by Muhammad al-Muwaylihi.* Oxford: Ithaca, 1992.

Almowanes, A.."History of Computing in Saudi Arabia: A Cultural Perspective." *International Journal of Social Science and Humanity* 7/7 (2017), pp. 437–441.

Alterman, Jon B. "The Middle East's Information Revolution." *Current History* 99 (2000): 21–26.

Altin, Ersin. *Rationalizing Everyday Life in Late Nineteenth Century Istanbul, c.1900.* PhD Thesis, New Jersey Institute of Technology, 2013.

Amin, Ahmad. *Qamus al-'Adat wal-Taqalid wal-Ta'abir al-Misriyya.* Cairo: al-Majlis al-A'la lil-Thaqafa, ed.1999.

Amin, Camron Michael. *The Making of Modern Iranian Women, Gender, State Policy, and Popular Culture, 1865–1946.* Gainesville, University Press of Florida, 2005.

Amin, Galal. *Whatever Happened to the Egyptians?* Cairo: American University in Cairo Press, 2001.

Ammon, D. *Crafts of Egypt.* Cairo: American University in Cairo Press, 1991.

Anastassiadou, Meropi. "Livres et 'Bibliothèques' dans les Inventaires après Décès de Salonique au XIXe Siècle", *REMMM* 87–88 (1999): 111–141.

Anid, Nada. *Les Très Riches Heures d'Antoine Naufal, Librairie à Beyrouth.* Paris: Calman-Levy, 2012.

Anker, Richard. *Gender and Jobs, Sex Segregation of Occupations in the World.* Geneva: ILO, 1998.
Appadurai, Arjun. *Modernization at Large.* Cultural Dimensions of Globalization, Minneapolis, University of Minnesota Press, 1996.
Araci, Emre. "Giuseppe Donizetti at the Ottoman Court: A Levantine Life." *The Musical Times* 143 (1880).
Arbel, Benjamin. "The Last Decades of Venice's Trade with the Mamluks – Importations into Egypt and Syria." *Mamluk Studies Review 8/2 (2004)*: 37–86.
Armbrust, Walter. *Mass Culture and Modernism in Egypt.* Cambridge: Cambridge University Press, 1996.
Arnaud, J.-L. *Le Caire, Mise en Place d'une Ville Moderne, 1867–1907.* Paris: Sindbad, 1998.
Arnold, David. "Global Goods and Local Usages: The Small World of the Indian Sewing Machine, 1875–1952." Journal of Global History 63 (2011): 407–429.
Arnold, David. *Everyday Technology, Machines, and the Making of India's Modernity.* Chicago: Chicago University Press, 2013.
Ashtor, Eliyahu and Guidobaldo Cevidalli. "Levantine Alkali Ashes and European Industries." *Journal of European Economic History* 12/3 (1983): 475–522.
Ashtor, Eliyahu. *Social and Economic History of the Near East in the Middle Ages.* London: Collins, 1976.
Ashur, R. *Qit'a min Uruba, Riwaya.* Cairo: Dar al-Shuruq, 2003.
Awad, Dana. "The Evolution of Arabic Writing due to European Influence. The Case of Punctuation." *Journal of Arabic and Islamic studies* 15 (2015), pp. 117–136.
Ayalon, Ami. *Reading Palestine, Printing and Literacy (1900–1948).* Austin: University of Texas Press, 2004.
Ayalon, Ami. *The Arabic Print Revolution, Cultural Production and Mass Readership.* Cambridge: Cambridge University Press, 2016.
Ayalon, Ami. *The Press in the Arab Middle East.* Oxford: Oxford University Press, 1995.
Ayubi, Nazih N.M. *Bureaucracy and Politics in Modern Egypt.* London: Ithaca 1980.
Azzam, Nabil Salim. *Muhammad 'Abd al-Wahhab in Modern Egyptian Music.* PhD Thesis, University of California, 1990.
Bachtin, Piotr. "The Royal Harem of Nasser al-Din Shah Qajar (r.1848–96): The Literary Portrayal of Women's Lives by Taj al-Saltana an Anonymous 'Lady from Kerman'." *Middle Eastern Studies* 51/6 (2015): 986–1009.
Badeau, John et al., *Genius of Arab Civilization, Source of Renaissance.* Oxford: Phaedon 1976.
Baldwin, A.H. *Optical-goods Trade in Foreign Countries.* Washington: Government Printing Office, 1911.
Bali, Rifat N. *Avram Benaroya: Un Journaliste Juif Oublié Suivi de ses Mémoires.* Istanbul : Isis, 2004.
Barak, On. *On Time, Technology and Temporality in Modern Egypt.* Berkeley: University of California Press, 2013.
Barakat, Magda. *The Egyptian Upper Class between Revolutions, 1919–1952.* Reading: Ithaca Press, 1998.
Baron, A. and S.E. Klepp, "'If I Didn't Have My Sewing Machine...': Women and Sewing Technology." In Joan M. Jensen and S. Davidson (eds.), *A Needle, a Bobbin, a Strike, Women Needle Workers in America.* Philadelphia, Temple University Press, 1984.
Baron, Beth. *Egypt as a Woman, Nationalism, Gender, and Politics.* Berkeley: University of California Press, 2005.
Baron, Beth. *The Women's Awakening in Egypt, Culture, Society, and the Press.* New Haven: Yale University Press, 1994.

Batheit, K. "Nachrichten über den Fezexport Oesterreichs nach dem Orient im 19. und Beginnenden 20. Jahrhundert", *Vierteljahrschrift für Sozial- und Wirtschaftsgeschichte*, 29 (1938): 296–303.
Baudrillard, Jean. *La Société de Consommation.* Paris: Folio, 1970.
Bauer, Arnold J. *Goods, Power, History: Latin America's Material Culture.* Cambridge: Cambridge University Press, 2001.
Beckles Willson, Rachel. *Orientalism and Musical Mission, Palestine and the West.* Cambridge: Cambridge University Press, 2013.
Beeching, Wilfred A. *The Century of the Typewriter.* Bournemouth: British Typewriter Museum, new ed. 1990.
Bellman, Jonathan (ed.). *The Exotic in Western Music.* Boston: NEU Press, 1998.
Ben-Bassat, Yuval. *Petitioning the Sultan, Protests and Justice in Late Ottoman Palestine.* London: Tauris 2013.
Benfeghoul, Farid. "Through the Lens of Islam: A Note on Arabic Sources on the Use of Rock Crystals and Other Glass as Vision Aids." In Cynthia Hahn and Avinoam Shalem (eds.), *Seeking Transparency, Rock Crystals Across the Medieval Mediterranean.* Berlin: Gebr. Mann Verlag, 2020.
Benfeghoul, Farid. "Through the Lens of Islam: the Pre-History of Eye-glasses according to Arabic Sources." In *Zeitschrift für Geschichte der Arabisch-Islamischen Wissenschaften* 23 (Sonderdruck 2022): 259–315.
Benjamin, Walter. *On Photography.* [1931, transl. Esther Lesly]: London: Reaktion Books, 2015.
Benjamin, Walter. *The Arcades Project.* Cambridge, Mass.: Belknap Press,1999.
Berger, Morroe. *Bureaucracy and Society in Modern Egypt, a Study of the Higher Civil Service.* Princeton: Princeton University Press, 1957.
Berque, Jacques. *Egypt, Imperialism and Revolution.* London: Faber & Faber, transl., J. Stewart, 1972.
Bier, Laura. *Revolutionary Womanhood, Feminisms, Modernity, and the State in Nasser's Egypt.* Stanford, Stanford University Press, 2011.
Birtle, Andrew James. *US Army Counterinsurgency and Contingency Operations Doctrine 1944–1946.*Washington D.C: Center for Military History US Army 2007).
Blind, A. *L'Orient vu par un Médecin, Egypte, Palestine, Syrie.* Paris: APM, 1913.
Bloch, Marc. *The Historian's Craft.* New York: Knopf, 1949 (1953).
Blumberg, Arnold (ed.). *A View from Jerusalem 1859–1858, the Consular Diary of James and Elizabeth Anne Finn.* Cranbury NJ: Associated University Presses,1980.
Bohne, W. *Handbook for Opticians.* New Orleans: Griswold 1895.
Boorstin, Daniel J. *The Americans, the Democratic Experience.* New York: Vintage Books, 1974.
Botz-Bornstein, Thorstein. *Veils, Nudity, and Tattoos: The New Feminine Aesthetics.* Lanham: Lexington, 2015.
Boulos, Issa I. *The Palestinian Music-Making Experience in the West Bank, 1920s to 1959: Nationalism, Colonialism, and Identity.* PhD Thesis, Leiden University, 2020.
Bourdieu, Pierre. *La Distinction.* Paris: Minuit, 1979.
Bowen, Donna Lee and Evelyn A. Early (eds.). *Everyday Life in the Muslim Middle East.* Bloomington: Indiana University Press, 2nd ed. 2002.
Brakha, Eliyahou. *Riding the Waves of My Life.* Tel Aviv: Teper, 2005.
Brandon, R. *Singer and the Sewing Machine, a Capitalist Romance.* Philadelphia: Lippincott, 1977.
Brian Jewell, Brian. *Antique Sewing Machines.* Timbridge Wells, Kent 1985.
Briggs, Asa and Anne MacCartney, *Toynbee Hall, the First Hundred Years.* London: Routledge, 1984.
Brookes, Douglas Scott. *The Concubine, the Princess, and the Teacher; Voices from the Ottoman Harem.* Austin: University of Texas Press, 2008.

Brummett, Palmira. *Image & Imperialism in the Ottoman Revolutionary Press, 1908–1911.* Albany: State University of New York Press 2000.

Brunwasser, Matthew. "Western Classical Music Struggles to Find Audience in Turkey." *The World* [public radio program in US], 11 June 2013.

Bulut, Damla and Ferit Bulut. "The Problems of Piano Teachers in Fine Arts and Sports High Schools and the Solution Offers." *Procedia* 51 (2012): 152–156.

Burke, Edmund III (ed.). *Struggle and Survival in the Modern Middle East.* Berkeley: University of California Press, 993.

Burman, Barbara (ed.). *The Culture of Sewing, Gender, Consumption and Home Dressmaking.* Oxford: Berg 1999.

Buruma, Ian. *Inventing Japan 1853–1964.* London: Weidenfeld and Nicolson, 2003.

Busch, Moritz. *Eine Wallfahrt nach Jerusalem. Bilder ohne Heiligenscheine.* Leipzig: Grunow, 3rd ed., 1881.

Byrn, Edward W. *The Progress of Invention in the Nineteenth Century.* New York: Munn & Co, 1900.

Cabasso, Gilbert, et al. *Juifs d'Egypte, Images et Textes.* Paris: Scribe, 1984.

Cachia, Anna and Pierre. *Landlocked Islands, Two Alien Lives in Egypt.* Cairo: American University in Cairo Press, 1999.

Çakır, Serpil. *Osmanli Kadin Hareketi.* Istanbul, Metis Yayınlari, 1993.

Carnoy, N. *La Colonie Française du Caire.* Paris: Presses Universitaires de France, 1928.

Carstensen, Frederick Vernon. *American Multinational Corporations in Imperial Russia. Chapters in Foreign Enterprise and Russian Economic Development.* PhD Thesis, Yale University, 1976.

Castelo-Branco, Salwa. "Western Music, Colonialism, Cosmopolitanism, and Modernity in Egypt." In Virginia Danielson et.al. (eds), *The Garland Encyclopedia of World Music,* vol. 6: The Middle East. New York and London: Routledge 2002.

Çelik, Burce "Cellular Telephony in Turkey: A Technology of Self-produced Modernity", *European Journal of Cultural Studies* 14/2 (2011):147–161.

Cevdet, A. *Ma'azurat.* Istanbul: Çağrı Yayınları, 1980.

Chadirji, Rifat. *The Photography of Kamil Chadirji.* Surbiton, 1991.

Chandler, A.D. *The Visible Hand, The Managerial Revolution in American Business.* Cambridge, Mass.: Belknap Press, 1995.

Chatty, Dawn. *Mobile Pastoralists.* New York, Columbia University Press, 1996.

Cinar, E. Mine. "Unskilled Urban Migrant Women and Disguised Employment: Home-working Women in Istanbul, Turkey", *World Development* 22 (1994): 369–390.

Coffin, Judith G. "Credit, Consumption, and Images in Women's Desires", *French Historical Studies* 18/3 (1994): 749–783.

Cohn, Herman. "Untersuchungen über die Sehleistungen der Ägypter", *Berliner Klinische Wochenschrift,* 16 May 1898.

Çorapçıoğlu, Yavuz (ed.), *International Ottoman Postal History Exhibition.* Ankara 1999.

Cresson, William Penn. *Persia: The Awakening East.* Philadelphia: Lippincott, 1908.

Crossick, G. and S. Jaumain (eds.). *Cathedrals of Consumption: The European Department Store 1850–1939.* Aldershot: Ashgate, 1998.

Cruz Fernandez, Paula A. de. *Gendered Capitalism: Sewing Machines and Multinational Business in Spain and Mexico, 1850–1940.* London: Routledge, 2013.

Curli, Barbara. "*Dames Employées* at the Suez Canal Company: The 'Egyptianization' of Female Office Workers, 1941–56", *International Journal of Middle East Studies* 46/3 (2014): 553–576.

d'Aumale, J. *Voix de l'Orient, Souvenirs d'un Diplomate.* Montreal: Editions Variétés, 1945.

Dalachanis, Angelos. *The Greek Exodus from Egypt, 1937–1962.* New York: Berghahn, 2017.
Davaz-Mardin, Asli. *Hanimlar Alemi'nden Roza'ya, Kadın Sureli Yayinlari Bibliografyasi; 1928–1996.* Istanbul: Babil, 1998.
David, Paul. "Understanding the Economics of QWERTY: The Necessity of History" in W.N. Parker (ed.), *Economic History and the Modern Economist.* New York: Blackwell, 1986.
Davies, Margerie M. *Women's Place is at the Typewriter, Office Work and Office Workers 1870–1930.* Philadelphia: Temple University Pess, 1982.
Davies, R.B. "'Peacefully Working to Conquer the World:' the Singer Manufacturing Company in Foreign Markets, 1854–1889", *Business History Review* 43/3 (1969): 299–325.
Davis, Fanny. *The Ottoman Lady, A Social History from 1718 to 1918.* New York: Greenwood, 1986.
Deleon, Jak. *The White Russians in Istanbul.* Istanbul: Remzi Kitabevi Publications, 1995.
Demirakın, Nahide Isik. *The City as a Reflecting Mirror: Being an Urbanite in the 19th Century Ottoman Empire.* PhD Thesis, Bilkent University, Ankara 2015.
Desjeux, Dominique et al., *Anthropologie de l'Électricité. Les Objects Électriques dans la Vie Quotidienne en France.* Paris: L'Harmattan, 1996.
Deuchar, Hanna Scott. "A Case of Multiple Identities: Uncanny Histories of the Arabic Typewriter." *International Journal of Middle East Studies* 52/3 (2023): 1–22
Diamond, Jared. *Guns, Germs, and Steel.* New York: Norton, 1997.
Dikötter, Frank. "Objects and Agency, Material Culture and Modernity in China." In Karen Harvey (ed.), *History and Material Culture: A Student's Guide to Approaching Alternative Sources.* London: Routledge, 2009: 58–172.
Dikötter, Frank. *Things Modern, Material Culture and Everyday Life in China.* London: Hurst, 2007.
Doctor, Vikram. "History of Typewriters in India: MK Gandhi's Love-hate Relationship with the Machine", *The Economic Times* (9 March 2012).
Douek, Raymond Ibrahim (ed.). *La Voie Égyptienne vers le Socialisme.* Cairo: Dar al-Maaref,1966.
Douglas, Allan and Fedwa Malti-Douglas. *Arab Comics, Politics of an Emerging Mass Culture.* Bloomington: Indiana University Press, 1994.
Douglas, Diane A. "The Machine in the Parlor: A Dialectical Analysis of the Sewing Machine." *Journal of American Culture* 5/1 (1982): 20–29.
Doumato, Eleanor Abdella. "Women and Work in Saudi Arabia: How Flexible Are Islamic Regimes?" *Middle East Journal* 53/4 (1999):20–29.
Doyle, Arthur Conan. "A Case of Identity." *The Strand Magazine* Sept. 1891.
Dreyfus, John. "The Invention of Spectacles and the Advent of Printing." *The Library* 10/2 (1988): 93–106.
Dudden, Arthur P. and Theodore H. von Laue. "The RSDLP and Joseph Fels: A Study in Intercultural Contact". *American Historical Review* 61/1 (1955):. 21–47.
Dumont, Paul. "Said Bey- The Everyday Life of an Istanbul Townsman at the Beginning of the Twentieth Century." In Albert Hourani et al. (eds), *The Modern Middle East.* London: Tauris, 1993.
E.A. Wallis Budge, *E.A. Cook's Handbook for Egypt and the Egyptian Sudan.* London: Cook and Son, 1911.
Early, E.A. *Baladi Women of Cairo. Playing with an Egg and a Stone* Cairo: American University in Cairo Press, 1993.
Eban, Suzy. *A Sense of Purpose, Recollections.* London: Halban, 2009.
Edgerton, David. *The Shock of the Old.* London: Profile Books, 2006.
Eğecioğlu, Ömer. "The Liszt–Listmann Incident." *Studia Musicologica* 49/3–4 (September 2008): 275–293.

Egypt Almanac. Wilmington: Egypto-file, 2003.
Ehrlich, Cyril. *The Piano, a History.* Oxford: Clarendon Press, rev ed. 1990.
Eldem, Edhem. *A 135-Year-Old Treasure, Glimpses from the Past in the Ottoman Bank Archives.* Istanbul: Osmanlı Bankası,1998.
Eldem, Edhem. *Bankalar Caddesi.* Istanbul: Osmanlı Bankası, 2000.
Emarah, M.H.M. "Arabic Eye Types for the Determination of Visual Acuity." *British Journal of Ophthalmology* 52 (1968), pp. 489–491.
Empire Ottoman. *Coup d'Oeil Général sur l'Exposition Nationale Constantinople.* Istanbul 1863.
Engler, George Nichols, *The Typewriter Industry: The Impact of a Significant Technological Innovation.* PhD Thesis, UCLA 1969.
Erol, Merih. "The 'Musical Question' and the Educated Elite of Greek Orthodox Society in Late Nineteenth- Century Constantinople." *Journal of Modern Greek Studies* 32/1 (2014): 133–163.
Fahmy, Ziad. "Media-Capitalism: Colloquial Mass Culture and Nationalism in Egypt, 1908–18', *International Journal of Middle East Studies* 42:1 (2010): 83–103.
Fahmy, Ziad. *Ordinary Egyptians Creating the Modern Nation through Popular Culture.* Stanford: Stanford University Press, 2010. Press, 1989.
Finn, Elizabeth Anne. *Reminiscences of Mrs. Finn.* London and Edinburgh: Marshal Morgan and Scott, 1929.
Floor, Willem. "Les Premières Règles de Police Urbaine à Téhéran', in Chahryar Adele and Bernard Hourcade (eds), *Téhéran: Capitale Bicentenaire.* Paris and Teheran: Institut Français en Iran, 992: 173–198.
Floor, Willem. *The History of Theater in Iran.* Washington DC: Mage 2005.
Fontaine, Laurence. *Histoire du Colportage en Europe, XVe-XIXe Siècle.* Paris: Albin Michel 1993.
Fontanier, V. *Voyages en Orient Entrepris par Ordre du Gouvernement Français de l'Année 1821 à l'Année 1829.* Paris, Librairie Universelle 1829.
François-Levernay (eds.). *Guide Annuaire d'Égypte, Année 1872–1873.* Cairo & Alexandria.
Fricke, Timothy Fricke et al. "Global Prevalence of Presbyopia and Vision Impairment from Uncorrected Presbyopia." *American Academy of Ophthalmology.* San Francisco, 2008.
Frishkopf, Michael (ed.). *Music and Media in the Arab World.* Cairo: American University of Cairo Press, 2010.
Frugoni, Chiara. *Le Moyen Âge sur le Bout du Nez: Lunettes, Boutons et Autres Inventions Médiévales* Paris: Belles Lettres (transl from Italian, 2001), 2011.
Fuhrmann, Malte. *The Port Cities of the Eastern Mediterranean.* Cambridge: Cambridge University Press, 2021.
Gaboda, Peter. "Back, Fülöp." *Magyar Múzeumi Arcképcsarnok.* Budapest: Pulszky Társaság, 2002: 29–30.
Gaboda, Peter. "Gamhudi Asásatás, 1907." *Múzsák Kertje, a Magyar Múseumok Születése.* Budapest: Pulszky Társaság, 2002: 46–47.
Gabriele Mitidieri, Gabriele. "'Un Autómata de Fierro': Máquinas de Coser, Ropa Hecha y Experiencas de Trabajo en la Ciudad de Buenos Aires en la Segunda Mitad de Siglo XIX." *Historia Crítica*, 85 (1 July 2022): 26–49.
Gallo, Ruben. *Mexican Modernity: The Avant-Garde and the Technological Revolution.* Boston: MIT Press 2010.
Gardey, Delphine. "The Standardization of a Technical Practice: Typing (1883–1930)", *Reseaux, The French Journal of Communication* 6/2 (1998): 255–281.
Gargour, Edouard. *Die Presse in Aegypten, Internationale Press-Ausstellung [sic].* Cologne, 1928.

Gaspar, Veronica. "History of a Cultural Conquest: The Piano in Japan." *Acta Asiatica Varsoviensa* 27 (2014): 1–15.

Gaspari, A.C. et al.. *Volkstümliches Handbuch der Neuesten Erdbeschreibng*. Weimar: Hassel, 1821.

Gendzier, Irene L. *The Practical Visions of Ya'qub Sanu'*. Cambridge Mass: Harvard University Press, 1966.

Gerholm, Tomas. "Economic Activities of Alexandrian Jews on the Eve of World War I." In Shimon Shamir, (ed.), *The Jews of Egypt, a Mediterranean Society in Modern Times* (Boulder 1987): 94–107.

Gershoni, Israel and James P. Jankowski, *Redefining the Egyptian Nation 1930–1945*. Cambridge: Cambridge University Press 1995.

Ghrab, Anas. "The Western Study of Intervals in 'Arabic Music' from the Eighteenth century to the Cairo Congress", *The World of Music* 47/3 (2005): 55–79.

Gil-Marchex, Henri. "L'Institut Musical de Beyrouth." *Le Monde Musical* (April 1929): 138

Gilbar, Gad G. *Trade and Enterprise, The Muslim Tujjar in the Ottoman Empire and Qajar Iran, 1860–1914*. London and New York: Routledge, 2023.

Giraud, Ernest. *La France à Constantinople*. Constantinople: Imprimerie Française, 1907.

Godfrey, Frank P. *An International History of the Sewing Machine*. London: Trans-Atlantic Pubs.,1982.

Godins de Soushesmes, G. des. *Au Pays des Osmanlis*. Paris: Victor-Havard, 1894.

Godley, Andrew. "Homeworking and the Sewing Machine in the British Clothing Industry 1850–1905" in B. Burman (ed.), *The Culture of Sewing*. Oxford: Berg,1999: 255–268.

Godley, Andrew. "Selling the Sewing Machine Around the World: Singer's International Marketing Strategies, 1850—1920." *Enterprise & Society* 7 / 2 (2006): 266–314.

Goehre, P. *Das Warenhaus*. Frankfurt a.M: Ruetten & Loening, 1907.

Goldschmidt, Arthur. *Biographical Dictionary of Modern Egypt*. Boulder: Lyne Rienner, 2000.

Golia, Maria., *Photography and Egypt*. London: Reaktion, 2010.

Göncü, T. Cengiz, *Belgeler ve Fotoğraflarla Meclis-i Mebûsân (1870–1920)*. Istanbul: TBMM Milli Saraylar, 2010.

Good, Edwin M., *Giraffes, Black Dragons, and Other Pianos, a Technological History from Cristofori to the Modern Concert Grand*. Stanford: Stanford University Press, 2nd. ed., 2001.

Gordon, Andrew. *Fabricating Consumers: The Sewing Machine in Modern Japan*. Los Angeles: University of California Press, 2012.

Gordon, Leland J. *American Relations with Turkey, 1830–1930*. Philadelphia: University of Pennsylvania Pess, 1932.

Gorman, Anthony. "Anarchists in Education: The Free Popular University in Egypt (1901)." *Middle Eastern Studies* 413 (2005): 303–320.

Gorman, Lyn and David McLean. *Media and Society in the Twentieth Century, a Historical Introduction*. Oxford: Wiley & Sons, 2003.

Gould, R.T. "The Modern Typewriter and its Probable Future Development." *Journal of the Royal Society of Arts* 76/3940 (1928).

Graham-Brown, Sarah. *Images of Women*. London: Quartet Books, 1988.

Grange, Daniel J.. *L'Italie et le Méditerranée (1896–1911)*. Rome: École Française, 1994.

Grazia, Victoria de and Ella Furlough (eds). *The Sex of Things*. Berkeley: University of California Press, 1996.

Grazia, Victoria de. *Irresistible Empire: America's Advance through Twentieth-Century Europe*. Cambridge Mass.: Belknap Press, 2005.

Green, Nile. "Journeymen, Middlemen: Travel, Trans-culture and Technology in the Origins of Muslim printing", *International Journal of Middle East Studies*.41:2 (2009): 203–224.

Green, Nile. "Persian Print and the Stanhope Revolution: Industrialization, Evangelicalism, and the Birth of Printing in Early Qajar Iran." *Comparative Studies of South Asia, Africa, and the Middle East* 30/3 (2010): 473–490.
Grehan, James. *Everyday Life & Consumer Culture in 18th Century Damascus*. Seattle: University of Washington Press 2007.
Grippo, James R. "What's Not on Egyptian Television and Radio! Locating the 'Popular' in Egyptian Sha'bi" in Frishkopf, Michael (ed.). *Music and Media*. Cairo: American University in Cairo Press 2010: 137–142.
Gronow, Pekka. "The Record Industry Comes to the East." *Ethnomusicology* 25/2 (1981): 251–284.
Groot, Rokus de. "Perspectives of Polyphony in Edwards Said's Writings." *Alif: Journal of Comparative Poetics* 25 (2002): 219–220.
Gündoğdu, A. Orkun. *Osmanlı/Türk Müzik Kültüründe Avrupa Müziği'nin Yayginlaşması Süreci ve Levanten Müzikçiler*. PhD Thesis, Başkent Üniversitesi, Ankara 2016.
Gurevich, D. *Foreign Trade of the Middle East* 1930–1931. Jerusalem: Jewish Agency, 1933.
Haan, Francisca de. *Gender and the Politics of Office Work, the Netherlands 1860–1940*. Amsterdam: Amsterdam University Press, 1998.
Hajjar Halaby, Mona. "School Days in Mandate Jerusalem at the Dames de Sion." *Jerusalem Quarterly* 31/07: 40–71.
Halevi, Leor. *Modern Things on Trial: Islam's Global and Material Reformation in the Age of Rida, 1865–1935*. New York: Columbia University Press, 2021.
Hanioğlu, M. Şükrü. *A Brief History of the Late Ottoman Empire*. Princeton: Princeton University Press, 2010.
Harbison, Frederick and Ibrahim Abdelkader Ibrahim. *Human Resources for Egyptian Enterprise*. New York: McGraw Hill, 1958.
Hasan, Aziz Hasan. *In the House of Muhammad Ali: A Family Album 1805–1952*. Cairo, American University in Cairo Press, 2000.
Hausen, Karin. "Technical Progress and Women's Labour in the Nineteenth Century, The Social History of the Sewing Machine." In Georg Iggers (ed.), *The Social History of Progress*. Leamington Spa: Dover 1985: 259–281.
Hauser, Julia. *German Religious Women in Late Ottoman Beirut, Competing Missions*. Leiden: Brill, 2015.
Hayut (Hayutman), Zekharya. *'Im Yitshaq Hayutman, Meyyased Metulla we-Tel-Aviv*. Haifa published by author, 5728 [1967–8].
Headrick, Daniel R. *Power over Peoples, Technology, Environments, and Western Imperialism, 1400 to the Present*. Princeton: Princeton University Press, 2010.
Headrick, Daniel R.. *The Tentacles of Progress*. Oxford, Oxford University Press, 1988.
Heilbroner, Robert. L. "Do Machines Make History?" and "Technological Determinism Revisited" in M.R. Smith and L.Marx (eds.), *Does Technology Drive History? The Dilemma of Technological Determinism*. Cambridge, Mass., MIT Press, 1994: 53–78.
Hemsi, Alberto. *La Musique Orientale en Egypte*. Alexandrie: Edition Orientale de Musique, 1930.
Heyworth-Dunne, James. *Introduction to the History of Education in Modern Egypt*. London: Luzac,1939.
Hilan, Rizkallah. *Culture et Développement en Syrie et dans les Pays Retardés*. Paris: Anthropos, 1969.
Hillel, H. *'Yisrael' beQahir, Iton Tzioni beMitzrayim haLe'umit 1920–1939*. Tel Aviv: Am Oved, 2004.
Hilton, M.L. "Retailing the Revolution: The State Department Store (GUM) and Soviet Society in the 1920s", *Journal of Social History* 37/4 (2004): 939–964.
Hirschberg, Julius. *History of Ophthalmology*. 10 vols. Bonn: Wayenborgh Verlag, 1991
Hobsbawm, Eric J. *Industry and Empire*. London: Pantheon Books, 1969.

Hobsbawm, Eric. *Age of Empire.* London: Weidenfeld & Nicolson, 1987.
Hoffman, R.E. "A Medical Missionary's Journey to Persia in War Time." *Cleveland Medical Journal* 15 (1916).
Hoke, Donald. "The Woman and the Typewriter: A Case Study in Technological Innovation and Social Change." *Business and Economic History* 8 (1979): 76–88.
Hondai, Susumi. "Organizational Innovation and the Development of the Sewing-Machine Industry." In R. Minami et al. (eds.), *Acquiring, Adapting and Developing Technologies, Lessons from the Japanese Experience.* London, 1995: 191–214
Hopwood, Derek. *Tales of Empire.* London: Tauris 1989.
Hout, Bayan Nuwayhed al- "Evenings in Upper Baqʻa: Remembering Ajaj Nuwayhed and Home." *Jerusalem Quarterly* 46 (2011): 15–22.
Huff, Toby E. *Intellectual Curiosity and the Scientific Revolution.* Cambridge: Cambridge University Press, 2011.
Hunter, Archie. *Power and Passion in Egypt: Life of Sir Eldon Gorst 1861–1911.* London: Tauris, 2007.
Ilardi, Vincent. *Renaissance Vision from Spectacles to Telescopes.* Philadelphia: American Philosophical Society, 2007.
İşiyağ, Ahmet Nadir. *Orosdi-Back Efsanesi (Horozdibeği) Adana.* Ankara: Akedemisyen Kitabevi, 2019.
Issawi, Charles (ed.). *The Economic History of the Middle East, 1800–1914.* Chicago and London: Chicago University Press, 1966.
Jagailloux, Serge. *La Médicalisation de l'Égypte au XIXe Siècle (1798–1918).* Paris: Synthèse 1986.
James, Jeffrey. "Leapfrogging in Mobile Telephony: A Measure for Comparing Country Performance." *Technology Forecasting & Social Change* 26 (2009): 991–998.
Järvi, Elisa. *The Keyboard Expands – Working on the New Quarter-Tone Piano.* PhD Thesis, Sibelius Academy, Helsinki 2019.
Jennings, A.M. "Nubian Women and the Shadow Economy." In R.A. Lobban (ed.), *Middle Eastern Women and the Invisible Economy.* Gainesville: University of Florida Press, 1998: 45–59.
Jessup, Henry. *Fifty-Three Years in Syria.* New York: Revell, c.1910.
Joseph, Suad (ed.). *Encyclopedia of Women & Islamic Cultures.* Leiden: Brill 2003.
Kal'a, Ahmet. *İstanbul Esnaf Birlikleri ve Nizamları 1.* Istanbul, İstanbul Araştırmaları Merkezi 1998.
Kamal. Amr Tawfik, *Empires and Emporia: Fictions of the Department Stores in the Modern Mediterranean.* PhD Thesis University of Michigan, 2013.
Kamalkhani, Z..*Women's Islam, Religious Practice Among Women in Today's Iran.* London, New York, Kegan Paul, 1998).
Kanaaneh, Moslih et al. (eds.). *Palestinian Music and Song Expression, and Resistance since 1900.* Bloomington & Indianapolis: Indiana University Press, 2013.
Kandil, Doaa Adel Mahmoud. "Abu Naddara, the Forerunner of the Egyptian Satirical Press." *Journal of the Association of Arab Universities for Tourism and Hospitality* 13/1 (2016): 9–40.
Karababa, Eminegül. "Investigating Early Modern Ottoman Consumer Culture in the Light of Bursa Probate Inventories." *Economic History Review* 65/1 (2012): 194–219.
Karanasou, Flora. *Egyptianization: The 1947 Company Law and the Foreign Communities in Egypt.* PhD Thesis, Oxford University,1992.
Karpat, Kemal. "Kossuth in Turkey. The Impact of Hungarian Refugees in the Ottoman Empire,1849–1851." In his (ed.), *Studies on Ottoman Social and Political History, Selected Articles and Essays.* Leiden: Brill, 2002: 194–219.
Katz, Ruth. *'The Lachmann Problem', an Unsung Chapter in Comparative Musicology.* Jerusalem: Magnes Press, 2003.

Kay, Shirley. "Social Change in Saudi Arabia." In T. Niblock (ed.), *State, Society and Economy in Saudi Arabia*. London, Routledge, 1982.

Khabbaz, Hana (ed.). *Mukhtar al-Muqtataf*, Cairo: Matba'at al-Muqattam, 1930.

Khalidi, Walid al-. *Qabl al-Shatat*. Beirut: Institute for Palestine Studies, 1987.

Khatabi, S. Al- and A.O. Danuton. "Arabic Visual Acuity Chart for Vision Examinations." *Ophthalmic and Physiological Optics* 14/3 (1994): 314–316.

Khuri, Shakir al- (Chaker Khouri). *Kitab Sihhat al-'Ayn*. Beirut: al-Matba'a al-'Umumiyya al-Kathulikiyya, 1890.

Kilincoğlu, Deniz. *Economics and Capitalism in the Ottoman Empire*. London: Routledge. 2015.

Kittler, Friedrich, *Gramophone, Film, Typewriter* (Stanford: Stanford University Press, transl. 1999 from German ed. 1986).

Kneip, James Robert. *A.S. Griboeder: His Life and Works as a Russian Diplomat, 1817–1829*. PhD Tthesis, Ohio State University, 1976.

Kocaman, Meltem. "Scientific Instrument Retailers in Istanbul in the Nineteenth Century, and Verdoux's Optical Shop." In Neil Brown et al. (eds.). *Scientific Instruments between East and West*. Leiden: Brill, 2020: 240–256.

Koloğlu, Orhan. *Reklamcılığımızın İlkyüzyılı 1840–1940*. Istanbul: Reklamcılar Derneği ,1999.

Kopytoff, Igor. "The Cultural Biography of Things: Commoditization as Process". In Appadurai, Arjun (ed.). *The Social Life of Things, Commodities in Cultural Perspective*. Cambridge: Cambridge University Press, 1986: 64–91.

Kosal, Vedat. *Western Classical Music in the Ottoman Empire*. Istanbul: Stock Exchange, 1999.

Köse, Yavuz. "Basare der Moderne von Pera bis Stamboul und ihre Angestellten." In his (ed.) *Istanbul: vom Imperialen Herschersitz zur Megapolis*. Munich: Meidenbauer, 2006: 314–350.

Köse, Yavuz. "Bicycling into Modernity in the Late Ottoman Empire: Ahmed Tevfik and his Bicycle Travelogue." In Ebru Boyar and Kate Fleet (eds.), *Entertainment Among the Ottomans*. Leiden: Brill, 2019: 183–207.

Köse, Yavuz. "Vertical Bazaars of Modernity: Western Department Stores and their Staff in Istanbul (1899–1921), *International Review of Social History* 54 (2009), Supplement: 91–114.

Köse, Yavuz. *Westlicher Konsum am Bosporus, Warenhäuser, Nestlé & Co im Späten Osmanischen Reich (1855–1923)*. München: Oldenburg, 2010.

Kotzebue, Moritz von. *Narrative of a Journey into Persia in the Suite of the Imperial Russian Embassy in the Year 1817*. London: Longman, 1819.

Krämer, Gudrun.*Minderheit, Millet, Nation? Die Juden in Ägypten 1914–1952*. Wiesbaden: Harrassowitz, 1982.

Kraus. Richard Kurt. *Pianos and Politics in China. Middle-Class Ambitions and the Struggle over Western Music*. New York: Oxford University Press, 1989.

Krüger, K. *Kemalist Turkey and the Middle East*. London: Allen & Unwin, 1932.

Krystall, Nathan. "The Fall of the New City." In Salim Tamari (ed.). *Jerusalem 1948: The Arab Neighbourhoods and Their Fate in the War*. Jerusalem: Institute of Jerusalem Studies, 2002: 84–139.

Küçükkalay, A. Mesud and Numan Elibol. "Ottoman Imports in the Eighteenth Century: Smyrna (1771–72)", *Middle Eastern Studies* 42/2 (2006): 723–740.

Kupferschmidt, Uri M. "On the Diffusion of 'Small' Western Technologies and Consumer Goods in the Middle East during the Era of the First Modern Globalization." In Liat Kozma et al.(eds), *A Global Middle East, Mobility, Materiality and Culture in the Modern Age, 1880–1940*. London: Tauris, 2015: 229–260.

Kupferschmidt, Uri M. "The Social History of the Sewing Machine in the Middle East." *Welt des Islams* 44:2 (2004): 1–19.
Kupferschmidt, Uri M. *European Department Stores and Middle Eastern Consumers: The Orosdi-Back Saga.* Istanbul: Ottoman Bank Archives and Research Centre, 2007.
Kupferschmidt, Uri M., "Who Needed Department Stores in Egypt? From Orosdi-Back to Omar Effendi." *Middle Eastern Studies* 43/2 (2007): 175–192.
Kurpershoek, P.M. *The Short Stories of Yusuf Idris, a Modern Egyptian Author.* Leiden: Brill, 1981.
Kurz, Otto. *European Clocks and Watches in the Near East.* London: Warburg Institute 1975.
Kutlay, Evren. "A Historical Case of Anglo-Ottoman Musical Interactions: The English Autopiano of Sultan Abdulhamid II." *European History Quarterly* 49/3 (2019): 386–419.
Kutlay, Evren. *Osmanlı'nın Avrupalı Müzisyenleri.* Istanbul: Kapı Yayınları, 2010.
Lagrange, Frederic. *Musiciens et Poètes en Egypte au Temps de la Nahda.* PhD Thesis, Paris, 1994
Lancaster, W. *The Department Store, A Social History.* London: Leicester University Press, 1995.
Landau, Jacob M. "An Insider's View of Istanbul: Ibrahim al-Muwaylihi's *Ma Hunalika, Welt des Islams* 27 (1987): 70–81.
Landes, David S. *The Wealth and Poverty of Nations, Why Some Are So Rich and Some So Poor.* New York: Norton, 1998.
Landes, David. *The Unbound Prometheus.* Cambridge: Cambridge University Press,1969.
Larson, B.K. "Women, Work, and the Informal Economy in Rural Egypt." In R.A. Lobban (ed.). Middle Eastern Women and the Invisible Economy. Gainesville: University Press of Florida, 1998: 148–165.
Latour, Bruno. *Pandora's Hope.* Cambridge Mass.: Harvard University Press, 1999.
Laura Vorachek, Laura, "'The Instrument of the Century': the Piano as an Icon of Female Sexuality in the Nineteenth Century", in *George Eliot-George Henry Lewes Studies*, vol.38/39 (2000): 26–43.
Leibowitz, J. and Stephen E. Margolis," The Fable of the Keys." *Journal of Law and Economics* 33 (1990): 1–25.
Lewis, Bernard. *The Middle East, 2000 Years of History.* London: Simon and Schuster, 1995.
Lewis, Bernard. *What Went Wrong? The Clash Between Islam and Modernity in the Middle East.* Oxford: Oxford University Press, 2002. New York: Harper, 2003.
Locke, Ralph P. *Music, Musicians, and the Saint-Simonians.* Chicago and London: Chicago University Press, 1986.
Loesser, Arthur. *Men, Women and Pianos.* London: Dover 1954.
Longuenesse, E. "L'Industrialisation et sa Signification Sociale." In A. Raymond (ed.) *La Syrie d'Aujourd'hui.* Paris: CNRS, 1980: 327–358.
Longuenesse, Elisabeth. "Système Educatif et Modèle Professionel: le Mandat Français en Perspective, l'Exemple des Comptables en Liban." In N. Meouchy and P. Sluglett (eds.), *The British and French Mandates in Comparative Perspectives.* Leiden: Brill, 2004: 544–545
Luthi, Jean-Jacques. *Lire la Presse d'Expression Française en Égypte, 1798–2008.* Paris, l'Harmattan 2009.
M. Molyneux, M. *State Policies and the Position of Women Workers in the People's Democratic Republic of Yemen, 1967–1977.* Geneva: ILO 1984.
Maalouf, Shireen, "Mikha'il Mushaqa: Virtual Founder of the Twenty-Four Equal Quartertone Scale", *Journal of the American Oriental Society*, vol. 123/4 (2003): 835–840.
Mackenzie, John M. *Orientalism, History, Theory, and the Arts.* Manchester and London: Manchester University Press, 1995.

Macleod, Arlene Elowe. "Transforming Women's Identity, the Intersection of Household and Workplace in Cairo." In Diane Singerman and Homa Hoodfar (eds.). *Development, Change, and Gender in Cairo, A View from the Household.* Bloomington: Indiana UP 1969: 27–50.
Mak, Lanver. *The British Community in Occupied Cairo, 1882–1922.* PhD Thesis, SOAS, London 2001.
Malcolm, Napier. *Five Years in a Persian Town.* New York: Murray 1908.
Maldonado, Tomás. "Taking Eyeglasses Seriously." *Design Issues* 17/4 (2001): 32–43.
Manguel, Alberto. *A History of Reading.* New York: Viking,1996.
Mardin, Serif. "Super-westernization in Urban Life in the Last Quarter of the Nineteenth Century." In P. Benedict *et al.* (eds.), *Turkey: Geographical and Social Perspectives.* Leiden: Brill, 1974: 403–445.
Mardin, Şerif. *Super Westernization in Urban Life in the Ottoman Empire in the Last Quarter of the Nineteenth Century.* Leiden: Brill 1974.
Margolin, Jean-Claude. "Des Lunettes et des Hommes ou la Satire des Mal-Voyants au XVIe Siècle." *Annales. Economies, Sociétés, Civilisations* 30/ 2–3 (1975): 375–393.
Marvin, Carolyn. *When Old Technologies Were New.* New York: Oxford University Press, 1990.
Marx, Karl. *Das Kapital.* Berlin: Dietz Verlag,1951.
Masson, Paul. *Histoire du Commerce Français dans le Levant au XVIIe Siècle.* Thesis, Faculté des Lettres, Paris 1896.
Matthews, R.D. and M. Akrawi, *Education in the Arab Countries of the Near East.* Washington, American Council on Education, 1949.
Mayer, S. *Ein Juedischer Kaufmann, 1831–1911.* Leipzig: Duncker & Humbolt, 1911.
Mazor, A. and K. Abbou Hershkovits, "Spectacles in the Muslim World: New Evidence from the Mid-Fourteenth Century." *Early Science and Medicine* 18/3 (2013): 291–305.
McConnell, Anita. *A Survey of the Networks Bringing Knowledge of Optical Glass-Working to the London Trade, 1500–1800.* The Whipple Museum of Science, 2016.
McKibben, Gordon. *Cutting Edge: Gillette's Journey to Global Leadership.* Boston, HBS Press, 1998.
Meeker, Michael E. *A Nation of Empire, the Ottoman Legacy of Turkish Modernity.* Berkeley: University of California Press, 2002.
Mehrez, Samia (ed.). *The Literary Atlas of Cairo: One Hundred Years on the Streets of the City.* Cairo: American University in Cairo Press, 2010.
Messick, Brinkley. *The Calligraphic State, Textual Domination and History in a Muslim Society.* Berkeley: University of California Press, 1993.
Messiri, Sawsan El-. *Ibn al-Balad, a Concept of Egyptian Identity.* Leiden: Brill 1978.
Mestyan, Adam. 'A Garden with Mellow Fruits of Refinement', Music Theatre and Cultural Politics in Cairo and Istanbul 1867–1921. PhD Thesis, Central European University, Budapest, 2011.
Meyerhof, Max." A Short History of Ophthalmia During the Egyptian Campaigns of 1798–1807." *The British Journal of Ophthalmology* 16/3 (March 1932): 129–152.
Michael, B.J. "Baggara Women as Market Strategists", in R.A. Lobban (ed.), *Middle Eastern Women and the Invisible Economy.* Gainesville: University of Florida Press 1998: 60–73.
Micklewright, Nancy. "Alternative Histories of Photography in the Ottoman Middle East." In Ali Behdad and Luke Gartlan (eds.), Photography's Orientalism: New Essays on Colonial Representation, Los Angeles: Getty Research Institute, 2013: 75–92.
Micklewright, Nancy. "An Ottoman Portrait", *International Journal of Middle East Studies* 40/3 (2008): 372–373.
Micklewright, Nancy. "London, Paris, Istanbul, and Cairo: Fashion and International Trade in the Nineteenth Century." *New Perspectives on Turkey* 7 (1992): 125–136.

Mikhail, Mona N. *Seen and Heard: A Century of Arab Women in Literature and Culture.* Northampton: Interlink Books, 2004.
Miller, M.B. *The Bon Marché, Bourgeois Culture and the Department Store, 1869–1921.* Princeton, NJ: Princeton University Press, 1981.
Millingen, Edwin van. *Bericht über die Jahren 1880 und 1881.* Salzburg: Selbstverlag, 1883.
Millingen, Edwin van. *Centralblatt für Augenheilkunde* 7 (1883)
Millingen, Edwin van. *Tabular Analysis of 1,118 Cases Treated at the Imperial Naval Hospital.* Constantinople, 1880 [?].
Millingen, Edwin van. *Tri-annual Report of 5703 Cases of Eye Diseases Seen and Treated in Private Practice at Constantinople in 1877,1878 & 1879.*Constantinople, 1880 [?].
Miranda, Manuel N. "The Environmental Factor in the Onset of Presbyopia." In Lawrence Stark and Gerard Obrecht (eds.), *Presbyopia.* Haiti: Third International Symposium on Presbyopia, 1985.
Mishaqa, Mikhail, transl. by Eli Smith, *Journal of the American Oriental Society* 1849: 171–217.
Mitchell, Timothy. *Colonising Egypt.* Berkeley: University of California Press,1988.
Modelsky, Sylvia. *Port Said Revisited.* Washington: Faros, 2000.
Moghadam, V.M. "Manufacturing and Women in the Middle East and North Africa: a Case of the Textiles and Garments Industry", *CMEIS Occasional Paper no. 49 (1995),* University of Durham
Mole, John. *The Diary of Thomas Dallam, 1599, London to Constantinople and Adventures on the Way.* London: Fortune, 2012.
Monaco, Cynthia. 'The Difficult Birth of the Typewriter", *Invention & Technology Magazine* 4/1 (Spring/Summer 1988).
Montague, Lady Mary Wortley. *The Complete Letters.* Oxford: Clarendon Press 1988.
Moore, Geoffrey A., *Crossing the Chasms.* New York: Harper Collins, 1991.
Moore, Lawrence S. "Some Phases of Istanbul Life' in Clarence Richard Johnson (ed.), *Constantinople To-Day or the Pathfinder Survey of Constantinople, A Study in Oriental Social Life.* New York: MacMillan, 1922.
Moosa, Matti. *The Origins of Modern Arabic Fiction.* Boulder: Lynne Rienner, 2nd ed. 1997.
Moreh, Shmuel. *Modern Arabic Poetry: 1800–1970: the Development of its Forms and Themes.* Leiden: Brill, 1976.
Morier, James. *A Second Journey Through Persia, Armenia, and Asia Minor.* London: Longman, 1818.
Morrison, Heidi. "Nation Building and Childhood in Early Twentieth Century Egypt." In Benjamin Fortna (ed.). *Childhood in the Late Ottoman Empire and After.* Leiden: Brill, 2016: 73–90.
Most van Spijk, Marileen van der. *Eager to Learn, an Anthropological Study of the Needs of Egyptian Village Women.* Leiden Research Center Women and Development, 1982.
Mozzato, Andrea. "Luxus und Tand: Der Internationale Handel mit Rohstoffen, Farben, Brillen, und Luxusgüttern in Venedig des 15. Jahrhunderts am Beispiel des Apothekers Agostino Altucci." In Christof Jeggle et al. (eds.) *Luxusgegenstände und Kunstwerke vom Mittelalter bis zur Gegenwart, Produktion, Handel, Formen der Anneignung.* München: UVK Verlag Konstanz, 2015.
Mubarak, 'Ali. *'Alam al-Din,* Alexandria: Matba'at Jaridat al-Mahrusa, 1882.
Müge Göçek, Fatma. *Rise of the Bourgeoisie, Demise of Empire: Ottoman Westernization and Social Change.* Oxford: Oxford UP 1996.
Müge Göçek, Fatma. *East Encounters West: France and the Ottoman Empire in the Eighteenth.* New York: Oxford University Press,1987.
Mülayim, Ahmet, *Dünden Bugüne Istanbul Ansiklopedisi,* "Ticaret Liseleri', vol. 7: 268–269.
Mullaney, Thomas. "Controlling the Kanjisphere: The Rise of the Sino-Japanese Typewriter and the birth of CJK." *Journal of Asian Studies* 75/3 (2016), pp. 725–753.

Mullaney, Thomas. *The Chinese Typewriter, a History.* Cambridge, Mass.: MIT Press, 2017.
Myntti, Cynthia. *Paris along the Nile, Architecture in Cairo from the Belle Epoque.* Cairo: American University in Cairo Press, 1999.
N.N. *al-Uslub lil-Sahih fi al-Darb 'ala al-Ala al-Katiba al-'Arabiyya, 'Ilmi – 'Amali bi-Tariqat al-Lams.* Beirut: al-Khayat, 1966.
Naor, Mordechai. *Ha-Rishonim: Sippurim min ha-'Aliya ha-Rishona le Eretz-Yisrael.* Tel Aviv: Am Oved, 1983.
Nelson, Elif C. Nelson, "Advertisement at Izmir Press During the Early 20th Century", *Tarih İncelemeleri Dergisi*, 34/1 (2019): 161–177.
Nemeth, Titus. *Arabic Type-Making in the Machine Age, The Influence of Technology on the Form of Arabic Type, 1908–1993.* Leiden: Brill, 2017.
Nevine Miller, Nevine. *The Ivory Cell.* Bloomington: Author House, 2013.
Nieuwkerk, Karin van. "Of Morals, Missions, and the Market: New Religiosity and 'Art with a Mission' in Egypt." In her ed., *Halal Soaps, and Revolutionary Theater: Artistic Developments in the Muslim World.* Austin: University of Texas Press, 2011.
Nowill, Sidney E.P. *Constantinople and Istanbul, 72 Years of Life in Turkey.* Kibworth, 2011.
Okkenhaug, Inger Marie. "She Loves Books & Ideas & Strides Along in Low Shoes Like an Englishwoman: British Models and Graduates from the Anglican Girls' Secondary Schools in Palestine, 1918–48." *Islam and Christian-Muslim Relations* 13/4 (2002): 461–479.
Os, Nicole A.N.M. van. "From Conspicuous to Conscious Consumers: Ottoman Muslim Women, the Mamulat-i Dahiliya Istihlaki Kadınlar Cemeyet-i Hayriyesi, and the National Economy." *Journal of Ottoman and Turkish Studies* 6/2 (2019): 113–130.
Os, Nicole van. "'Müstehlik Değil Müstahsil', Not Consumers but Producers; Ottoman Muslim Women and Milli İktisat'", in *The Great Ottoman-Turkish Civilisation.* Ankara: Yeni Türkiye Yayınları, 2000, vol. II: 269–275.
Osmanoğlu, Ayşe. *Babam Sultan Abdülhamid (Hâtıralarım).* Istanbul, Momo Sahaf. Yayınevi, 1984.
Overholt, C. et al. (eds.), *Gender Roles in Development Projects, a Case Book.* West Hartford: Kumarian Press, 1985.
Özbek, Nadir."Policing the Countryside", *International Journal of Middle Eastern Studies* 40 (2008): 47–67.
Palmer, Monte et al. *The Egyptian Bureaucracy.* Syracuse: Syracuse University Press, 1988.
Pamuk, Orhan. *The Innocence of Objects.* New York: Abrams 2012.
Papastefanaki, Leda. "Sewing at Home in Greece, 1870s to 1930s, a Global History Perspective." In Malin Nilsson et al. (eds). *Home-Based Work and Home-Based Workers (1800–2021).* Leiden: Brill, 2021: 74–95.
Parakilas, James. *Piano Roles, Three Hundred Years with the Piano.* New Haven: Yale University Press, 1999.
Parry, V.J. and M.E. Yapp (eds.), *War, Technology and Society in the Middle East.* Oxford: Oxford University Press, 1975.
Pasdermadjian, H. *The Department Store.* London: Newman, 1954.
Patai, Raphael (ed.). *The Complete Diaries of Theodor Herzl.* New York: Herzl Press 1960.
Peters, Ruud. "Religious Attitudes Towards Modernization in the Ottoman Empire: A Nineteenth Century Pious Text on Steamships, Factories, and the Telegraph." *Welt des Islams*, 26/1–4 (1986): 76–105.

Petersen, Sonja. "Piano Manufacturing Between Craft and Industry: Advertising and Image Cultivation of Early 20th [Century] German Piano Production."' *ICON, Journal of the International Committee for the History of Technology* 17 (2011): 12 – 30.
Petricioli, Marta. *Oltre il Mito, l'Egitto degli Italiani (1917 – 1947.* Milan: Mondadori, 2007.
Petropoulos, Elias. "Ah Allegra." In *Salonique 1850 – 1948, Autrement* (Paris 1992).
Petroski, Henry. *The Evolution of Useful Things.* New York: Vintage Books, 1994.
Phillips, Amanda. *Everyday Luxuries, Art and Objects in Ottoman Constantinople, 1600 – 1800.* Berlin: Vetler (app. 2015).
Pirker, Michael. "Janissary Music." *Grove Music Online* (2001).
Poché, Christian. "Vers une Musique Libanaise de 1850 a 1950." *Les Cahiers de l'Oronte* 7 (1960): 77 – 100
Powell, I. *Disillusion by the Nile, What Nasser has Done to Egypt.* London: Solstice, 1967.
Prausnitz, Gotthold. *Das Augenglas in Bildern der Kirchlichen Kunst im XV. und XVI. Jahrhundert.* Strassburg: Heitz & Mindel, 1915.
Puig, Nicolas. "Egypt's Pop-Music Clashes and the 'World-Crossing', Destinies of Muhammad 'Ali Street Musicians." In Diana Singerman and Paul Ammar (eds.) *Cairo Cosmopolitan.* Cairo, American University in Cairo Press, 2006: 513 – 515.
Puig, Nicolas. "Le Long Siècle de l'Avenue Muhammad 'Ali: d'Un Lieu et des Publics Musiciens", l'Égypte dans le Siècle 1901 – 2000." *Égypte Monde Arabe* 1 – 2 / 4 – 5 (2001): 207 – 223.
Purkhart, Markus. *Die Österreichische Fezindustrie.* PhD Thesis, University of Vienna, 2006.
Qasimi, Jamal al-Din al- (et al.). *Qamus al-Sina'at al-Shamiyya / Dictionnaire des Métiers Damascains.* Paris: Mouton 1960.
Qassim Hassan, Sheherazade. "Musical Instruments in the Arab World." In *The Garland Encyclopedia of World Music*, vol. 6: The Middle East. New York and London: Routledge 2002.
Quataert, Donald. *Manufacturing and Technology Transfer in the Ottoman Empire.* Istanbul and Strasbourg: Gorgias Press, 1992.
Quataert, Donald. *Workers, Peasants, and Economic Change in the Ottoman Empire 1730 – 1914.* Istanbul: Isis Press,1993.
Raafat, Samir. "'From Mag-Arabs to al-Magary", *Egyptian Mail* (13 April 1996).
Raafat, Samir. "'The House of Cicurel." *al-Ahram Weekly (*15 Dec. 1994).
Raafat, Samir. "Davies Bryan & Co. of Emad el Din Street." *Egyptian Mail (*27 May 1995).
Raafat, Samir. *Cairo, the Glory Years.* (Alexandria: Harpocrates, 2003.
Raby, Julian. "The Serenissima and the Sublime Porte: The Art of Diplomacy, 1453 – 1600." In Stephano Carboni (ed.), *Venice and the Islamic World, 828 – 1797.* New York: Metropolitan Museum of Art 2007.
Racy, A.J. *Making Music in the Arab World.* Cambridge: Cambridge University Press, 2003.
Racy, Ali Jihad Racy. "The Record Industry and Egyptian Traditional Music, 1904 – 1932." *Ethnomusicology* 20/1 (1976): 23 – 48.
Racy, Ali Jihad. "Historical Worldviews of Early Ethnomusicologists: An East-West encounter in Cairo, 1932. In Steven Blum, Philip Bohlman and Daniel Neumann (eds.), *Ethnomusicology and Modern Music History*, Urbana: University of Illinois Press,1991: 68 – 94.
Racy, Ali Jihad. "Words and Music in Beirut: A Study of Attitudes." *Ethnomusicology* 30 (1986): 413 – 427.
Raff, Wolfgang. *Deutsche Augenärtzte in Ägypten -von Franz Ignaz Pruner bis Max Meyerhof (1831 – 1945)*, MD Thesis, Technische Universität München.

Ramsay, W. M. *The Revolution in Constantinople and Turkey: A Diary.* London: Hodder and Stoughton, 1909.
Rasim, Ahmed. *Şehir Mektuplari.* Istanbul: Oğlak Yayıncılık, 2005.
Rasmussen, Anne K. "Theory and Practice at the 'Arabic Org': Digital Technology in Contemporary Arab Music Performance" *Popular Music* 15/3 (1996): 345–365.
Ravndal, G. Bie. *Turkish Markets for American Hardware,* Special Consular Reports, no. 77. Washington, 1917.
Raymond, André et al. *Le Caire.* Paris: Citadelles, 2000.
Raz, Adam. *Bizat haRekhush ha'Aravi beMilhemet haAtzma'ut.* Jerusalem: Carmel 2020.
Regev, Motti. *'Ud ve-Gitara.* Tel Aviv: Beth Berl, 1993.
Reid, Donald M. "Syrian Christians, the Rags to Riches Story, and Free Enterprise."' *International Journal of Middle East Studies* 1 (1970): 358–367.
Reinhard, Ursula. "Turkey an Overview", in *The Garland Encyclopedia of World Music,* vol. 6: The Middle East. New York and London: Routledge 2002.
Reynolds, Nancy. "Sharikat al-Bayt al-Misri: Domesticating Commerce in Egypt, 1931–1956." *Arab Studies Journal* 7/8 (1999–2000): 75–107.
Reynolds, Nancy. *A City Consumed, Urban Commerce, the Cairo Fire, and the Politics of Decolonization in Egypt.* Stanford: Stanford University Press, 2012.
Reynolds, Nancy. *Commodity Communities, Interweavings of Market Cultures, Consumption Practices and Social Power in Egypt, 1907–1961.* PhD Thesis, Stanford University, 2003.
Riad, H. *L'Egypte Nasserienne.* Paris: Minuit, 1964.
Rigler, Lorenz. *Die Türkei und seine Bewohner in ihren Naturhistorischen, Physiologischen, und Pathologischen Verhälltnissen von Standpunkt Constantinopel's.* Vienna: Gerold, 1852.
Rihani, Amīn al-. *Tārīkh Najd al-Hadīth.* Beirut, al-Maṭbaʿah al-ʿIlmīyah li-Yūsuf Ṣādir, 1928.
Rihani, Amin. *Ibn Saʿoud of Arabia, His People and his Land.* London: Constable, 1928.
Roberts, Sophie. *The Lost Pianos of Siberia.* New York: Grove Press, 2020.
Robson, Laura. "A Civilizing Mission? Music and the Cosmopolitan in Edward Said." *Mashriq & Mahjar* 2/1 (2014): 461–479.
Rogers, Everett M. *The Diffusion of Innovations* 4th ed. New York:,Simon & Schuster, 1985.
Rogers, Everett M. *The Diffusion of Innovations* 5th ed. New York: Free Press, 2003.
Rogers Mary Eliza. *Domestic Life in Palestine.* London: Bell and Dalby 1862.
Rosen, Edward. "The Invention of Eyeglasses." *Journal of the History of Medicine and Allied Sciences,* part. I: 11/1 (Jan. 1956): 13–46 and part II: 11/2:183–213.
Rosenberg, E.S. *Spreading the American Dream.* New York: Hill and Wang, 1982.
Rosenthal, J. William. *Spectacles and Other Vision Aids, a History and Guide to Collecting.* San Francisco: Norman, 1996.
Rowbotham, S. "Feminist Approaches to Technology, Women's Values or a Gender Lens?" In S. Mitler and S. Rowbotham (eds.), *Women Encounter Technology, Changing Patterns of Employment in the Third World.* London and New York, 1995: 44–67.
Rowland, David. *A History of Pianoforte Pedaling.* Cambridge: Cambridge University Press, 1993.
Rugh, Andrea B. *Reveal and Conceal, Dress in Contemporary Egypt.* Cairo: American University in Cairo Press, 1986.
Ruppin, Arthur. *Syrien als Wirtschaftsgebiet.* Berlin, Vienna: Beniamin Harz, 2nd. ed. 1920.
Russel, Mona L. *Creating the New Egyptian Woman, Consumerism, Education, and National Identity, 1863–1922.* New York: MacMillan, 2004.

Russell, Mona L. "Creating al-Sayida al-Istikhlakiyya: Advertising in Turn-of-the Century Egypt." *Arab Studies Journal*, 8/9 (2000/01): 61–96.

Russell, Mona L. "Modernity, National Identity and Consumerism: Visions of the Egyptian Home, 1805–1922." In R. Shechter (ed.), *Transitions in Domestic Consumption and Family Life in the Modern Middle East*. New York: Palgrave, 2003: 37–62.

Ryan, James D. *The Republic of Others. Opponents of Kemalism in Turkey's Single Party Era, 1919–1950*. PhD Thesis, University of Pennsylvania, 2017.

Ryzova, Lucie. "My notepad is My Friend. Effendis and the Act of Writing in Modern Egypt." *Maghreb Review* 32 (2007): 323–348.

Sabra, A.I. "Manazir, or 'Ilm al-Manazir." In *Encyclopaedia of Islam*. Leiden: Brill, 2nd. ed., 1960–2007.

Sadid, Mohammed. *L'Alphabet Arabe et la Technologie*. Rabat, ISESCO, 1993.

Said, Edward. "The Imperial Spectacle." *Grand Street* 6/2 (1987): 82–104.

Said, Edward. *Culture and Imperialism*, New York: Vintage Books, 1994.

Said, Edward. *Musical Elaborations*. London: Chatto and Windus, 1991.

Said, Edward. *Out of Place, a Memoir.* London: Granta Books, 1999.

Said, Edward. *Reflections on Exile & Other Literary and Cultural Essays*. Cambridge, Mass.: Harvard University Press, 2000.

Sakorafas, Argyrios. "The Great Civilizer": The Global Diffusion of the Sewing Machine and its Impact on Greece during the Late 19th-Early 20th Century." Global History Blog, The Scottish Center for Global History, 29 March 2021.

Salamandra, Christa. *A New Old Damascus, Authenticity and Distinction in Urban Syria*. Bloomington: Indiana University Press, 2004.

Sanayi Devrimi Yıllarında Osmanlı Saraylarında Sanayi ve Teknoloji Araçları. Istanbul: Yapı ve Kredi Bankasi 2004.

Sanudo (Sanuto), Marin. *I Darii*, Venice, 1887.

Saul, Samir. *La France et l'Égypte de 1882 a 1914, Intérêts Économiques et Implications Politiques*. Paris : Ministère de l'Economie, 1997.

Sawaie, Mohammed. "Rifa'a Rafi al-Tahtawi and his Contribution to the Lexical Development of Modern Literary Arabic", *International Journal of Middle East Studies* 32/3 (2000): 395–410.

Saz, Leyla Hanımefendi, *The Imperial Harem of the Sultans, Daily Life at the Çirağan Palace during the 19th Century*. Istanbul: Hil Yayın 1955.

Scarce, Jennifer. *Domestic Culture in the Middle East*. Edinburgh: National Museum of Scotland, 1996.

Schacht, Joseph. "Max Meyerhof", *Osiris* 9 (1950): 6–32.

Schaefer Davis, S. "Working Women in a Moroccan Village", in L. Beck and N. Keddie (eds.), *Women in the Muslim World*. Cambridge, Mass.: Harvard University Press, 1978.

Schor, Juliet B. "In Defense of Consumer Critique: Revisiting the Consumption Debates of the Twentieth Century." *Annals AAPS*, 611 (May 2007): 16–30.

Schumacher, E.F. *Small is Beautiful, Economics as if People Mattered*. New York: Harper, 1973.

Scott, J. *Genius Rewarded, the Story of the Sewing Machine,* New York: Caulon,1880.

Sednaoui, Selim. "Western Classical Music in Umm Kulthum's Country" in Sherifa Zuhur (ed.). *Images of Enchantment, Visual and Performing Arts in the Middle East*. Cairo: American University in Cairo Press, 1998: 123–134.

Serjeant, R.B. and R. Lewdock (eds.). *Sana'ā', an Arabian Islamic City*. Cambridge World of Islam Festival Trust, 1983.

Sha'rawi, Huda al-. *Harem Years: The Memoirs of an Egyptian Feminist* (transl. Margot Badran), New York: SUNY, 1986.

Shafik, Viola. *Arab Cinema, History and Cultural Identity.* Cairo: American University in Cairo Press, 2000.
Shannon, Jonathan Holt. *Among the Jasmine Trees, Music and Modernity in Contemporary Syria.* Middletown: Wesleyan University Press, 2006.
Shatzmiller, Maya. *Labour in the Medieval Islamic World.* Leiden: Brill, 1994.
Sheehi, Stephen. *The Arab Imago, A Social History of Portrait Photography. 1860–1910.* Princeton and Oxford: Princeton University Press, 2016.
Sheehi, Stephen. "A Social History of Early Arab Photography or a Prolegomenon to an Archaeology of the Lebanese Imago", *International Journal of Middle East Studies* 39 (2007): 177–208.
Sheil, Lady. *Glimpses of Life and Manners in Persia.* London: Murray,1856.
Shermer, Michael. "Exorcising Laplace's Demon, – Chaos, and Antichaos, History and Metahistory." *History and Theory* 34 (1995): 74–75.
Sheva, Shelomo and Dan Ben-Amotz, *Eretz Tzion Yerushalayim.* Jerusalem: Weidenfeld, 1973.
Shiloah, Amnon. *Music in the World of Islam: A Socio-Cultural Study.* Detroit: Wayne State University Press, 1995.
Shinder, Joel. "Mustafa Efendi: Scribe, Gentleman, Pawnbroker", *International Journal of Middle East Studies* 10/3 (1979): 415–420
Sid Ahmad,N.A. *Al-Hayat al-Iqtisadiyya wal-Ijtima'iyya lil-Yahud fi Misr, 1947–1957.* Cairo: Hay'at al-Kitab,1991.
Siegelbaum, Lewis H. (ed.). *Cars for Comrades: Automobility in the Eastern Bloc.* Ithaca: Cornell University, 2011.
Singer Manufacturing Company. *Sewing Machines*, The Columbian Exposition, May 1893.
Singerman, Diana. *Avenues of Participation, Family, Politics, and Networks in Urban Quarters of Cairo.* Princeton: Princeton UniversityPress, 1995.
Siniora, Randa George. "Palestinian Labor in a Dependent Economy: Women Workers in the West Bank Clothing Industry." *Cairo Papers in Social Studies* 12/3 (1989).
Sirgany, Dorothee Greans El. *Les Brevets d 'Invention en Égypte.* Cairo, 1978.
Sladen, Douglas B. W. *Oriental Cairo, The City of the Arabian Nights.* 1911, repr. Altenmünster: Jazzybee Verlag, 2022.
Sladen, Douglas. *Queer Things About Egypt.* London: Hurst & Blacket, 1910.
Smith Allen, James. *In the Public Eye: A History of Reading in Modern France, 1800–1940.* Princeton: Legacy Library, 1991.
Snouck Hurgronje, Christian. *Mekka in the Latter Part of the Nineteenth Century.* Leiden: Brill, 1931; repr. 1970.
Sönmez, M. *Statistical Guide to Istanbul in the 1990s.* Istanbul, n.d.
Sreberny-Mohammedi, Annabelle and Ali Mohammedi, *Small Media, Big Revolution.* Minneapolis: University of Minnesota Press, 1994.
Stearns, Peter N. *Consumerism in World History, Global Transformation of Desire.* Abingdon: Routledge, 2nd ed. 2006.
Stein, Abrevaya. *Making Jews Modern, the Yiddish and Ladino Press in the Russian and Ottoman Empires.* Bloomington: Indiana University Press, 2004.
Stoichiţă, A. "Trois Continents, Une Passion. Entretien Avec Salwa El-Shawan Castelo-Branco." *Cahiers d'Ethnomusicologie* 26 (2013): 241–254.
Stone, Christopher. *Popular Culture and Nationalism in Lebanon, the Fairouz and Rahbani Nation.* Abingdon: Routledge 2008.
Storrs, Ronald. *Orientations.* London: Nicholson, ed. 1945.

Strom, Sharon Hartman. *Beyond the Typewriter, Gender, Class, and the Origins of Modern American Office Work, 1900–1930.* Urbana and Chicago: University of Illinois Press, 1992.
Stuart, Donald. *The Struggle for Persia.* London: Methuen, 1902.
Szyliowicz, Joseph S. Political *Change in Rural Turkey, Erdemli.* The Hague, Paris: Mouton,1966.
Tabatabaei, Seyyed Hadi. "The Introduction of the Telescope into Iran before the Nineteenth Century." In Neil Brown et al. (eds.) *Scientific Instruments Between East and West.* Leiden: Brill 2020: 142–153.
Tamari, Salim and Isam Nassar (eds.). *The Storyteller of Jerusalem, the Life and Times of Wasif Jawhariyyeh, 1904–1948.* Jerusalem: Institute for Palestine Studies, 2014.
Tamilia, R.D. "The Wonderful World of the Department Store in Historical Perspective: A Comprehensive Bibliography Partially Annotated". Working Paper, Centre de Recherche en Gestion, March 2002.
Tanatar-Baruh, Lorans."Tracing the Painted-Tray Dealers in Istanbul, a Commercial and Spatial Reading." In Flavia Nessi and Myrto Hatzaki (eds), *Rituals of Hospitality, Ornamented Trays of the 19th Century in Greece and Turkey.* Athens: Melissa, 2013.
Tanatar-Baruh, Lorans. *A Study in Commercial Life and Practices in Istanbul at the Turn of the Century: The Textile Market.* MA-thesis, Boğazici University, 1993.
Tansuğ, Feza. "Ottoman Elites and their Music: Music Making in the Nineteenth Century Istanbul." *Musicology Today* 39/3 (2019): 201–209.
Tarazi, Rima. "The Palestinian National Song, a Personal Testimony", *This Week in Palestine*, 27 June 2010.
Taylor, Jean Gelman. "The Sewing-Machine in Colonial-Era Photographs: A record from Dutch Indonesia." *Modern Asian Studies* 46/1 (2012): 71–95.
Tekçe, B., L. Oldham and F. Shorter, *A Place to Live, Families and Child Care in a Cairo Neighborhood.* Cairo: American University in Cairo Press, 1994.
Thévenot, Jean de. *The Travels of Monsieur de Thévenot into the Levant*, transl. pt. I. London: Faithorne et al., 1687.
Thompson, Elisabeth. *Colonial Citizens, Republican Rights, Paternal Privilege, and Gender in French Syria and Lebanon.* New York, Columbia University Press, 2000.
Tignor, R.L. *Capitalism and Nationalism at the End of Empire.* Princeton: Princeton University Press, 1998.
Tilly, Charles. "Retrieving European Lives." In O. Zunz (ed.). *Reliving the Past, the Worlds of Social History.* Chapel Hill,1985.
Tobi, Josef. Yisrael Tzubeiri [Israel Subeiri] *Yehudi beSherut haImam.* Tel Aviv: Afiqim, 2001–2002.
Trentmann, Frank. *Empire of Things: How we Became a World of Consumers, from Fifteenth Century to the Twenty-First.* New York: Harper Collins 2016.
Trietsch, Davis. *Levante-Handbuch.* Berlin, Orient Verlag, 1914.
Trietsch, Davis. *Levante-Handbuch.* Berlin, Orient Verlag,1909.
Trietsch, Davis. *Palästina-Handbuch.* Berlin, Orient Verlag, 1910.
Trimbur, Dominique. "Fortune and Misfortune of a Consul of France in Jerusalem: Amédée Outrey, "1938–1941." *Bulletin du Centre de Recherche Français:* 1998/2.
United Nations Development Program, *Arab Human Development Report.* New York: 2003.
US Department of Commerce. *World Trade in Typewriters 1948–1958.* Washington, 1959.
Vaujany, H. de. *Le Caire et les Environs.* Paris : Plon, 1883.
Veyrat, Nicolas et al. "Social Embodiment of Technical Devices: Eyeglasses over the Centuries and According to Their Uses." *Mind, Culture, and Activity* 15 (2008):185–207.

Vincent, David. *The Rise of Mass Literacy, Reading and Writing in Modern Europe.* Cambridge: Polity 2000.

Vlam, Grace A. H. "The Calling of Saint Matthew in Sixteenth Century Flemish Painting." *The Art Bulletin* 59/4 (1977): 561–570.

Volait, Mercedes. *Le Caire-Alexandrie, Architectures Européennes, 1850–1950.* Cairo: IFAO, 2001.

Vorachek, Laura. "'The Instrument of the Century': the Piano as an Icon of Female Sexuality in the Nineteenth Century." In *George Eliot-George Henry Lewes Studies* 38/39 (2000): 26–43.

Vrolijk, Arnoud. *Een Turks Alfabet op Latijnse Grondslag.* Leiden: Universiteitsbibliotheek 1998.

Wagemans Marianne and O. Paul Van Bijsterveld, "The French Egyptian Campaign and its Effects on Ophthalmology." *Documenta Ophthalmologica* 68/1–2 (1988): 135–144.

Wahid, A. al-Din Wahid. *Masrah Muhammad Taymur.* Cairo: Hay'at al-Kitab, 1975.

Wali, Najm. *Tel Al Laham,* London: Dar Alsaqi, 2001.

Walker, Bethany J. "The Late Ottoman Cemetery in Field L, Tell Hisban." *Bulletin of the American Schools of Oriental Research* 322 (May 2001).

Wallis Budge, E.A. *Cook's Handbook for Egypt and the Egyptian Sudan.* London: Cook

Walters. D.M. "Invisible Survivors, Women and the Diversity in the Transitional Economy of Yemen." In R.A. Lobban (ed.), *Middle Eastern Women and the Invisible Economy.* Gainesville 1998: 74–95.

Weakly, E. *Report on the Conditions and Prospects of British Trade in Syria.* London: Board of Trade, 1911.

Weber, Max, *The Rational and Social Foundations of Music* (1921 posthumous, 1958).

Wills, C.J. *In the Land of the Lion and the Sun, Modern Persia.* London: Ward, Lock & Co, 1891.

Wills, C.J. *Persia as It Is.* London: Sampson Low 1886.

Witkam, J.J. "Scenes of Learning in the Hotz Photographic Collection." In L.A. Ferydoun Bariesteh van Waalwijk van Doorn (ed.), Qajar Era Photography". *Journal of the International Qajar Studies Association* 1 (2001): 49–55.

Woodard, Kathryn. "Music in the Imperial Harem and the life of Composer Leyla Saz (1850–1936)" *Sonic Crossroads* (2011).

Woodsmall, Ruth Francis. *Moslem Women Enter a New World.* New York: Round Table Press 1936.

Wright, Arnold. and H. A. Cartwright, *Twentieth Century Impressions of Egypt.* London: Lloyds 1908.

Yağız, Burcu and Aygül Ağir. "XIX. Yüzyıl Sonu İstanbulu'nda Batılı Tüketici Ürünlerinin Dolaşıma Girdikleri Kanallar ve Yarattıkları Hareketlenmeler; Şark Ticaret Yıllıkları Üzerinden Bir Araştırma", *Aralık* (2017): 31–35.

Yılmaz, Asım Egemen. and Emrah Çiçek. "Optimized Rearrangement of Turkish Q and F Keyboards by Means of Language-Statistics and Simple Heuristics." *Hittite Journal of Science and Engineering* 3/1 (2016): 23–28.

Yılmaz, Ela. "Osmanlı Topraklarında Uluslararası bir Kuruluş: Orosdi-Back ve Adana Mağazası." *Ankara Üniversitesi Sosyal Bilimler Dergisi* 13/1 (2022): 28–44.

Yusuf Salah, Rima. *The Changing Roles of Palestinian Women in Refugee Camps in Jordan.* PhD Thesis SUNY,1986.

Zaatari, Akram (ed.). *The Vehicle.* Beirut: The Arab Image Foundation, 1999.

Zaatari, Akram. *The Vehicle. Picturing Moments of Transition in a Modernizing Society.* Beirut: Arab Image Foundation, 1999.

Zachs, Fruma. "Mikhail Mishaqa: The First Historian of Modern Syria." *British Journal of Middle Eastern Studies* 28/1 (2001): 67–87.

Zachs, Fruma. "The Beginning of Press Advertising in 19th Century Beirut: Consumption, Consumers and Meanings. In G. Prochazka-Eisl, and M. Strohmeier (eds.), *The Economy as an Issue in the*

Middle Eastern Press. Neue Beihefte zur Wiener Zeitschrift für die Kunde des Morgenlandes; 3 2008): 187–202.
Zdafee, Keren. *Cartooning for Modern Egypt.* Leiden: Brill, 2019.
Zeidan, Joseph T. *Arab Women Novelists, The Formative Years and Beyond.* New York, SUNY, 1995.
Zeuge-Buberl, Uta. *The Mission of the American Board in Syria, Implications of a Transcultural Dialogue.* Stuttgart: Steiner, 2017.
Zirinski, Roni. *Ad Hoc Arabism. Advertising, Culture, and Technology in Saudi Arabia.* New York: Lange, 2005.
Zophy, A.H. and F.M. Kavenick (eds.), *Handbook of American Women's History.* New York: Garland Press, 1990.
Zoughi, Saleem. "Salvador Arnita", *This Week in Palestine,* January 2021.
Zwemer, S. M. "'The Clock, the Calendar, and the Koran." *The Moslem World* 3 (1913).

Some periodicals

Al-Lata'if al-Musawwara (Cairo)
La Revue Commerciale du Levant , Chambre de Commerce Français, Istanbul
Al-Muqtataf (Beirut, Cairo)
Misr al-Mahrusa
l'Egypte Nouvelle
Le Monde Musical
Cairo Times
al-Ahram Weekly
The Daily Star
The Levant Herald
The Levant and Eastern Express
Le Monde Musical
Le Ménestrel

Diverse yearbooks and reports

L'Indicateur Ottoman
Indicateur des Professions Commerciales for Izmir (1896).
E.J. Blattner (ed.), Le Mondain Égyptien (Cairo)
Annuaire Orientale
Great Britain, Department of Overseas Trade, Report on the Economic and Financial Situation of Egypt (June 1925), pp. 32–33; idem for June 1939, p. 83; Report on Economic and Social Conditions in Turkey, April 1939, p. 73; idem for September 1947, p. 196; idem for April 1950, p. 157.
Papers Relating to the Foreign Relations of the United States, 1902 (Washington 1903, idem for 1905 (Washington 1906), pp. 883–885; idem for 1910 (Washington 1910), pp. 1073–1074; idem for 1908 (Washington 1912), p. 756.

Archives

Ministère des Affaires Étrangères (Brussels) (MAE)
Public Record Office, today National Archives (Kew)
Centre des Archives Diplomatiques (Nantes) (CAD)
Centre des Archives du Monde de Travail (Roubaix) (CAMT)

Selected online sources

Igrec, Mario, *Pianos Inside Out*, partly summarized as *"Marketing History of the Piano"* http://www.cantos.org/Piano/History/marketing.html (accessed 27 December 2011).
Farhi, A. 'Back to the Nile', http://www.farhi.org/Documents/backtothenile.htm
V.D. Sanua, 'A Return to the Vanished World of Egyptian Jewry', http://www.sefarad.org/diaspora/egypt/vie/001/0.html.
V.D. Sanua, 'The Sepharadim in Egypt', http://www.sefarad.org/publication/lm/033/6.html; G. Mizrahi, 'Ma vie au 19ème siècle', http://www.monimiz.com/giacomo.html.
G. Mizrahi, 'Ma vie au 19ème siècle', http://www.monimiz.com/giacomo.html
http://www.ajoe.org/grandmag/gm.htm.
'Des Lunetiers Moreziens', p. 10, in www.hal.archives-ouvertes.fr/
.../Des_lunetiers_a_l_echelle_du_monde.pdf (accessed 1 July 2013).
Lunettiers Moreziens à l'Échelle du Monde: Les Fils d'Aimé Lamy (1889–1914)
Ryzova, Lucie. "I have the picture: Egypt's photographic heritage between digital reproduction and neoliberalism', part 1 and 2", www.photography.jadaliyya.com/pages/index/8297 and 9825 (accessed 1 July 2013).
Zeina Dowidar and Ahmed Ellaithy, "The Invention of the Arabic Typewriter", https://medium.com/@kerningcultures/the-invention-of-the-arabic-typewriter-a6d26e0554a (accessed 8 Aug. 2022).
Robert Messenger, https://oztypewriter,blogspot.co.il/2014/10/the-arabic-typewriter (accessed 11 May 2018).
Robert Messenger. https://oztypewriter.blogspot.com/2015/05/ataturk-sultan-his-harem-remington-7.html (first accessed 1 June.2018)
Gürmen Y. et al., 'Higher Education Institutions and the Accounting Education in the Second Half of the XIXth Century in the Ottoman Empire' [from web]
Eamonn Fitzgerald's Rainy Day, "The Crime? Possession of a Typewriter", http://www.eamonn.com/2004/10/the_crime_possession.
http://www.sampiyon-kurslari.com.tr/kurucumuz_eng.htm.
Tuncay Kayaoğlu, "Keyboard Warriors: Battle for Turkish Typists Heats Up" (3 March 2015) http://aa.com.tr/en/turkey/keyboard-warriors-battle-for-turkis... , accessed 15 Dec. 2017.
http://maviboncuk.blogspot.com/2011/09/p-foyagian-constantiople-galata.htm-
Peter Thoegersen, "Charles Ives's Use of Quartertones: Are They Structural or Expressive?', https://.academia.edu.
See also his "Quartertones Are Not Out of Tune: You are!", https://www.academia.edu/2562791/Quartertones_Are_Not_Out_of_Tune_You_AreDes
Arab Republic of Egypt, Ministry of Communication, *Measuring the Digital Society in Egypt: Internet at a Glance* (2015)
Goes, Frank Joseph. *The Eye in History*. New Delhi; Jaype Bros 2013. [internet].

Antique Spectacles, curator David A. Fleishman, https://www.collectorsweekly.com/hall-of-fame/view/antiquespectacles-com.

Lunettiers Moreziens à l'Échelle du Monde: Les Fils d'Aimé Lamy (1889–1914

Lutfallah Gari, "The Invention of Spectacles between the East and the West", *Muslim Heritage* (first published 12 November 2008), https://muslimheritage.com/people/authors/9725-2/ (accessed 16 December 2021).

https://www.sothebys.com/en/buy/auction/2021/arts-of-the-islamic-world-india-including-fine-rugs-and-carpets-2/a-pair-of-mughal-spectacles-set-with-emerald.

http://maviboncuk.blogspot.com/2011/09/p-foyagian-constantiople-galata.htm-.

Bilim Tarihi, "The Essad Ophthalmoscope", http://www.bilimtarihi.org/OBA/2007-08_9-1-2.htm

British Optical Association Museum, https://www.college-optometrists.org/the-british-optical-association-museum

Recep Uslu, "Western Music Theory Works Among the Ottomans, 19th Century", Turkish Music Portal, http://www.turkishmusicportal.org/en/articles/western-music-theory-works-among-the-ottomans-19th-century

Ömer Eğecioğlu, internet site *History of Istanbul*, vol.7,

"Iranian Piano", *Wikipedia*.

Recep Uslu, "Western Music Theory Works Among the Ottomans", http://www.turkishmusicportal.org/en/articles/western-music-theory-works-among-the-ottomans-19th-century, accessed 4 August 2022.

Jan Altaner,"Kaps – a Grand Piano from Dresden in the Chouf Mountains", Goethe Institut, Lebanon, https://www.goethe.de (last accessed 3 August 2022)

Farhat, "Piano in Persian Music", *Encyclopedia Iranica*, https:// iranicaonline.

Yuval Shaked, "On Contemporary Music", https://www.searchnewmusic.org.

Ersin Altin," Baker Mağazalari: Göstere Göstere Tüketmek", *Manifold* (https://manifold.press/baker-magazalari-gostere-gostere-tuketmek)

Samir Raafat, "Ignace Tiegerman, Could He Have Dethroned Horowitz?", Egy.Com-Judaica {internet], originally published *Egyptian Mail*, 20 September 1997, and Allan Evans, an audio restoration pioneer," Ignace Tiegerman: The Lost Legend of Cairo", *Arbiterrecords* (on internet 1999). Tiegerman also taught the princes Hasan 'Aziz Hasan (grandson of Khedive Ismai'il) and the young prince Faruk. See further Samir Raafat, "Pianissimo: Henri Barda, an Egyptian-born Virtuoso", *Egyptian Mail*, 1 April 1995 and Egy.com.on the web, as well as "'Cairo Condidential' – Une Rencontre avec le Pianiste Henri Barda", *Concerts de Monsieur Croche* , accessed 11 November 2019.

Gary Graffman "Visiting the Pyramids and Pianos of Egypt", 4 February 1979, https://www.nytimes.com/1979/02/04/archives/visiting-the-pyramids-and-pianos-of-egypt-visiting-egypt.html (accessed 4 August 2022).

Pirker, Michael. "Janissary Music." *Grove Music Online* (2001).

Sabra, A.I., "Manāẓir, or 'Ilm al-Manāẓir", in: *Encyclopaedia of Islam*, Second Edition, Edited by: P. Bearman, Th. Bianquis, C.E. Bosworth, E. van Donzel, W.P. Heinrichs. Consulted online on 26 February 2023 http://dx.doi.org.ezproxy.haifa.ac.il/10.1163/1573-3912_islam_SIM_4911

List of Figures

Fig. 1:	Orosdi-Back building, Beirut, postcard *(author's collection)*.
Fig. 2:	Orosdi-Back, Istanbul, poster *(Bibliothêque Nationale de France,* https://gallica.bnf.fr/ark:/12148/btv1b53185410m.item *(by permission)*.
Fig. 3:	Établissements Orosdi-Back, map of commercial network *(by author)*.
Fig. 4:	Ramses Perfumes *(courtesy: Mrs. Grace Hummel,* see her website https://cleopatrasboudoir.blogspot.com/2014/01/ramses-perfumes.html (accessed 30 September 2022).
Fig. 5:	Orosdi-Back, Istanbul interior *(University of Istanbul, Orosdi Back Şirketi Mağazaları,* no. 91539–0017)
Fig. 6:	Interior of Orosdi-Back in Baghdad *(courtesy: Dr. Amer Hanna Fatuhi, USA)*.
Fig. 7:	Orosdi-Back Istanbul, music department *(University of Istanbul, Orosdi Back Şirketi Mağazaları,* no. 91539–0007.
Fig. 8:	Orosdi-Back Istanbul, clock department *(University of Istanbul, Orosdi Back Şirketi Mağazaları,* no.91539–1113.
Fig. 9:	Trade mark Établissements Orosdi-Back as registered on 7 January 1891 *(permission: ProMark Paris)*.
Fig. 10:	First article on Arabic typewriter, *al-Hilal,* vol. 13 (15 May 1904).
Fig. 11:	Omar Effendi building in Cairo *(photograph by author)*
Fig. 12:	Orosdi-Back Cairo souvenir fan *(courtesy: Mrs. Hélène Alexander, The Fan Museum, London)*.
Fig. 13:	Alois Hába at Főrster quartertone piano *(courtesy: Wolfgang and Annekatrin Főrster, Löbau)*.
Fig. 14:	Electric fan, advertisement, *al-Lata'if al-Musawwara,* 26 June 1922.
Figs. 15 and 16:	Cartoons, "car scare", *al-Lata'if al-Musawwara,* 10 July 1922.
Fig. 17:	Singer sewing machine manual *(courtesy: Prof. Ekmeleddin İhsanoğlu)*.
Fig. 18:	Woman with sewing machine in Kuwait, photo by Jean-Philippe Charbonnier, 1955 *(Getty Images, iStock)*.
Figs. 19 and 20:	Ottoman typewriter *(courtesy: Rahmi M. Koç Müzesi, Ankara)*.
Fig. 21:	A "Misr" Arabic typewriter, *al-Lata'if al-Musawwara,* 24 January 1927.
Fig. 22:	Sydney Nowill invoice for Remington typewriters, 1929 *(author's collection)*.
Fig. 23:	Textbook for typists *Al-Uslub al-Sahih fi'l Darb 'ala al-Ala al-Katiba al-'Arabiyya* (Beirut 1966) (Nazarian Library, University of Haifa).
Fig. 24:	*Sekreter Daktilograf* magazine, first issue1956 *(author's collection)*.
Fig. 25:	Street-based public scribes, Istanbul *(photograph by author)*.
Fig. 26:	Miniature portrait of Riza Abbasi, 1673 *(Garret Collection, Princeton University Library,* public domain*)*.
Fig. 27:	Foyagian optician's store trade card *(credit: Ara the Rat, Los Angeles,* also https://www.aratherat.com/blogs/home-page/ottoman-armenian-business-letterhead-designs)*.
Fig. 28:	Lawrence and Mayo opticians, advertisement, *al-Lata'if al-Musawwara,* 18 December 1922.
Figs. 29 and 30:	Photograph of Midhat Pasha, ca. 1877 *(Felix Nadar Archives, public domain)* and of an unknown high official, ca. 1910 *(Getty Research Institute, Los Angeles,* by permission).

Fig. 31:	Title page of magazine Abou Naddara 5/1 (1881) (BULAC, Paris)
Fig. 32:	al-Misri Effendi, cartoon, *Ruz al-Yusuf*, 12 June 1926.
Fig. 33:	Painting "Beethoven in the Palace" *(the İstanbul State Art and Sculpture Museum, public domain)*.
Fig. 34:	Comendinger, advertisement, *Annuaire Orientale du Commerce, 1889–1890 (courtesy: Salt Research, Yearbooks,* Istanbul*)*.
Fig. 35:	The 'Ariyan quartertone piano, *al-Lata'if al-Musawwara*, 11 September 1922.
Figs. 36 and 37:	Főrster quartertone piano in Prague *(photographs by Naama and Zwi Kupferschmidt, with permission of České Muzeum Hudby / Národní Muzeum Praha)*.
Fig. 38:	Front cover Zeina Abirachid, *Le Piano Oriental* (Brussels 2015) *(by permission of Editions Casterman)*.

Index of proper names

'Abd al-Nasir   66, 143, 234,
'Ali Mubarak   48
'Abd al-Haqq, Wajiha   236
'Abd al-Massih, Mathilda and Sophie   229
'Abd al-Wahhab, Muhammad   88, 224, 226
Abdülaziz, Sultan   197, 198
Abdülhamid II, Sultan   35, 80, 124, 182, 199, 200
Abdülmecid, Sultan   197, 198
Abirached, Zeina   236
Abou Hershkovits, Keren   160
Abou Naddara (Abu Naddara, Abu Nadhdhara)
   180, 181, 183
Adelson, Robert   192
Adorno, Theodor   188
Afghani, Jamal al-Din al-   36, 71
Aharoni, Israel   105
Akbal, Oktay   139
Alhazen, see Ibn Haytham
Ali Shah (Iran)   195
'Ariyan, Emile   220, 223, 224
Asmahan   227
Atatürk   232
Atrash, Farid al-   227
Avarino   66
Ayscough, James   181
Ayşe Osmanoğlu   199
Azud al-Dawla   201

Bach, Johann Sebastian   189
Back family   10, 38, 50 – 55, 61
Bacon, Roger   158
Baker, James   205, 206
Balatti, 213
Barda, Henri   229
Barth, Hans   218
Bartók, Béla   87, 221, 222
Bayramoğlu, Fuat   128
Beethoven   20, 198
Beger, Johann   168
Belefantis   211
Benaroya, Avram   134
Benfeghoul, Farid   160
Berger, Morroe   143

Berggrün, Joseph   228
Berque, Jacques   215
Bertero   211
Birbary, Leila   231
Bloch, Marc   90
Bluwstein, Rachel and Shoshanna   210
Bodenstein R.   211
Botta, Paul-Emil   204
Boulous   211
Bourdieu, Pierre   188
Brandon, Raoul   56
Busch, Moritz   204
Bustani, Salim al-   208

Cachia, Pierre   138
Calderon   211
Cantoni, Fortunato   226
Caro, Vincenze   228
Cemil, Mesut   226
Chahine (Shahin), 'Abdallah   236
Cheridjian, Miss   231
Chidiac   211
Chopin, Frederic   190
Clark, Edward   99
Cohn, Hermann   169
Collangettes, Xavier (Père)   192, 222
Cristofori, Bartelomeo   27, 85, 189
Czaczkes, Ludwig   133
Czerny, Carl   194

D'Arenda   197
Darwish, Sayyid   219 – 220
David, Félicien   195 – 196
Davies, Margarie   118
De Vaux, Carra   222
Donizetti Pasha, Giuseppe   196 – 197
Dumont, Paul   92
Dursunoğlu, Cevat   233
Dussap Pasha   197, 207

Edison   1, 27, 77, 92, 148
Ehrlich, Cyril   232
El Dabh (Dab'), Halim   229

Eldem, Ekrem   131
Elizabeth I, Queen   195
Érard, Sebastien   189
Eugénie, Empress   200

Farid, Muhammad   185
Farmer, Henri   226
Faruq, Prince, King   200, 229
Fathi, Muhammad Bey   224, 226
Fesch, Paul   91, 104
Firzan (Firzein), Bishara   224
Fontanier, Victor   166
Foucault, Michel   73
Foyagian   170
Freud, Sigmund   138
Fuad (Fouad), King   8, 66, 221

Galasso, Giuseppe   228
Gandhi, Mahatma   98
Gari, Lutfallah   160–161
Gattegno   5, 46, 48
Gendzier, Irene   183
Georgeon, François   92
Gergis, Sonia   234
Giraud, Ernest   131
Glidden, Carlos   117
Göçek, Fatma Müge   30, 77, 165
Gökalp, Ziya   232–233
Gottheil, Richard   81, 124
Granato   211
Greis, Yusef   229
Grünzel, Joseph   214
Guarino, Piero   228
Guatelli, Callisto   197
Gutenberg, Johannes   27, 156

Habá, Alois   87, 217, 220–221, 224, 226
Haddad, Shibli   81, 121, 123, 136
Hamdis, 'Abd al Mattar al-   160
Hampartsum   192
Harari, Victor   55
Hasan 'Aziz Hasan, Prince   200
Haykal, Muhammad Husayn   185
Hayutman, Yitshaq   105
Heizer, Oscar   171
Hemsi, Alberto   220
Herzl, Theodor   80, 124–126

Hickmann, Hans   228
Hifni, Mahmud al   217, 221
Hilan, Rizkallah   95
Hindemith, Paul   87, 221, 226, 233
Hobsbawm, Eric   4
Hornbostel, Erich von   217, 226
Hugh de St. Cher, Cardinal   155
Husayn, Ahmad   61
Husayn, Saddam   139
Husayni, Magdi al-   238
Hüttel, Joseph   214
Huygens, Constantijn and Christiaan   155

Ibn Firnas, 'Abbas   75, 157
Ibn Haytham (Alhazen)   75
Ibn Sa'ud   135
Ibn Sahl, Abu Sa'd al-'Ala   157
Ibrahim Adham Pasha   182
Ibrahim Hakki Pasha   176
Ilardi, Vincent   157, 162
Ileri, Selim   139
Isik, Eset   179
Isma'il, Khedive   54, 86, 181, 200, 230
Ismat al-Dawla   201
Ives, Charles   218

JamI, Nur al-Din 'Abd al-Rahman   161–162
Jarallah, Husam al-Din   210
Jessup, Henry   202, 219

Kantemiroglu (Cantemir), Dimitri   92
Karanasou, Flora   243
Keppler, Johannes   158
Khayrat, Abu Bakr   229
Khuri, Shakir al-   178
Koç family   128
Konguel, Arkadie   235
Köse, Yavuz   3, 6, 34
Kossuth, Layos   50

Lamy   76, 172
Landes, David   98
Lange, Paul   197, 228
Latife, Hanım   232
Latour, Bruno   25, 73
Lawrence and Mayo   172–173
Leeuwenhoek, Antonie van   155

Index of proper names

Leibniz, Gottfried 155
Lemaire, Alfred Jean Baptiste 201
Lewin, Yeshayahu 105
Lewis, Bernard 161
Liszt, Franz 190, 197

Mahfuz, Najib 138
Mahjubi, Morteza Khan-I 214
Mahmud II, Sultan 196, 198
Mahmud Sadiq Khan 201
Malcolm, Napier 167
Mardani, Shams al-Din 161
Marinovitz 55
Markowitz, George 233
Marrash, Maryana 207
Marx, Karl 4, 39–40, 79, 98
Mayer family 4, 49, 54–55
Mazor, Amir 160
Mehmed Said Effendi 195
Messenger, Robert 134
Meyerhof, Max 179
Micklewright, Nancy 30, 175
Midhat Pasha 175
Millingen, Edward van 179
Minyalawi, Yusuf al 20, 62, 77, 90
Mishaqa, Mikhail 192, 223
Misri Effendi, al 151
Möllendorff, Willy von 216
Morgenstern 55
Mozart, Wolfgang Amadeus 189, 193, 230
Mozzato, Andrea 162
Mubarak, 'Ali 48
Murad III, Sultan 195
Mustafa Effendi 165

Nahad, Najib 218
Nahhas, Najib (Naguib) 219, 223–224
Napoleon 24, 85, 176, 195
Nasim Pasha, Muhammad Tawfiq 184
Nasr-al-Din Shah 85, 202
Nasser. Nasirist, see 'Abd al-Nasir
Nazli Fadil, Princess 207
Neidlinger, Georg 101–102
Nemeth, Titus 121
Nevine Abbas Halim, princess – 201
Nietzsche, Friedrich 138

Nimetullah, Princess 200
Nuwayhad, Bayan al-Hout 132

Oliphant, Lawrence 210
Orosdi family 10, 50–55. 61

Papasian 211
Poche (Poché), Christian 218
Poliakine 211
Puglisi, Ettore 200, 228, 234

Qasimi, Jamal-al-Din 83, 178,
Qaytbay, Sultan 160
Quataert, Donald 103, 107

Raafat, Samir 67
Racy, Ali Jihad 203, 237
Raghib, Sirri Idris Bey 123
Rahbani brothers 235
Rasim, Ahmad 173
Ravndal, consul 172
Reiner, Karel 217
Renner, 236
Reşad V, Sultan 198
Richmond, Ernest Tatham 123
Richter, Joseph 228
Rida, Rashid 36, 83
Rigel, Henri-Jean 195
Rigler, Lorenz 168
Rihani, Amin 94, 135
Rogers, Everett 29, 31, 74, 117, 120, 148, 150, 154, 185, 204
Rogers, Mary 204
Rothschild 91
Ruppin, Arthur 105
Rusafi, Ma'ruf al 210

Saad El-Din, Mursi 138
Sabra, Wadi' (Wadia) 219, 223–224, 226
Sachs, Curt 217, 222–223, 226
Said Bey 7, 92
Said, Edward 81, 127, 229
Said, Wadie 127
Sakakini family 210
Sakhavi, Shams al-Din 161
Salib, Rachel 228
Sanu', Yaqub (Sanua) 180–181, 183

Sanudo, Marino   163
Saul, Samir   52, 63
Sawwa family   89, 227
Say, Fazil   233
Saygun, Ahmet Adnan   233
Sayyida Zaynab   67
Saz, Leyla   198
Schnabel (see also Orosdi)   10, 50
Schumann, Clara and Robert   190
Sednaoui, Selim   229
Sednaoui, Sidnawi family   5, 48, 67
Senefelder, Alois   130
Sha'rawi, Huda al   207
Shahin, see Chahine
Shawan, Salwa al-   200
Sheehi, Stephen   84, 175
Shinder, Joel   165
Sholes, Christopher Latham   27, 117
Shurbaji   61
Silbermann, Gottfried   189
Sladen, Douglas   174
Spinoza, Baruch   155
Sreberny, Annabelle   72
Stearns, Peter   43
Stein, Johann Andreas   189
Storrs, Ronald   205
Stufa, Luigi di Angelo della   162
Subeiri, Israel   135
Sureyyya, Musa   233
Surur al-Mulk   195, 201
Suyuti, Jalal al-Din   161
Szulcz. Joseph   228

Tabatabaei, Seyyed Hadi   162
Taha Husayn   185, 234
Tahtawi, Rifa'a al-   91
Takács, Jenő   214

Taymur, Ahmad   159
Taymur, Muhammad   37, 46
Tchouhadjian (Chukaajian), Dikran   207
Thévenot, Jean de   181
Tibi, Shams al-Din al   160
Tiegerman, Ignace   201, 228–229
Tilly, Charles   97
Tommaso da Modena   155
Totah, Khalil   210
Twain, Mark   80, 138

'Ukasha, Tharwat   234
Umm Kulthum   87, 89–90, 224, 227, 230–231, 235
Urmawi, Safi al-Din   191
Uryanzade, Cemil Molla   91

Vahanian, Srpouhi   207
Veblen, Thorstein   14, 37, 39, 73
Vefik, Ahmed   205
Vergilius   155

Wakid, Philippe   81, 121, 123
Wällisch, Soma   126
Weakly, Ernest   102, 112
Woodsmall, Ruth   143
Wornum, Robert   190
Wyschnegradsky, Ivan   87, 216–217

Yekta Bey, Rauf   223
Yener, Ihsan Sitki   140, 144

Zen, Pietro   163
Zivy, Cesar   76
Zola, Emile   55
Zuckmayer, Edward   233
Zumpe, Johannes   189

Index of geographical names

Adana 5, 12, 52
Aden 109, 136, 172
Afghanistan 163
Aleppo 4, 12, 30, 50, 52, 11, 134, 207, 218
Alexandria 4, 12, 30, 47, 49 – 50, 52 – 53, 55, 81, 106, 130, 132 – 134, 136, 172, 188, 211, 213, 219 – 220, 224, 228
America(n) 3 – 4, 6, 9, 14, 22 – 23, 26, 29, 34, 42 – 44, 47 – 48, 63, 68, 70, 74, 77, 80 – 8 – 1 – 82, 90, 97,100, 104 – 105, 111, 113, 115 – 119, 121 – 124, 126 – 128, 131, 134 – 136, 138, 140, 142 – 143, 148, 153, 156 – 157, 163, 169 – 171, 187, 189 – 190, 202 – 203, 218 – 219, 232
Amman 201
Anatolia 191, 205, 220
Ankara 128, 233
Argentina 41, 228
Asia Minor 103
Aswan 106
Asyut 49
Austria(ns) 1, 4, 15, 37,54 – 55, 59, 85, 113, 168, 170, 177, 190, 205, 214 – 215

Baghdad 5, 6, 52, 103, 134, 157, 237
Balkan 1, 40, 49, 89, 227
Bani Suwayf 213
Barcelona 12
Basra 52, 103, 157
Beirut 4 – 6, 9 – 10, 12 – 13, 30, 40, 49, 52, 62, 77, 84, 90, 101 – 102, 105, 112, 123, 133 – 134, 162, 172, 178, 188, 192, 196, 223 – 224, 236
Bir Zeit 203
Birmingham 212
Bizerte 12, 52
Blankenloch 99
Bohemia 61 – 62, 113
Bombay (Mumbay) 172
Bonnières 99
Bosnia 100
Bradford 12
Britain, Great Britain 3, 24, 43, 47, 59, 65, 71, 85, 106, 125, 135, 190, 203

Bucharest 12, 52
Bulgaria 6, 102
Bursa 103
Byzantine 158

Cairo, Cairene 4 – 6, 8, 10, 12, 14, 18, 30, 33, 35, 47 – 53, 56, 58 – 60, 63, 67, 69, 76, 87, 102, 106, 121, 123, 127, 134 – 136, 138, 141, 144, 157, 172, 174, 179, 181, 188, 195, 205, 207, 211, 213 – 215, 217, 220 – 221, 226, 228, 230 – 231, 234
Campinas 99
Casablanca 13, 52
Chaudry 12
Chemnitz 12
Chicago 77, 100, 121
China, Chinese 14, 26, 38, 41, 101, 116,, 121, 153 159
Clydebank 99, 115
Constantinople, *see* Istanbul
Cordoba 157
Costa Rica 54

Damascus, Damascene 83, 104 – 107, 162, 178, 207
Damiette 106
Dayr al-Zur 12, 52
Diyarbakir 103
Dresden 127, 211

East Asia 153
Egypt 3 – 4, 10, 15, 24, 34, 37, 46 – 48, 50, 52, 54, 57, 59, 61, 64, 67 – 69, 82, 86 – 87, 89, 93, 96, 106, 109, 113, 116, 119, 123, 125, 135 – 137, 139, 141 – 143, 150, 160, 169, 172, 176, 180 – 182, 184 – 185, 193, 194 – 195, 200, 203 – 205, 214 – 215, 219, 221, 227 – 229, 234 – 235, 238
El Salvador 153
Elaziğ 103
Eskişehir 116
Europe 1 – 4, 6, 9 – 10, 14, 23 – 24, 26, 29 – 30, 43, 47 – 48, 52, 63, 65, 70, 74 – 76, 79 – 81, 87, 90, 92 – 93, 98, 101, 115, 118 – 119, 121, 125,

Index of geographical names

130, 134–135, 142, 148, 152–158, 161, 163, 166–170, 172, 177, 183, 187–188, 190, 193, 205–207, 210, 214, 216–217, 219, 232, 235

Fayum 213
Florence 154, 162
France, French 1–4, 6, 8–10, 13–15, 17, 28–29, 38, 43, 47, 49, 51, 54–56, 58–59, 61, 65–66, 68–69, 76, 81, 85, 87, 91, 102, 104, 131–133, 136, 139, 144, 155, 166, 170–172, 177, 181–182, 190, 195–198, 201, 204–205, 207, 211, 214–215, 218, 222, 234
Frankfurt 139, 154

Gablonz (Jablonec) 12
Galgocz 51
Germany 40, 43, 47, 52, 54, 61, 85–86, 106. 113, 118, 135, 140, 154–155, 169, 171, 177, 190, 203, 211, 217
Giza 234
Glasgow 99
Greece, Greek 41, 143, 165, 170, 179, 210, 226, 234
Guatemala 153
Gulf, Arab/Persian 150, 185

Haifa 204
Hungary 113

Iran 3, 25, 37, 42, 72, 80, 96, 135, 167, 185, 188, 195, 201, 206, 214, 221
Iraq 10, 21, 93, 135, 139
Ireland 100
Israel 201
Istanbul 4, 6, 8, 10, 12, 14–15, 18, 20, 24, 29, 33, 35, 49–52, 59–62, 64, 78, 84, 91–92, 104, 107, 114, 116, 124–126, 131, 133, 142, 144, 162, 171–173, 179–180, 182, 188, 197–198, 200, 206–207, 213, 215, 219, 228
Italy, Italian 6, 48, 55, 61, 75, 85–86, 89, 105–106, 133, 136, 138, 154, 157–159, 162, 172, 190, 192, 196–197, 200–201, 204–205, 213–214, 220, 226, 234
Izmir 4, 12, 30, 48, 52, 103, 173, 211, 220, 232

Jaffa 105, 205, 210
Japan 12, 26, 38, 52, 62, 101, 126, 159

Jerusalem 105, 127, 132, 162, 180, 204–205, 230
Jordan 212, 238

Karak 105
Kharput 103
Kilbowie 99
Kobe 12
Kolkata 172
Konya 103

La Chaux-de-Fonds 12
Lebanon, Lebanese 66, 81, 105, 123, 127, 135, 144, 192, 200, 202–203, 208, 218–219, 222, 224, 226, 235–236, 238
Leipzig 211
London 100, 154–155
Luxor 166
Lyon 12, 195

Mainz 156
Malta 172, 181
Manchester 12
Mansura 49
Mediterranean 162, 172
Meknes 13, 52
Metula, 105
Mexico, Mexican 41, 100
Milan, 12, 220
Minya 49, 213
Moldavia 192
Monza 99
Morez 76, 172
Morocco(can) 10, 136, 227
Mosul 103

Nabi Rubin 180
Naples 197
Netherlands, Dutch 26, 101, 196, 204
New York 77, 81, 100, 124, 127, 218
North America 48, 74, 97, 116, 134, 153, 156
Nuremberg 154

Odessa 196
Oman 172
Ottoman Empire 4–5, 15, 30, 48, 50, 77, 107, 119, 162, 165, 168, 172, 211, 214

Index of geographical names

Padua   189
Palestine   80, 91, 93, 105, 123–124, 132–133, 135, 143, 180, 188, 204, 210, 235
Paris, (see also articles de Paris, „Paris along the Nile")   6, 10, 14, 47–48, 53–54, 58, 73, 100, 179, 181, 196, 201, 205, 210, 217, 219, 223–224, 226
Persia(n) (see also Iran)   48, 54, 80, 85, 89, 97, 119, 123, 126, 158, 191, 195, 227
Pisa   27, 151, 160
Plovdiv (Philippopoli)   6, 12, 52
Podolsk   99
Poland, Polish   193, 228
Port-Said   12, 49, 52–53, 106, 113
Prague   217
Pressburg (Bratislava)   49
Prussia(n)   193, 196–197, 204

Regensburg   154
Rishon le-Zion   91
Riyad (Riyadh)   95, 135
Rosetta   106
Roubaix   12
Rusçuk (Rusu)   12, 52
Russia, Russia(n)   87, 100–101, 119, 171, 180, 193, 204, 215–216, 228, 235

Salonika (Thessaloniki)   4, 12, 50, 52, 107, 171, 200
Samsun   12, 52
Saudi-Arabia   68–69, 95, 113–114, 160
Scotland, Scots   180
Serbia, Serbs   100
Sfax   12, 52
Silahtarağa   92
Sivas   103
Smyrna, see Izmir
South-Africa   153
Soviet Union, see Russia
Spain   41, 119

Strakonice   61
Stuttgart   200, 236
Syria(n)   5, 48–49, 61, 65, 77, 93, 95, 102–103, 105–107, 109, 111, 133, 135, 158, 160, 206, 210, 236

Tabriz   13, 52
Tanta   12, 49, 52–53
Tehran   13, 52, 116, 202, 205
Tirebolu   209
Trabzon   166
Tripoli (Leb.)   105, 134, 162
Tunis   6, 10, 12, 17, 48, 52, 159
Turgutlu   220
Turkey   5–6, 9, 20, 28–29, 37–38, 42, 48, 52, 61, 64, 71, 80, 82, 89, 93, 96, 101, 109, 114, 116, 121, 123–127, 130, 133–134, 135, 137–140, 142–145, 148, 150, 167–168, 173–174, 176, 179–180, 183, 185, 191, 197–198, 202, 221, 223, 230, 232–234

United Kingdom (see also Britain)   106, 135
United States (see also America)   1, 42, 47, 79, 85, 93, 98–99, 117–118, 135, 203

Varna   12, 52
Venice, Venetian(s)   14, 76, 154, 158, 162–164
Veria (Karaferia)   165
Vienna   8, 12, 49, 51, 61, 100, 126, 171–172, 193–194, 197, 216, 236

Wales, Welshmen   5, 49
Wittenberg   99
Wuppertal   12

Yazd   167
Yemen   112, 135–136, 166
Yokohama   12

Zaqaziq   12, 49, 52–53

Index of subjects

accordion 188, 226–227
Ades, department store 5, 48
advertising, advertisements 9, 13–14, 28, 33, 40, 59, 66, 74–76, 80, 83, 87, 89, 95, 98, 102, 111, 120, 125, 136–137–138, 140, 144, 167, 170
'alim, see 'ulama
alla turca 85, 193, 230
Alliance Israélite Universelle 134
Anglican(s) 133
Anwal Co. 68–69
Aq Quyunlu 162
Arab Music Congress 87, 221
Arab Spring 69
Arabesque music 233
Arabic 6, 9, 38–39, 41, 48, 59, 66, 80–82, 121–128, 130–141, 144, 148–149, 159, 167, 169, 219, 238
Armenian(s) 111, 143, 170, 12, 207, 213
arms, weapons 25, 73, 80
Articles de Paris 14, 58, 73
artisans 152, 158
arzuhalci, see street scribes
atelier, *see* workshops
'Ataba square 18, 49, 56, 63, 66
Au Bon Marché 4, 6, 47, 49, 56, 59–60
Au Printemps 6, 49
automobile 23, 25, 28, 33–35, 37, 64, 72, 75, 93–96
azan 76
Azhar 64, 77
Bağlama 234

Bağlarbaşi Palace 198
Bahçe Kapı 18, 50, 173
bailo 76, 163
Baker, department store 5, 206
banks 12, 131
bazaar 18, 51, 60, 68, 107, 112, 175
Bechstein, pianos 190,200, 211, 215
Bedouins 169
Benzion, department stores 5, 46, 48
Berliet, automobile 64

Beyoğlu quarter, *see* Pera
Bijouterie (see also jewelry) 15, 62
binoculars 21
Blickensderfer, typewriters 148
Blüthner, pianos 190, 205, 210, 215
bonneterie 15, 60, 73
bourgeois, bourgeoisie 10, 46, 66, 84, 86, 111, 187, 206–207, 230, 234
Bureaucracy 142

calligraphy 128
camera 27. 84. 157
car, (motorcar) *see* automobile
Catholic (-ism) 54, 156, 210
cello 227
Chamber of Commerce 29, 131, 133
chasm (*see also* time lag) 30, 34, 120, 134
Chavanne, pianos 195
Chemla, department stores 5, 48, 56, 66
Chickering, pianos 218
Christians 5, 24, 49, 64, 83, 106, 118, 123, 143, 203, 206, 210–211, 235
Christmas 135
Cicurel, department stores 5, 46, 48, 62–66
Cirağan Palace 196, 198
civil service 143, 173
civilization, civilized 14, 73, 100, 157, 205
clarinet 227
Clark & Co 62
class(es) 4, 8, 24, 26, 31, 43, 58, 60–61, 63, 68, 72, 84, 86, 89, 98, 102, 114, 173, 187–188, 208–211, 213, 228, 231–232
clerks, clerical 42, 105, 118–119, 142–144, 146–148,
clocks (*see also* watches) 6, 17, 20, 30, 75, 77, 165, 194–195
cloth 62, 113–114
clothing 1, 2, 6–7, 10, 14–15, 18–19, 21–22, 28, 33, 37, 43, 49, 51–52, 59–60, 62–64, 73, 79, 97–98, 100, 107, 109, 111–114, 206
Club Syrien d'Alexandrie 220
Colonial (-ism) 23, 26, 42, 58, 71, 74,119, 202, 205

https://doi.org/10.1515/9783110777222-009

Index of subjects

Columbian press   77
Comendinger, music store   9, 211, 213
Commodities   9, 14, 23, 26, 33, 37, 48, 51–53, 55, 58, 60, 63, 70–71, 73, 93, 142, 170, 211
computer   39, 42, 44, 82, 97, 138, 140, 146, 148–150
concerts, concert halls   189, 198, 202, 211, 234
confection (*see also* ready-to-wear)   1, 49, 51, 112, 114
conservatory/-ies   201, 214, 219, 221, 223–224, 228, 233–235
consuls, consul(-ar), consular reports   51, 59, 91, 170–172, 204–205, 211
consumer durables   18, 23, 28, 72, 79
consumerism, consumer culture   2–3, 5, 9, 14, 30, 33, 39, 169
consumption   2, 3, 14, 31, 38–39, 46, 48, 51, 58–60, 63–64, 66, 68, 73, 112
Copts   143
courts (royal etc.)   9, 35, 44, 86, 102, 119, 164, 188, 193, 195–201, 206–207, 211, 214, 227–228, 232, 234
craftsmen, *see* artisans
credit   7, 39, 60, 64, 99, 103, 127, 150
Crédit Foncier credit   65
Crimean War   49, 59, 211
crystal   154, 160, 162, 164, 182
cutlery (spoons, forks)   16, 78

dactylography, *see* typewriting
Dar al-Fonun   201
Darülelhan   233
da*stgah* mode   201, 214
Davies Bryan, department store   5, 48
Defterdar   75, 163
diplomats   56, 128, 163, 202, 205, 210
Dolmabahçe Palace   196, 200
Dominican   151
dress   26, 58, 60, 113–114, 197, 227
drums   85, 193–194, 234, 238
dulcimer   189

Eastern piano, *see* Oriental piano
Effendi   61, 67, 141, 184, 234
Egypt, revolution   1952 46, 66–67, 69, 234

electricity (electric implements, light)   8, 21, 23, 25, 28, 34, 39, 56, 71, 73, 90–93, 148, 215, 238
electronic keyboard   39, 44, 238–239
elevators   6, 56
embroidery   109
Eminönü, quarter   50
en *gros, en détail*   10, 52
entrepreneur (-ial), enterprise   1, 3–4, 10, 12, 18, 23–24, 47, 49, 71, 121
Érard, pianos   189, 198, 200, 201, 217
Établissements Orosdi Back, *see* Orosdi-Back
ethnomusicology   217
excavations   53, 166

Farsi (Persian)   48, 80, 89, 121–123, 149, 160, 191, 201, 227
fashion   1, 2, 14, 21, 59–60, 62, 65, 98, 111, 113–114, 168, 172, 176, 207
fatwa   36, 83, 167
female, feminization (*see also* women)   42, 90, 98, 107, 109, 118, 142–143, 146, 213, 231
fez (tarbush)   15, 38, 61–62, 183
field glasses, *see* binoculars   165
folk music   217, 220, 222, 232–233
folklore   217
Ford, automobiles   73, 94
Förster, pianos   117–224
fountain pens   73, 77–78
Four C's   35
Franciscans   158
Frankfurt School   39, 73
Franks, Frankish   106, 119, 124, 161
Free Officers   46
French Revolution   155
Fridgidaire, frigidaire   73
furniture   8, 15, 58–59, 63, 97, 197, 206

galabiyya   113
Galata, quarter   10, 50, 173
Galeries Lafayette, department store   49, 56
garments, *see* clothing   1, 114
Gaveau, pianos   224
gender   5, 7, 26, 36, 40, 97, 108–109, 114, 142
General Electric   73
Gillette   73, 78
girls, *see* women

glass, glass making   6, 51, 75, 152, 155, 157–161, 171, 179, 231–232
globalization   1, 52, 70, 82, 90, 93, 149–150, 215
Golestan Palace   202
gramophone   20, 62, 89–90, 208, 214, 232
grand magasin   5, 47, 49, 51, 56, 58

Hadassa   180
Hammond, typewriters   117, 124
Hanımlara Mahsus Gazete   209, 213
Hannaux, department stores   5, 47–48, 65
harmonium   20, 199, 216
harmony   202, 218–219, 222
Hebrew   122, 132, 149
Herberger, piano manufacturer   224
Hermes, typewriters   127
Hi*lal, al-* (journal)   33, 121, 159, 167
Hilal, typewriter   123
Ho*da, al-*, newspaper   127
Hoffman, piano manufacturer   236
Holy Sepulcher   162
household goods   16, 19. 23. 25, 28, 34, 43, 58, 63, 90
Howe-Singer, *see* Singer   78
Huguenots   200

IBM Selectric   148
ib*n al-balad*   184
imperialism   1, 2, 24, 26, 42, 44, 71, 230
Industrial Revolution   4, 70, 190
Infitah   67, 68, 109, 114
installments, payment, *see* credit
Institut d'Égypte   220
insurance companies   119, 143
internet   5, 10, 17, 116, 149, 152
invention(s), inventor   1, 27, 28, 41, 44, 47, 74, 79, 92, 98, 102, 119–120, 123, 126, 154, 156, 159–160, 175, 189, 219

Janissary/ -ies   85, 193, 196–197
jewelers, jewelry   35, 73, 171–172
Jews, Jewish   4–5, 10, 24, 47–48, 51–53, 55–56, 64–66, 68, 76, 83, 132, 135–136, 142, 170, 172, 179, 210–211, 213,220, 228, 233–235
journals (magazines, newspapers, periodicals)   14, 28, 33, 70–71,75, 95, 102, 111, 121, 136, 142, 144, 156, 167, 175, 209, 211, 213, 218, 220

Kadıköy, quarter   133, 198
kahhal (kuhl)   178
Kaps, pianos   211
ke*mence, kaman*   89, 227
keyboard (*see also* piano keyboard)   39, 80–82, 122–123, 126–128, 140, 144, 149
Khedive   14, 54, 81, 86, 102, 119, 181–182, 185, 200, 230
Kodak   1, 73, 75, 84
Kredit-Anstalt   61
Kurds, Kurdish   233

labor force   109, 118, 142
Lata'if al-Musawwara, al-   33, 34, 173, 184, 220
Latin-script, *see* Roman-script
leapfrogging   38, 39, 42, 44, 82, 89, 116
lenses   152–155, 157, 159, 161–162, 170, 179–181, 183
Levant(ine)   4, 24, 48, 66, 211
light bulbs   23, 28, 73, 90, 92
Linotype   82, 127
literacy, illiteracy   36, 43, 76, 82, 119, 142, 145, 149–150, 172–173, 186
locks   78, 165
lunetiers, *see* opticians

magnifying glass   152, 157–159
Mamluks   158, 160, 162, 166
Maqam   44, 191–192, 194, 222, 224, 227, 232–233, 238
marasme (depression)   52
marches, musical   196–198, 219
Masonic   230
mass production   1, 8, 70, 98, 101, 108, 153
Mayflower   156
McCormick, reapers   73
meh*ter, mehterhane*   193, 196
Ménestrel, le   196–197
microscopes   155
middle class (*see also* classes)   4, 31, 43, 60, 63, ,65, 68, 84, 87, 114, 185, 187–188, 190, 209, 231
Midmar, al   220,
mirrors   48, 157–158, 161

Index of subjects

Misr al-Fatat 141
„Misr" typewriter 136
mission (-missionaries) 180, 202–203, 210
Mission Civilisatrice 131, 204
missionary/ies 44, 134, 167, 203, 219
Moghul 75, 162
Mondain Égyptien, le 185
monks 151, 155
monocle 43, 176, 180
Morum's, department store 55
mosque 77, 90, 114
motorcar, *see* automobile
movies 94, 111, 113, 116, 146, 185,214–215, 224
Muqtataf, al- 33, 79–80, 91, 102, 120, 123, 167
Muski, quarter 18, 33, 51, 53, 60, 174
Muslim Brothers /-hood 46, 141
Muslims 75, 83, 106, 117, 143, 157, 160, 175, 204
myopia, myopic 152, 153, 168, 179

nationalization(s) 5, 13, 65–67, 69, 141, 143
Naum Theater 196
nay, *ney* (flute) 198, 227, 234
Nizam- i Jadid 200
nomads 105, 108
notation (of music) 123, 187, 191–192, 194, 214, 218, 222
novelties, *nouveautés* 2, 6, 14, 19, 28, 34, 58, 63, 73
Nubians 169

offices, office culture 36, 38, 42,81–82, 117, 119, 124, 130, 132, 136, 142–143, 145–147, 149
Olivetti, typewriters 127, 138
Omar Effendi (Ömer Efendi), department stores 10, 46, 50–51, 67–69
opera 86, 196, 200, 207, 213, 217, 226, 230
ophthalmology, ophthalmologists 75, 158, 159, 168, 172, 176, 178–179
ophthalmoscope 177, 179
opticians 76, 169–172, 181
optics, optical 8, 75, 151–152, 155, 157, 159, 170, 179
organ (-ist) 195, 197, 219, 228, 238
Oriental piano 216, 219, 218–221, 223
Orientalism 229–230

Orosdi-Back, department stores 4, 6, 9–10, 12–13, 15, 17–18, 20–21, 37–38, 50–56, 59–65, 67, 105, 113
Ottoman Bank 12, 131

„Paris along the Nile" 14, 54
patents 27–28, 123–124
peddlers, peddling 33, 156, 169, 174
Pera (quarter), Rue de Pera 8, 18, 33, 162, 172, 211
Pera Palace Hotel 91
perfume 15, 53, 62
periodicals, *see* journals
photographs, photography 20, 35,55, 83–84, 175–176, 216
piano keyboard 189, 217–218, 231, 236, 238
pianos, *see* various manufacturers
Piastre Plan 61
pin*ce-nez* 43, 163, 176
Pleyel, pianos 205, 211, 217, 219, 224
polyphony 194, 218, 222, 231
port cities 12, 30, 52
presbyopia 76, 151–153, 155, 158, 167–169
printing press 14, 117, 121, 125
Protestant 133, 156, 202–203, 218–219
publicity 5, 9, 13, 29, 33, 70, 75, 95, 101, 119, 214

qadi 160, 210
Qajars 201, 214
qanun 88, 189, 194, 209, 227, 234
Quaker 210
quartertone pianos 87, 217–218, 220, 224, 226, 236
quartertones 39, 87, 89, 192, 199, 217, 220–221, 224, 227
quartz 152, 157
Qur'an 128
QWERTY 82, 122, 139–140, 148

rababa, rebeb 227
radio 25, 39, 90, 150, 211, 214, 232
railways 12, 23, 31, 36, 42, 52, 71–72, 94, 119
Ramadan 64, 77
Rawdat al-Balabil 220
razor blades, *also* Gillette 78

Index of subjects

ready-to-wear clothing 1, 10, 14–15, 18, 49–51, 59–60, 107
Remington, typewriters etc. 1, 41, 44, 73, 80–81, 117, 119, 124–126, 128, 131, 136, 138
Renaissance 43, 75, 156
Roma („gypsies") 198, 227
Roman script 42, 80, 82, 122, 125, 130, 135–136, 140, 143, 150, 169
Royal, typewriters 81, 117, 127
Ruz al-Yusuf 183

Safawids 162–163
Saint-Simon(ians) 195
Salamander, department store 54
san*tur* 88, 189, 195, 201, 234
Schaeffer, fountain pens 73
Schmidt's College 210
scribes, scribal occupation 82, 152, 175
seamstress (*see also* female) 52, 79, 98, 108, 111, 113–114
secretary/-ies (*see also* female) 79, 135, 143–144, 146–147
Seidel and Naumann, typewriters 128, 137
Şeyhülislam 83
Shari'a 167
Shepheard's Hotel 172
shoes 15, 61–62, 77–78, 107
shopkeepers 156, 166
shopping, shopping malls/streets 6–8, 33, 48, 60, 68
shorthand 42, 132–133, 143
sika mode (E half flat tone) 222
Singer, sewing machines 1, 18, 35–36, 40–41, 62, 73, 75, 78–79, 98–103, 105–106, 111–112, 114, 118, 127
Smith Corona, Smith Premier typewriters 81, 117, 136
Snellen card 177
socks (stockings) 15, 60–61, 107
spectacles (*see also* eyeglasses) 76, 151–152, 156, 158, 160, 162–163, 165, 167–168, 174–175, 178–183
St John's Hospital 180
St. Joseph University 192
Stanhope printing press 24, 77
statistics 29, 42, 59, 74, 76, 78, 134, 149, 155, 163, 169–170, 211, 214

steam power 97
Stein, department stores 4, 54–55
Steinway pianos 190
stenographer(s), stenography (*see also* shorthand) 79, 82, 118, 132–134, 138, 142–144, 146
street scribes 145, 175
Stross, department stores 4, 50, 54
Sublime Porte 81, 130
Suez Canal Company 143
sultans 86, 193, 196–198
suq(s) 6, 9, 114
Sütterlin script 118
synagogue 51, 53
synthesizer 39, 44, 89, 238
Syrian Protestant College (AUB) 133

tadbir al-manzal 15, 209
tailors 51, 98, 105–107, 152
tar (lute) 201
tarbush, see fez
technologies (big) 23, 35–36, 71–72, 97
technologies (small) 18, 23–29, 31, 34–35, 38–41, 43–44, 72–73, 77–78, , 83, 85, 90, 96, 98, 114, 128, 134, 136, 149–151, 159, 187, 189–191, 216
technology (general) 22–23
technoscapes 70
telegraph 23–24, 42, 71, 143
telephone 27, 39, 45, 119, 143, 150, 176
textile(s) 59, 62
taht 227
„Time is Money" 42, 80, 116, 121, 126
time keeper 76–77
time lag 23, 74
Tiring, department stores 4, 49, 54–56
tombola 6, 64
Topkapi Palace 77, 194
toys 16, 73
Transuntino (harpsichord) 192
types, typography, typesetting 27, 121–122, 127, 128–130, 136
typewriters, *see* different manufacturers
typewriting, typing (speed, skills) 121, 132–134, 138, 140, 144, 149
typist(s) 42, 79, 118, 135, 138, 142–144, 146–147

Index of subjects

'ud 88, 198, 209, 218, 227, 234
'ulama 36, 71, 83, 106, 163, 210
umbrellas, parasols 15, 52, 62, 73, 107
Underwood, typewriters 117
urbanization 1, 135, 168

Verdoux, store 9, 172–173, 182
village(s) 111–112, 166, 210
violin (*see also kaman, kemence*) 44, 89, 198–199, 209, 215, 217, 227, 234

watches (*see also* clocks) 12, 16, 20, 62, 73, 76–77, 172, 174
Waterman, fountain pens 73
Wertheim, department stores 52
window displays 56
women (*see also* female) 1, 6–7, 15, 18, 34, 36, 38, 41–43, 48, 55, 60, 62, 64–65, 79, 86, 98–100, 103, 106–109, 111–115, 118–119, 133, 142–144, 146–147, 149, 152, 168–169, 185, 187, 196, 198, 203, 207–210, 231
word processor 148
workshops, sweatshops 41, 52, 62, 79, 98, 102, 107, 112–114
World War I (WWI) 40, 49, 53–55, 76, 81, 90, 93, 130, 142, 203–204, 216, 228
World War II (WWII) 47, 65, 84, 109, 134, 139, 216

Yildiz palace 81, 91
Young Turks 37, 126
Yüksek Kalderım, street 8

za*j*al 89
Zeiss, optics 155
Zionism 80, 105, 124, 210
Zoroastrians 167
Zulus 100